Hans G. Völz

Industrial Color Testing

© VCH Verlagsgesellschaft mbH, D-69451 Weinheim (Federal Republic of Germany), 1995

Distribution:
VCH, P.O. Box 10 11 61, D-69451 Weinheim (Federal Republic of Germany)
Switzerland: VCH, P.O. Box, CH-4020 Basel (Switzerland)
United Kingdom and Ireland: VCH (UK) Ltd., 8 Wellington Court, Cambridge
    CB1 1HZ (England)
USA and Canada: VCH, 220 East 23rd Street, New York, NY 10010-4604 (USA)
Japan: VCH, Eikow Building, 10-9 Hongo 1-chome, Bunkyo-ku, Tokyo 113 (Japan)

ISBN 3-527-28643-8

Hans G. Völz

Industrial Color Testing

Fundamentals and Techniques

Translated by Ben Teague

Weinheim · New York · Basel · Cambridge · Tokyo

Dr. Hans G. Völz
Deswatines-Straße 78
(formerly: Bayer AG, Krefeld-Uerdingen)
D-47800 Krefeld

This book was carefully produced. Nevertheless, author, translator and publisher do not warrant the information contained therein to be free of errors. Readers are advised to keep in mind that statements, data, illustrations, procedural details or other items may inadvertently be inaccurate.

Editorial Directors: Karin Sora, Dr. Ulrike Wünsch
Production Manager: Peter J. Biel

Library of Congress Card No. applied for.

A catalogue record for this book is available from the British Library.

Die Deutsche Bibliothek – CIP-Einheitsaufnahme
Völz, Hans G.:
Industrial color testing : fundamentals and techniques / Hans G. Völz. – Weinheim ; New York ; Basel ; Cambridge ; Tokyo : VCH, 1995
Dt. Ausg. u.d.T.: Völz, Hans G.: Industrielle Farbprüfung
  ISBN 3-527-28643-8

© VCH Verlagsgesellschaft mbH, D-69451 Weinheim (Federal Republic of Germany), 1995

Printed on acid-free and chlorine-free paper.

All rights reserved (including those of translation into other languages). No part of this book may be reproduced in any form – by photoprinting, microfilm, or any other means – nor transmitted or translated into a machine language without written permission from the publishers. Registered names, trademarks, etc. used in this book, even when not specifically marked as such, are not to be considered unprotected by law.
Composition: Hagedornsatz GmbH, D-68519 Viernheim. Printing: Druckhaus Diesbach, D-69469 Weinheim.
Bookbinding: Industrie- und Verlagsbuchbinderei Heppenheim GmbH, D-64630 Heppenheim
Printed in the Federal Republic of Germany

Preface to the English Edition

The English version of my book is appearing just four years after the successful German edition entitled *Industrielle Farbprüfung*. It is closely based on the German text. Revisions have been necessary in only a few passages to correct errors and update (and sometimes expand) the literature citations. Furthermore, equations that are essential for theoretical understanding and practical application have been enclosed in boxes for emphasis.

New material in the colorimetric part includes the CIELAB 92 color-difference system (Section 2.3.3) and the RAL Design color-order system (Section 2.3.4). A greatly simplified and clearer description has been given for the proposition (important in the determination of tinting strength) that equality of the CIE tristimulus values Y in lightness matching implies equality of the Kubelka-Munk functions, that is, that $\Delta Y = \Delta F_Y = 0$ (Sections 4.6.4 and 4.7.1). In connection with tests of significance, it is shown how the error of preparation can be separated into specific error components (Section 6.3.3). To aid the understanding of confidence levels in relation to acceptability, an explanatory paragraph has been inserted in Section 6.5.3, and the same chapter has been expanded by the new Section 6.5.4, "Acceptability and Test Errors."

The index of standards (Appendix 1) is completely new. Organized through keywords, it lists standards of the International Organization for Standardization (ISO), the European Committee for Standardization (CEN), and several national standards bodies: the American Society for Testing and Materials (ASTM), the British Standards Institution (BSI), the Association Française de Normalisation (AFNOR), the Japanese Industrial Standards (JIS), and the Deutsches Institut für Normung (DIN). An important role in the preparation of this table was again performed by Mr. E. Fritzsche of DIN, Berlin, to whom I would like to express my gratitude.

I am also grateful to my colleague and successor, Dr. H. Heine, Krefeld, for his generous assistance, to Mr. Ben Teague for a sympathetic translation, and to Mrs. Karin Sora of VCH Publishers for her helpful collaboration.

July 1994   Hans G. Völz

Preface

This book is directed chiefly to all technical personnel who are involved with pigments and dyes, whether as manufacturers, converters, or consumers. It is often not enough to be familiar with the procedure used in a test; more must be known about the quantity being measured and the way in which it varies as a function of other parameters – in brief, about its physicochemical habit, which is embedded in the well-known fields of chemistry and physics. This is especially important when test methods are to be rationalized; rationalization requires computers, and all computer use is based on applied mathematics. Effective rationalization is possible only when the mathematical and physical background of the quantities being measured is kept fully in view. The lack of suitable worked-out examples to illustrate test methods often creates difficulties in practice as well. This book should help close the gap by providing at least one practical example for each of the test methods described.

The book should, however, also offer some utility to the scientist engaged in research and development work on pigments and dyes. Discussion of principles makes up a substantial part of the text, and many suggestions and hints are included to further the scientist's work. On the other hand, anyone working in research or development but lacking a knowledge of both the principles and the state of the art will scarcely be in a position to advance the colorimetric properties of these substances.

A third audience – so the author hopes – will find the book beneficial: university lecturers and students in the natural sciences. The study of colorimetry in the testing of coloring material is a perfect example of the interdisciplinary effect of chemistry, physical chemistry, physics, theoretical physics, mathematics, and cybernetics (informatics). It goes without saying that one must understand the principles of each discipline before one can understand their interaction. There is, however, another point: The integrative components of the knowledge assembled in this book are hardly taught (if at all) in our universities and colleges of technology – colorimetry and Kubelka-Munk theory being just two examples. Even beyond the field of coloring materials, both these bodies of knowledge have significant roles to play in many other technical areas, and it is hard to comprehend that people with university-level diplomas even today are entering the colorant manufacturing and converting industries without ever having come in contact with either topic.

In choosing the contents of this book from the disciplines named, the author has adopted a standard that he considers trustworthy, one that has continually aided him in distinguishing key topics from less important ones: He wanted to write the book he wishes had been available when he entered the field of pigment testing. Naturally, such a wish is a bit of a fantasy, for it is only now that this book could be written in its present form. Some three-quarters of the test methods described here did not even exist then. The same holds, in modified form, for the principles; while certain branches of knowledge had already been developed 30 years ago, they were not yet relevant for our field – the statistical analysis of color coordinates is one of many topics that might be mentioned. Finally, the introduction of the computer in colorimetric testing has led to major changes in recent years.

To avert misunderstandings, some special points relating to text selection should be stated in advance. The title of the book should be read as referring to the colorimetric examination of pigments and dyes *in media*; thus, all tests not carried out in mixtures with a lacquer, paint, plastic, or other coating are excluded (e. g., on textile fibers or in printing inks). The chapter on "Colorimetry" has been kept short even though it threatened to grow into another whole volume. In other words, this book does not wish to be mistaken for a textbook on colorimetry. The only aspects of that field that are covered are topics essential for the context of the book. Similarly, one will search in vain for instructions on color formulation and correction calculations. These operations do not come under the heading of pigment and dye testing, even if they must be preceded by testing. Similarly, the application of Mie theory in the determination of particle-size distributions is not covered, for it is based not on colorimetric methods but on general spectroscopic methods.

In Chapter 2, on the other hand, a comparatively large proportion has been devoted to the statistical analysis of color coordinates. The reasons for this are that these applications are relatively novel in practice and that key topics such as acceptability (tolerances) are related to them. New in Chapter 3 are the four-flux equations for specular substrates, which may prove to be of interest in the future. Chapter 4 offers – for the first time, as far as the author is aware – a theoretical mathematical justification for the universal practice of treating wide-band measurements (e. g., reflectometer values) or color coordinates (e. g., CIE tristimulus values) as spectral values in Kubelka-Munk analysis and of matching not the tristimulus values but their Kubelka-Munk functions. In Chapter 5, instructions are given for extending Mie calculations to the particle-size distribution of anisometric (i. e., bladed or rod-shaped) pigments.

While the text was in preparation, a number of colleagues expressed willingness to read it and offer vital suggestions and corrections. The entire text was read by Dr. H. Heine (Krefeld-Uerdingen), Dr. S. Keifer (Krefeld-Uerdingen), and A. van Leendert (Krefeld). The following experts made themselves available to read portions of the text: Dr. L. Gall (Ludwigshafen), Dr. W. Herbst (Frankfurt-Höchst), Prof. G. Kämpf (Leverkusen), Prof. J. Krochmann (Berlin), Dr. H. Pauli (Basel), Dr. K. Sommer (Krefeld-Uerdingen), Dr. J. Spille (Leverkusen), Dr. J. Thiemann (Leverkusen), Dr. J. Wiese (Krefeld-Uerdingen), and Dr. K. Witt (Berlin). Mrs. E. Lechner and Mr. E. Fritzsche of DIN, Berlin, have prepared the exemplary listing of the standards cited. The author would like to take this opportunity to thank all these colleagues for their helpfulness and their valuable advice and suggestions.

I am grateful to Bayer AG for its continuing interest and assistance in the writing of this book, and to VCH Publishers for its intelligent collaboration and valuable advice.

Krefeld, November 29, 1989   Hans G. Völz

Contents

Part I Principles   1

1 Introduction   3

   1.1 Coloring Materials   3
   1.2 Color Properties   9
   1.3 Summary   12
   1.4 Historical and Bibliographical Notes   12

2 How Colors Depend on Spectra (Colorimetry)   15

   2.1 Introduction   15
      2.1.1 Concerns and Significance of Colorimetry   15
      2.1.2 Reflection and Transmission   17
   2.2 CIE Standard Colorimetric System   20
      2.2.1 Spectral Distribution and Color Stimulus   20
      2.2.2 Trichromatic Principle   21
      2.2.3 CIE System   23
   2.3 Sensation-Based Systems   26
      2.3.1 Lightness, Hue, Saturation   26
      2.3.2 Physiologically Equidistant Systems   27
      2.3.3 CIELAB System   28
      2.3.4 Color Order Systems   37
   2.4 Mathematical Statistics of Color Coordinates   40
      2.4.1 Normal Distribution in Three Dimensions   40
      2.4.2 Standard Deviation Ellipsoid   44
      2.4.3 Standard Deviations   49
      2.4.4 Color Measurement Errors and Significance   53
      2.4.5 Acceptability   56
   2.5 List of Symbols Used in Formulas   59
   2.6 Summary   61
   2.7 Historical and Bibliographical Notes   63

3 How Spectra Depend on the Scattering and Absorption of Light (Phenomenological Theory)   67

   3.1 Introduction   67
      3.1.1 Phenomenological Theory and Its Significance   67
      3.1.2 Many-Flux Theory   68
      3.1.3 Surface Phenomena   71
   3.2 Four-Flux Theory   75
      3.2.1 The Differential Equations and Their Integration   75
      3.2.2 Transmittance and Transmission Factor   79
      3.2.3 Reflectance and Reflection Factor   83
      3.2.4 Limiting Cases of Reflection   87
      3.2.5 Determination of the Coefficients   90
   3.3 Kubelka-Munk Theory   93
      3.3.1 Significance and Formalism   93
      3.3.2 Limiting Cases of Reflection   95
      3.3.3 Determination of the Absorption and Scattering Coefficients   99
   3.4 Hiding Power   101
      3.4.1 General   101
      3.4.2 Achromatic Coatings   102
      3.4.3 Scattering and Absorption Components   105
   3.5 Transparency   107
      3.5.1 Description and Definition   107
      3.5.2 Coloring Power   108
      3.5.3 Achromatic Coatings   110
   3.6 Principle of Spectral Evaluation   112
      3.6.1 Description and Significance   112
      3.6.2 Application to Hiding Power   113
      3.6.3 Application to Transparency and Coloring Power   118
   3.7 List of Symbols Used in Formulas   120
   3.8 Summary   122
   3.9 Historical and Bibliographical Notes   123

4 How Light Scattering and Absorption Depend on the Content of Coloring Material (Beer's Law, Scattering Interaction)   129

   4.1 Introduction   129
      4.1.1 Description and Significance of the Concentration Dependence   129
      4.1.2 Pigment Particle Size   129
      4.1.3 Dispersing of Pigments   131
      4.1.4 Measures of Pigment Content   133
   4.2 Absorption and Content of Coloring Material   137
      4.2.1 Dyes   137
      4.2.2 Pigments   137

4.3 Scattering and Pigment Content   139
    4.3.1 Scattering Interaction   139
    4.3.2 Experimental Test of an Empirical Formula   143
4.4 Systematic Treatment of Pigment/Achromatic Paste Mixing   145
    4.4.1 Standard Methods of Pigment/Paste Mixing   145
    4.4.2 Importance of the Methods   147
4.5 Kubelka-Munk Functions of Pigment/Paste Mixture   148
    4.5.1 General Formula   148
    4.5.2 Black Pigments Mixed with White Paste   150
    4.5.3 White Pigments Mixed with Black Paste   155
    4.5.4 Colored Pigments Mixed with White Paste   157
4.6 Tinting Strength   164
    4.6.1 Significance and Definition   164
    4.6.2 Coloristic Matching Criteria for Pigments   167
    4.6.3 Color Matching Studies   172
    4.6.4 Lightness Matching   181
    4.6.5 Color Depth Matching   187
4.7 Special Problems   189
    4.7.1 Lightening Power   189
    4.7.2 Color Differences after Matching   191
    4.7.3 Change in Tinting Strength   194
4.8 List of Symbols Used in Formulas   200
4.9 Summary   203
4.10 Historical and Bibliographic Notes   206

5 How Light Scattering and Absorption Depend on the Physics of the Pigment Particle (Corpuscular Theory)   211

5.1 Introduction   211
    5.1.1 Description and Significance of the Corpuscular Theory   211
    5.1.2 Particle-Size Distribution   212
    5.1.3 The Optical Constants: Refractive Index and Absorption Index   217
5.2 Mie Theory   222
    5.2.1 Integration of the Wave Equation   222
    5.2.2 Absorption and Scattering of the Particle Ensemble   224
    5.2.3 Scattering Behavior of Pigments   226
    5.2.4 Absorption Behavior of Pigments   229
5.3 List of Symbols Used in Formulas   231
5.4 Summary   232
5.5 Historical and Bibliographical Notes   234

Part II Test Methods 237

6 Measurement and Evaluation of Object Colors 239

  6.1 Reflection and Transmission Measurement 239
      6.1.1 Gloss Measurement and Assessment 239
      6.1.2 Measurement and Evaluation Conditions 244
  6.2 Practical Evaluation of Color Differences 248
      6.2.1 Preparation of Specimens 248
      6.2.2 Color Measuring 249
      6.2.3 Full Shade 254
      6.2.4 Special Problems 258
  6.3 Test Errors 262
      6.3.1 Calculation of the Standard Deviation Ellipsoid 262
      6.3.2 Visualization of the Ellipsoid 264
      6.3.3 Total Error and Its Components 266
  6.4 Significance 270
      6.4.1 Calculation of Standard Deviations 270
      6.4.2 Test of Significance 272
  6.5 Acceptability 275
      6.5.1 Specimen Preparation 275
      6.5.2 Matching and Color Measurement 277
      6.5.3 Evaluation and Example 279
      6.5.4 Acceptability and Test Errors 282

7 Determination of Hiding Power and Transparency 285

  7.1 Measurement of Film Thickness 285
      7.1.1 Selection of Method 285
      7.1.2 Gravimetric, Wedge-Cut, and Dial-Gauge/Micrometer Methods 287
      7.1.3 Pneumatic Method 289
  7.2 Scattering and Absorption Coefficients 291
      7.2.1 Kubelka-Munk Coefficients S and K of White Pigments 291
      7.2.2 Kubelka-Munk Coefficients S and K of Black and Colored Pigments 296
      7.2.3 Four-Flux Coefficients s^+, s^-, k' 300
  7.3 Transparency 304
      7.3.1 Single-Point and Multi-Point Methods 304
      7.3.2 Method Based on Principle of Spectral Evaluation 307
  7.4 Hiding Power 310
      7.4.1 General (Graphical Method) 310
      7.4.2 Achromatic Case 313
      7.4.3 Based on Principle of Spectral Evaluation 319
      7.4.4 Economic Aspects 326

8 Determination of Tinting Strength and Lightening Power 331

   8.1 Content of Coloring Material 331
      8.1.1 Dyes 331
      8.1.2 Pigments 332
   8.2 Relative Tinting Strength 333
      8.2.1 Dyes 333
      8.2.2 Inorganic Black and Colored Pigments (Lightness Matching) 335
      8.2.3 Organic Pigments ("FIAF" Method Based on Principle
          of Spectral Evaluation) 339
      8.2.4 Change in Tinting Strength 348
   8.3 Lightening Power 354
      8.3.1 Graphical Method 354
      8.3.2 Rationalized Method 357
      8.3.3 PVC Dependence from One Gray Mixture 360

Appendix: List of DIN, DIN ISO, and ISO Standards Cited 365

Subject Index 369

Name Index 375

Part I
Principles

1 Introduction

1.1 Coloring Materials

This book deals with color-imparting substances and the color phenomena which they produce. The general term for all coloring substances is "coloring materials";* and includes, in particular, pigments and dyes.

Pigments are substances consisting of particles that are practically insoluble in the application medium. They are used chiefly as coloring materials but can also be employed for other properties in special applications (e. g., anti-corrosion or magnetic pigments). The demarcation between pigments and fillers was formerly somewhat problematic as it cannot be stated in terms of a limiting refractive index. However they can be differentiated according to their application. In contrast to pigments, a filler is not employed to impart color but to impart certain technical properties to the coating or to increase the volume of the coating material. Such a convention does not rule out the possibility that certain substances can find use as both pigments and fillers. However, there is a large group of substances that act as pigments in the great majority of cases (e. g., TiO_2 is never used as a filler). When we speak of pigments, we are referring to this group. This statement also applies to the present book.

Pigments can be classified according to a variety of criteria; here we generally follow the standard scheme and speak of two classes:
– Inorganic pigments
– Organic pigments

The next step in classification is according to their origin:
– Natural pigments
– Synthetic pigments

Economically, synthetic pigments are very much more significant than natural pigments, and they will therefore be our exclusive focus in the following classification.

---

* For standards see "Coloring Materials, Terms" in Appendix 1.

According to standards*, inorganic and organic pigments are broken down as follows:
- White pigments. Their optical effect depends on the nonselective scattering of light.
  Examples, *inorganic*: titanium dioxide pigments, zinc sulfide pigments, lithopone, lead white, zinc white; *organic*: white resin powders.
- Black pigments. The optical effect of black pigments is based on the nonselective absorption of light.
  Examples, *inorganic*: carbon black, black iron oxide; *organic*: aniline black.
- Colored pigments. Colored pigments act by selective absorption of light, often combined with selective scattering.
  Examples, *inorganic*: iron oxide pigments, cadmium pigments, ultramarine pigments, chrome yellow, molybdate red, nickel rutile yellow, cobalt blue; *organic*: azo pigments, quinacridone pigments, dioxazine pigments, perylene pigments, phthalocyanine pigments, lake colors based on azo dyes and triphenylmethane dyes.
- Luster pigments. These are metallic particles or particles with high refractive indices. They are predominantly planar in structure and orientation, which produces gloss effects (see below for further classification).
- Luminescent pigments. These pigments fluoresce or phosphoresce (see below for further classification).

The last two groups require additional explanation. Luster (gloss) pigments include:
- Metal-effect pigments. These consist of lamellar metal particles (in contrast, powdered zinc and lead are referred to as anticorrosive "metallic pigments" and do not belong to this group).
  Examples, *inorganic* only: gold bronze, aluminum bronze.
- Nacreous pigments. These pigments consist of transparent lamellar particles, which produce a pearl-like (nacreous) gloss by multiple reflection when in a parallel orientation.
  Examples, *inorganic*: specially prepared basic lead carbonate, bismuth oxychloride, titanium dioxide precipitated on mica; *organic:* fish scale.
- Interference pigments. If the lamellae have the appropriate thickness, iridescence effects are produced by interference.
  Example, *inorganic*: iron oxide on mica.

Luminescent pigments include the following types:
- Fluorescent pigments. Their optical effect is based on selective absorption of light with simultaneous luminescence. Excitation is achieved by high-energy radiation such as ultraviolet or short-wavelength visible light.

---

*For standards see "Coloring Materials, Classification" in Appendix 1.

Examples, *inorganic*: pigments for display tubes and fluorescent lamps, radioactive luminescent pigments; *organic*: 2-hydroxy-1-naphthaldazine.
- Phosphorescent pigments. The optical effect of phosphorescent pigments is based on selective absorption and scattering of light plus delayed luminescence. Excitation is achieved by high-energy radiation such as ultraviolet or short-wavelength visible light.
Examples, *inorganic* only: zinc sulfide/alkaline-earth sulfides doped with heavy-metal ions.

The chemical makeup of pigments will only be dealt with briefly to discuss the hues which are obtainable. Inorganic pigments, with a few exceptions, are oxides, sulfides, oxide hydroxides, silicates, sulfates, and carbonates. In general, they consist of individual particles which are uniform in chemical composition (e.g., iron oxide red is α-Fe_2O_3). Organic pigments, in principle, are also powders of uniform composition; the two major groups are azo pigments and polycyclic pigments. The colors associated with individual pigments are listed in four tables:
- White and black pigments: Table 1-1
- Inorganic colored pigments: Table 1-2
- Organic colored pigments, azo pigments: Table 1-3
- Organic colored pigments, polycyclic pigments: Table 1-4

Table 1-1. White and black pigments.

	Chemical class	White pigments	Black pigments
Inorganic pigments	Oxides	Titanium dioxide Zinc white	Black iron oxide Iron-manganese black Spinel black
	Sulfides	Zinc sulfide Lithopone	
	Carbon and carbonates	White lead	Carbon black
Organic pigments	Perylene pigments		Perylene black
	Others	White resin powders	Aniline black Anthraquinone black

1 Introduction

Table 1-2. Inorganic colored pigments.

Chemical class	Green	Blue-green	Blue	Violet	Red	Orange, brown	Yellow	Yellow-green
Oxides and hydrated oxides								
Iron oxide pigments					Red iron oxide	Brown iron oxide / Mixed brown (spinel and corundum phases)	Yellow iron oxide	
Chromium oxide pigments	Chromium oxide green	Hydrated chromium oxide green						
Mixed phase oxides	Cobalt green		Cobalt blue				Nickel rutile yellow / Chromium rutile yellow	
Sulfides and sulfoselenides				Cadmium sulfoselenide / Cadmium mercury sulfide			Cadmium sulfide	Cadmium zinc sulfide
Chromate pigments	Chrome green				Molybdate red	Chrome orange	Chrome yellow / Zinc yellow / Alkaline-earth chromates	
Ultramarine pigments	Ultramarine green, Ultramarine violet		Ultramarine blue, Ultramarine red					
Iron blue pigments			Iron blue					
Others			Manganese blue	Cobalt and manganese violet				

Table 1-3. Organic colored pigments: azo pigments.

	Blue	Violet	Red	Orange, brown	Yellow	Yellow-green	Green
Monoazo pigments				Monoazo orange pigments (also as lakes*)	Monoazo yellow pigments (also as lakes**)		
Disazo pigments				Diaryl yellow pigments			
			Disazopyrazolone pigments		Bisaceto-acetic acid arylide pigments		
Others				β-Naphthol pigments (also as lakes*)		Metal complex pigments**	
			Naphthol AS pigments (also as lakes*)				
			Naphthalenesulfonic acid pigments in lake form				
			Benzimidazolone pigments				
			Disazo condensation pigments				
				Isoindoline pigments			
				Isoindolinone pigments			

* Pigments in lake form are Me salts between sulfonic acid and/or carboxylic acid groups of dyes and metal salts (Me = alkaline earth, Na, Mn).

** Me complexes with azo or azomethine groups (Me = Ni, Co, Fe, Cu, Zn).

8   1 Introduction

Table 1-4. Organic colored pigments: polycyclic pigments.

	Yellow-green	Green	Blue-green	Blue	Violet	Red	Orange, brown	Yellow
Aromatic and heterocyclic pigments	Phthalocyanine pigments				Quinacridone pigments			
						Perylene pigments		
						Perinone pigments		
					Thioindigo pigments			
Anthraquinone pigments	Triarylcarbonium pigments							Flavanthrone yellow pigments
						Pyranthrone pigments		Anthrapyrimidine pigments
					Dioxazine pigments	Anthanthrone pigments		Quinophthalone pigments

Dyes are coloring materials which are soluble in the application medium. Dyes used to be exclusively natural products, but these are of virtually no commercial importance today. It is not difficult to classify all dyes in a similar way to pigments. However, the main area of application for dyes is in textile coloring as although their ease of dispersion makes them advantageous for use in coatings and plastics, dyes are often less lightfast than pigments in these media. Special types are, however, significant in the plastics industry. We do not need to classify dyes as in Tables 1-3 and 1-4 because it would look the same.

Along with the classifications already mentioned, other ways of breaking down the classes of pigments and dyes are possible. For example, they might be listed by application.

The industrial synthesis of coloring materials is now subject to rigorous monitoring by qualitative and quantitative chemical analysis performed in plant laboratories, quality control laboratories and special research laboratories. The chemical analysis of raw materials, intermediates, end products, by-products, and waste products serves not only to meet quality requirements, but also to satisfy environmental protection standards. Final inspection, which also considers physical properties and application qualities, takes place in properly equipped testing laboratories; this process can no longer even be thought of without the aid of computers. The principles and methods of testing, as specifically applicable to coloristic pigment and dye properties, are the subject of this book.

The creation of international standards by the ISO (International Organization for Standardization) and of national standards by AFNOR, ASTM, BSI, DIN, JIS*, and other bodies has not only made it possible to state unambiguously what requirements pigments have to meet, but has also led to greater progress towards universally accepted test methods (see Appendix 1).

1.2 Color Properties

All color phenomena that we can observe in coatings have their origin in the interaction between the coating material and visible light, i.e., the electromagnetic radiation in the wavelength range of 400 to 700 nm. Special importance is attached to the fact that such coatings need not be homogeneous but can exhibit typical inhomogeneities due to the presence of discrete pigment particles. The physical processes that result from this interaction can be broken down into a few elementary processes.

When a photon enters into a dyed or pigmented coating, three processes can determine its subsequent fate (Fig. 1-1):
– The photon collides with a dye molecule or a pigment particle and is absorbed.
– The photon collides with a pigment particle and is scattered.
– The photon passes through the coating without collision (assuming that the surrounding medium does not absorb) and is scattered and/or absorbed by the substrate.

The fundamental physical and optical properties of coloring materials are thus their absorbing and scattering powers.

Inhomogeneity of the coating is not a necessary condition for *absorption;* even homogeneous coatings can absorb if they contain *light-absorbing* substances, such as dissolved dyes. The case is different for light scattering, which can occur only in a coating that has the following properties:
– The coating is heterogeneous in composition.**
– Its individual components differ in refractive index.

---

* AFNOR = *Association Française de Normalisation*
 ASTM = *American Society for Testing and Materials*
 BSI = *British Standards Institution*
 DIN = *Deutsches Institut für Normung*
 JIS = *Japanese Industry Standard*
** Light scattering on molecules can be ignored here because it is orders of magnitude less than scattering on multimolecular pigment particles.

10   *1 Introduction*

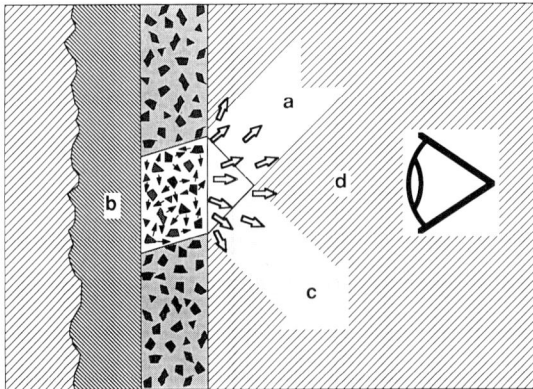

Fig. 1-1. Absorption and scattering of light in a coating with colored pigment particles.
a: Directionally incident white light
b: Gray substrate
c: Directly reflected light
d: Diffusely reflected light

In a somewhat simplified scheme, forward and backward scattering are distinguished by whether the light is scattered in the direction of incidence (into the front half-space) or in the opposite direction (into the back half-space).

White pigments have very little absorbing power in comparison to their scattering power. Black pigments, on the other hand, have very little scattering power relative to their absorbing power. Colored pigments, as a rule, feature selectivity in both absorbing and scattering power, i.e., they show wavelength-dependent behavior. This explains how the two elementary processes of absorption and scattering give rise to what we can measure: the reflection spectrum, the spectral reflectance $\varrho(\lambda)$ between wavelengths 400 and 700 nm.* Pigmented coatings, and thus indirectly pigments themselves, can be unambiguously characterized by their reflection spectra. But the human eye cannot see reflection spectra; it merely communicates color stimuli to the brain. The missing link in the chain is the conversion of reflection spectra to color stimuli; this is accomplished by colorimetry.

Modern testing of coloring materials is able to establish an almost perfect connection between the color stimulus (and hence the optical properties of the pigments) and fundamental physical quantities. The point of making such a connection is to be able to operate under well-known rules of physics. The schematic diagram shown in Figure 1-2 will aid in understanding these relationships. Key physical quantities are listed at the bottom left, while application properties of importance in practice are shown at the top left of the diagram. Particularly important are the boxed entries in between, which represent theoretical ideas that link the technical properties with the physical ones. The following relationships exist:

– *Colorimetry* relates the perceived color quality to the color stimulus **F** and this, in turn, to the reflection spectrum $\varrho(\lambda)$.

---

* The symbols employed in each subsequent chapter are listed at the end of that chapter. See also "Symbols for use in formulas" in Appendix 1.

1.2 Color Properties   11

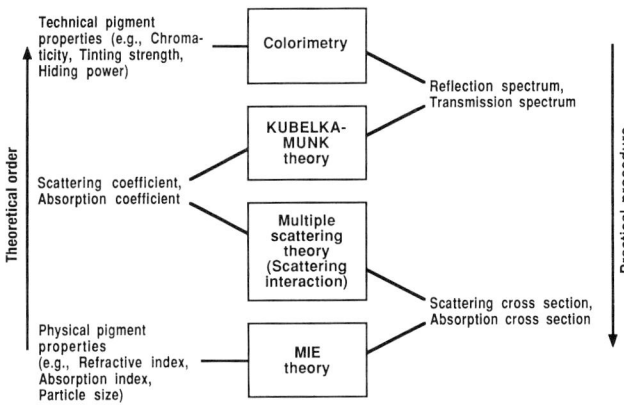

Fig. 1-2. Schematic diagram showing relationships between color properties and theoretical fundamentals.

- *Kubelka-Munk theory* relates $\varrho(\lambda)$ to the scattering coefficient $S(\lambda)$, absorption coefficient $K(\lambda)$, and coating thickness h.
- *Multiple scattering theory* relates S to the scattering cross section $Q_S(D,\lambda)$ of the individual particle and the pigment concentration σ. (The absorption coefficient K is directly proportional to the absorption cross section $Q_A(D,\lambda)$ and σ.)
- *Mie theory*, finally, relates $Q_S(D,\lambda)$ and $Q_A(D,\lambda)$ to the particle size D, the wavelength λ, the refractive index n, and the absorption index $\varkappa$; the last two optical properties are constants of the material.

In practice, either downward or upward motion is possible. For instance, to determine the hiding power it is first necessary to obtain the Kubelka-Munk coefficients by color measurements (moving downward) so that the expected color stimulus of the hiding layer can then be calculated (moving upward).

A separate chapter in this book is devoted to each of the four theoretical complexes listed above.

The ideas behind modern test methods are rather difficult, since they are based on the light source/sample/eye/brain configuration and the methods must reproduce it. The need to simulate this very complicated system leads to the insight that further progress is possible only if new problems with an even greater degree of difficulty can be handled. "Simple" test methods such as appear in the specialist literature from time to time cannot be expected to solve these problems. The only promising approach is to enhance our knowledge and understanding of the cybernetic and physico-chemical principles involved in the perception of optical stimuli by the eye and associated organs. The objection sometimes heard, that new and complicated methods entail excessive calculation effort, long ago lost its justification when powerful and economical computers became available, which meant that calculations could be done in quite a short time. This is one of the key conditions that have led to advances in this field, i.e., the remarkably fast development in computer technology. Only by virtue of this development have present-day colorimetric and testing laborato-

12   *1 Introduction*

ries been able to formulate such complex testing programs in an objective fashion and to perform tests such as those described in this book in a prompt and efficient manner.

1.3 Summary

The terms "coloring materials", "pigment", and "dye" are defined and explained in this chapter. The usual classification of pigments is according to chemical nature (inorganic, organic) and genesis (natural, synthetic). The subclasses of white, black, colored, luster, and luminescent pigments are also characterized. Inorganic pigments consist of oxides, oxide hydroxides, silicates, sulfates, sulfides, and carbonates; organic pigments are divided into two major groups, the azo pigments and polycyclic pigments. Dyes can be classified under criteria similar to those of organic pigments. The industrial synthesis of pigments and dyes today is subject to monitoring by analytical and physical-chemical methods, which would be inconceivable without the aid of computers. The existing international and national standards are of great importance for both specifications and test methods.

This book is concerned with colorimetric methods used for testing coloring materials in media. The starting point is therefore the interaction of light with a coating. Typical phenomena are described: the light can be scattered and/or absorbed. While absorption of light is apparent in the case of dyes and pigments, scattering is expected only when pigments are present. White pigments are characterized by a high scattering power with very little absorption; black pigments by a high absorbing power. Colored pigments feature selective absorption, and frequently selective scattering as well. For pigment and dye testing, the color can be related to fundamental physical quantities from color stimulus to the spectrum, to scattering and absorption, to refractive and absorption indices, and to wavelength and particle size. These relationships are used in an appropriate sequence when establishing test methods. The current state of the art in color test is chiefly due to advances in computer technology, which have offered ways to perform many testing jobs in an economical way.

1.4 Historical and Bibliographical Notes

It can be presumed that the use of coloring materials for coloring by human beings goes back to the origins of cultural history. The oldest artifacts that have come down to us are the well-known cave paintings, made with natural inorganic pigments. It must not be overlooked that the materials and the fundamental knowledge available at that time were limited even though artistic expressiveness and craft knowledge were already highly advanced. It was a long historical path [1], [2], [3] that led to the present high level of scientific and technological development in the areas of inorganic pigments [4], [5], [6], organic pigments [2], and dyes [3]. An important factor in this progress was the elucidation of the principles underlying the optical and colorimetric properties of dyes and pigments, together with the resulting advances in objective testing and evaluation methods. The methods in use today are the subject of this book. Stages of development that have occurred mainly in recent decades can be researched in survey articles [1], [7], [8]. The development of standardization in parallel with testing techniques is covered in [9].

Each chapter in Part I of this book contains a section of historical and bibliographical notes together with a list of references.

References

[1] H. Heine, H. G. Völz, "Pigments, Inorganic, Introduction", in *Ullmann's Encyclopedia of Industrial Chemistry*, 5. edit., VCH Weinheim 1992, Vol. A 20, 245.
[2] W. Herbst, K. Hunger, *Industrial Organic Pigments*, VCH, Weinheim 1993.
[3] G. Booth, H. Zollinger, K. McLaren, W. G. Sharples, A. Westwell, "Dyes, General Survey", in *Ullmann's Encyclopedia of Industrial Chemistry*, 5. edit., VCH Weinheim 1987, Vol. A 9, 73
[4] G. Buxbaum (Ed.), *Industrial Inorganic Pigments*, VCH Weinheim, 1993.
[5] G. Buxbaum, H. Heine, P. Köhler, P. Panek, P. Woditsch, Anorganische Pigmente, Winnacker/Küchler, *Chemische Technologie* 4. edit., Vol. 3, *Anorganische Technologie II*, Hanser, München 1983, 349.
[6] W. Büchner, R. Schliebs, G. Winter, K. H. Büchel, *Industrial Inorganic Chemistry*, VCH, Weinheim 1988.
[7] H. G. Völz, Progr. Org. Coat **1** (1973)
[8] H. G. Völz, *Angew. Chem.* Internat. edit. **14** (1975) 655
[9] E. Fritzsche, *farbe + lack*, **80** (1974) 731.

2 How Colors Depend on Spectra (Colorimetry)

2.1 Introduction

2.1.1 Concerns and Significance of Colorimetry

Colorimetry, the techniques based on it, and the colorimetric way of thinking have become the norm in all fields of technology that deal with colors in any way. The special uses of colorimetry can be summed up under the following headings:
- Color identification codes. Color names are verbal descriptions; they offer little help in the unique identification or labeling of colors. Scientists, however, are accustomed to describing a phenomenon objectively in terms of a measuring unit and number.
- Systematic description of color phenomena. Colorimetric concepts make it possible to construct logical and meaningful color systems based on easily understood principles.
- Evaluation of color differences. The quantification of color differences is one of the principal tasks of colorimetry. It is vital, for example, in production control of coloring materials, varnishes, paints, etc.
- Color rendering. Color rendering involves the evaluation of color differences between the object and the reproduction, as in color television or color photography.
- Colorimetric color formulation. Colorimetry makes it possible to fulfil an ancient wish, to calculate in advance the components and the proportions needed to obtain a desired color.

The color properties of coatings, which are the subject of this book, involve all of these points to a greater or lesser degree.

We will return to the meaning of *color difference* in this context. The testing of coloring materials involves measuring them between a given specimen ("reference sample") and a specimen to be evaluated ("test sample"). The objective determination of color differences is essential for the following areas:
- Development of new products. Key factors in technical progress are the continuous improvement of products which are already being produced and the creation of new coloring materialss. Indispensable to both, however, is a fundamental knowledge of colorimetric relationships and the testing and in-

spection methods derived from them. Analysis of optical performance can no longer be done without colorimetric testing methods based on exact scientific principles.
- Production control. Human beings use their visual apparatus, consisting of eye, nervous system, and brain, to match and assess coloring substances. This represents an information-processing system in the cybernetic sense, with the special feature that the output of this system gives rise to both physiological and psychological effects. It follows that assessment is generally not free of human prejudices, and so there is a pressing need for objective measurement techniques.
- Understanding between manufacturer and purchaser. Objective measurements are crucial so that the parties to a transaction understand each other clearly, and especially so in an area as emotionally charged as color. Thus it is easier to agree on acceptable deviations from rated values by referring to unimpeachable measuring methods than it would be to agree on how great a color difference can "just be seen".

When we raise the question of color deviations and measurement errors, we are in the domain of statistical analysis of color measurements. Such analysis is important for the following:
- Significance. The questioner wishes to know how precisely colors are being measured, that is, whether there is a significant difference between the measured color values for the two specimens being compared. A color comparison demands a knowledge of the significance level, which has to be tested in each case. Only when the significance level is known can one decide whether it is low enough for color comparisons to be made in a specific situation. This knowledge is needed even when color values for the same collection of samples, as measured with different types of instruments, are to be compared. What is called for in this situation is not a measure of the quality of color measurement, but a measure of the quality of the colorimeter. Given a suitable set of color samples covering the entire color space, it should be possible to compare the precisions of the colorimeters.
- Acceptability. This term refers to the limits within which measured color values must lie in order for the purchaser to classify them as "conforming to a standard." Acceptability must not be confused with the perceptibility of color differences, which is a threshold forming the upper limit of the region in which colors cannot be distinguished. Acceptability, in contrast, is based on a convention and on the experience of colorists. It has to do with the region within which specimens are assessed to "conform," that is, conform to a specified type or reference specimen. This means that some perceptible color differences may well still be acceptable.

Both the criteria of significance and acceptability are built up on the same mathematical principles and require the same mathematical apparatus, namely the extension of the Gaussian normal distribution from one to three dimensions. Because these applications of well-known mathematical principles are fairly new and unusual in everyday colorimetric practice, a section is devoted to them alone.

Colorimetry is discussed only as much as is necessary and sufficient in light of the theme and context of this book. Monographs and other literature on colorimetry are listed in Section 2.5.

2.1.2 Reflection and Transmission

The reflection spectrum mentioned in Section 1.1 is the spectral distribution typical of object colors, and it forms the starting point for any measurement of them. It seems necessary to look more closely at the widely-used terms "reflectance" and "reflection factor". First, what are "object colors"? The standards* give the following answer: They are "non-luminous perceived colors" i.e. colors of "non-self-radiating" media that require illumination for their visibility. Color perception thus only takes place with light which illuminates the non-self-radiating medium. Certain standard sources have been defined as illuminants for colorimetry*; those of interest here are
- Standard Illuminant A for incandescent light
- Standard Illuminant C for artificial daylight
- Standard Illuminant D65 for natural daylight at a color temperature of 6504 K

Reflection and transmission are purely physical phenomena; the related measurable quantities describe properties of how materials interact with radiation. These quantities are defined in the standards**. Of the many factors that influence them, the ones of interest here are those directly related to radiation: the direction of incidence, the direction of observation, and the aperture angles of the incident and observed beams (Fig. 2-1). Three configurations of the observed radiation are consistent with good technical realization:
- Directional (the beam consists of quasi-parallel rays)
- Conical (the beam has the shape of a narrow cone)
- Hemispherical (the beam is uniformly distributed over the half-space)

In *reflection measurements*, the name of the result depends on the following measuring conditions:
- Reflectance ϱ (hemispherical observation)
- Reflection factor R (conical observation)
- Radiance factor β (directional observation)

---

* For standards see "Colorimetry" in Appendix 1.
** For standards see "Photometric Properties of Materials" in Appendix 1.

2 How Colors Depend on Spectra (Colorimetry)

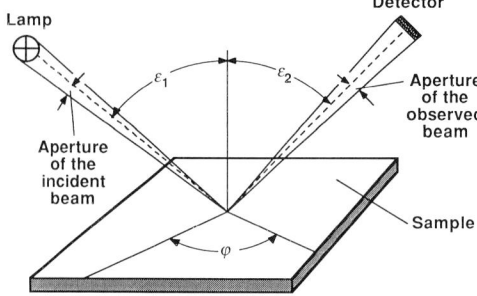

Fig. 2-1. Nomenclature of the parameters. ε_1, ε_2, angles of incidence and reflection; φ, azimuth angle.

If the illumination condition is included in the classification, nine geometries must be listed since there are three possibilities for illumination and three for viewing (each can be parallel or rather quasi-parallel, conical, or hemispherical). Each of the illumination configurations is represented by a subscript on the symbol ϱ, R, or β (Table 2-1): dif, diffuse (hemispherical case); g, directional; c, conical. Of the nine combinations, however, only three are common in colorimetry and realized in colorimeters. In the list below, the illumination geometry comes first and is followed by the observation geometry:
- Conical/hemispherical   Reflectance ϱ_c
- Hemispherical/conical   Reflectance factor R_{dif}
- Conical/conical   Reflection factor R_c

The preferred measuring conditions are as follows (Fig. 2-2):
- For ϱ_c, 0°/diffuse and 8°/diffuse (abbreviated 0/d and 8/d respectively)
- For R_{dif}, diffuse/0° and diffuse/8° (d/0 and d/8)
- For R_c, 45°/0° and 0°/45° (45/0 and 0/45)

Table 2-1. The nine types of reflectance measurement (after DIN 5036 Part 3).

Geometry		Quantity	Symbol
Illumination	Viewing		
hemispherical	hemispherical	reflectance	ϱ_{dif}
hemispherical	conical	reflection factor	R_{dif}
hemispherical	directional	radiance factor	β_{dif}
conical	hemispherical	reflectance	ϱ_c
conical	conical	reflection factor	R_c
conical	directional	radiance factor	β_c
directional	hemispherical	reflectance	ϱ_g
directional	conical	reflection factor	R_g
directional	directional	radiance factor	β_g

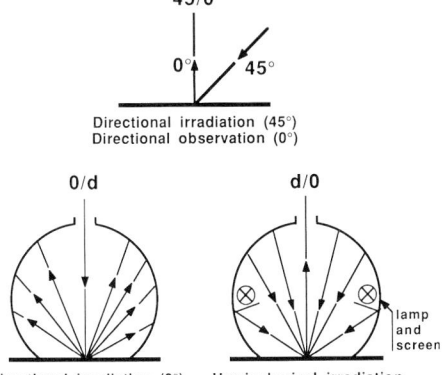

Fig. 2-2. Measuring conditions for colorimetry.

All measurements of surface colors in reflected light are referred to the ideal matte white surface, which reflects the incident radiation with a reflectance $\varrho = 1$ independent of the angle. Because this perfect white standard is not technically realizable, a real reflection standard (secondary standard) must be adopted for colorimetry. The standard features a set of fixed values and deviates as little as possible from the ideal. It consists of a circular tablet pressed from pulverized barium sulfate*. The rated values for tested commercial barium sulfate powder are specified in Germany by the *Physikalisch-Technische Bundesanstalt* PTB (Federal Physical-Technical Institute); a tablet pressed from such powder can be employed for instrument calibration, either directly or by means of a working standard.

A special point to note is that the surface of the specimen whose surface color is to be determined may exhibit peculiar reflectance behavior typical of the surface itself. If this point is not considered, measured color values may be falsified (see Section 6.1.2 for further detail).

Object colors can also be evaluated in transmitted light, as in the case of plastic films, dye solutions, and the like. The property sought in such cases is the transmission. There are again three configurations for *measuring the transmission* of object colors, and the terminology is analogous to that used in reflection measurements (with the following preferred measuring conditions):
– Transmittance τ_c, 0/d
– Transmission factor T_{dif}, d/0
– Transmission factor T_c, $0/\varepsilon_2$ (ε_2 is an arbitrary observation angle)

If both incidence and observation angles are 0° (geometry 0/0), the transmission factor is symbolized by T_r.

---

* For standards see "Colorimetry" in Appendix 1.

2.2 CIE Standard Colorimetric System

2.2.1 Spectral Distribution and Color Stimulus

In the context of physiological processes of color vision, colors are sensory impressions. If the psychological reactions associated with color vision are also considered, they are experiences. For this reason, it first appears contradictory to apply the term "metric", meaning measurement, to colors. It will be shown that the discussion is actually about a metrical arrangement of equality judgments, in which the human eye functions solely as a null indicator.

The starting point for the development is the well-known fact that any color stimulus can be prepared from the spectral colors by mixing. Thus a spectral distribution $\varphi(\lambda)$ can be measured with a spectrograph. The product $\varphi(\lambda)d\lambda$ is the intensity corresponding to the narrow wavelength interval between λ and $\lambda + d\lambda$ in the spectrum. If we use the term "color stimulus" and the symbol **F** for the color sensation registered by the eye, then there must be a correlation between the spectral distribution $\varphi(\lambda)$ and **F**. Determining how these quantities are related involves knowing not only the purely physical circumstances but also the laws describing the physiology of colors, which can be summarized roughly as follows: If three suitable spectral distributions are selected, other color stimuli can be perfectly reproduced through additive mixing of these three "reference stimuli" (also called "primaries"). The desired result is always achieved with a unique mixing ratio. It must be added that these mixtures are always prepared by the mixing of lights ("additive" color mixing), as distinct from the use of successive optical filters each having its own transmission behavior ("subtractive" color mixing). The physiological behavior just described is a summary of the rules called "Grassmann's laws".

If the addition of color stimuli is now defined as their mixing, the multiplication of a color stimulus by a number is also defined. For example, multiplication by the factor 2 corresponds to the addition of two identical color stimuli, etc. Now Grassmann's laws can be written out as follows:

$$\mathbf{F} = X \cdot \mathbf{X} + Y \cdot \mathbf{Y} + Z \cdot \mathbf{Z}$$

or

$$\mathbf{F} = m\,(x \cdot \mathbf{X} + y \cdot \mathbf{Y} + z \cdot \mathbf{Z}) \tag{2.1}$$

where $m = X + Y + Z$, $x = X/m$, $y = Y/m$, and $z = Z/m$. Here X, Y, and Z are called "tristimulus values"; x, y, and z are called "chromaticity coordinates"; m is a physical measure of intensity; and **X**, **Y**, and **Z** are the reference stimuli. From the above definition of addition, it also follows that changing from reference stimuli **X**, **Y**, **Z** to different reference stimuli **X'**, **Y'**, **Z'** involves a linear ("affine") transformation.

2.2.2 Trichromatic Principle

The task of colorimetry is now to determine m, x, and y; the determination of z is redundant, since $x + y + z = 1$. The chromaticity coordinates form a 3-manifold, and this fact suggests an analogy with the three dimensions of solid geometry. Indeed, in view of the addition rule, **X**, **Y**, and **Z** can be thought of as three-dimensional vectors forming a trihedral in which the color stimulus is a vector $\mathbf{F} = \{X, Y, Z\}$. The above equations also show that the wavelength dependence of **F** is wholly contained in X, Y, Z (or x,y,z), since the reference stimuli are fixed color stimuli. The spectral colors therefore constitute a one-dimensional manifold and lie on a three-dimensional curve ("spectral locus") whose points can be labeled with the associated wavelengths λ. Any color stimulus $\mathbf{F} = \{X, Y, Z\}$, by virtue of the intensity distribution $\varphi(\lambda)$, can be reproduced from the spectral colors by mixing, hence the sums taken over the spectral loci must give the tristimulus values exactly. This result can now be briefly derived:

A color equation as in eq. (2.1) can be written for the spectral color having the wavelength 400 nm:

$$\mathbf{F}_{400} = \bar{x}_{400} \cdot \mathbf{X} + \bar{y}_{400} \cdot \mathbf{Y} + \bar{z}_{400} \cdot \mathbf{Z}$$

where $\bar{x}_\lambda$, $\bar{y}_\lambda$, $\bar{z}_\lambda$ are the tristimulus values of the spectral colors. Now it must be remembered that this spectral color, in the color stimulus **F** being analyzed, is represented solely by the contribution φ_{400} given by the distribution function $\varphi(\lambda)$. The contributions of the spectral colors at 10 nm intervals are then

$$\varphi_{400} \cdot \mathbf{F}_{400} = \varphi_{400}\bar{x}_{400} \cdot \mathbf{X} + \varphi_{400}\bar{y}_{400} \cdot \mathbf{Y} + \varphi_{400}\bar{z}_{400} \cdot \mathbf{Z}$$

$$+ \varphi_{410} \cdot \mathbf{F}_{410} = \varphi_{410}\bar{x}_{410} \cdot \mathbf{X} + \varphi_{410}\bar{y}_{410} \cdot \mathbf{Y} + \varphi_{410}\bar{z}_{410} \cdot \mathbf{Z}$$

$$\vdots$$

$$+ \varphi_{700} \cdot \mathbf{F}_{700} = \varphi_{700}\bar{x}_{700} \cdot \mathbf{X} + \varphi_{700}\bar{y}_{700} \cdot \mathbf{Y} + \varphi_{700}\bar{z}_{700} \cdot \mathbf{Z}$$

$$\mathbf{F} = \sum_{\lambda=400}^{700} \varphi_\lambda \cdot \mathbf{F}_\lambda = \sum_{\lambda=400}^{700} \varphi_\lambda \bar{x}_\lambda \cdot \mathbf{X} + \sum_{\lambda=400}^{700} \varphi_\lambda \bar{y}_\lambda \cdot \mathbf{Y} + \sum_{\lambda=400}^{700} \varphi_\lambda \bar{z}_\lambda \cdot \mathbf{Z} \quad \text{(sum)}$$

$$= X \cdot \mathbf{X} + Y \cdot \mathbf{Y} + Z \cdot \mathbf{Z}$$

Taking the sum of the components gives the initial color stimulus; comparison of coefficients then yields the tristimulus values of the color stimulus:

$$X = \sum_{\lambda=400}^{700} \varphi_\lambda \bar{x}_\lambda; \quad Y = \sum_{\lambda=400}^{700} \varphi_\lambda \bar{y}_\lambda; \quad Z = \sum_{\lambda=400}^{700} \varphi_\lambda \bar{z}_\lambda \quad (2.2)$$

If the intervals are sufficiently small, the sums go over to integrals, so that

$$X = \int_{400}^{700} \varphi(\lambda) \bar{x}(\lambda) \, d\lambda$$

$$Y = \int_{400}^{700} \varphi(\lambda) \bar{y}(\lambda) \, d\lambda$$

$$Z = \int_{400}^{700} \varphi(\lambda) \bar{z}(\lambda) \, d\lambda \qquad (2.3)$$

These formulae give the sought relation between **F** and $\varphi(\lambda)$. For object colors, $\varphi(\lambda)$ is the spectral distribution of the reflectance $\varrho(\lambda)$ or of the reflection factor $R(\lambda)$ as measured under illumination with white* ("equal-energy") light (see Fig. 2-3 for examples). If light with a different spectral distribution $S(\lambda)$ is used for illumination, the reflected light will have a distribution $\varrho(\lambda) \cdot S(\lambda)$, and $\varphi(\lambda)$ in the integrals above must be replaced by this product. The "color matching functions" $\bar{x}(\lambda)$, $\bar{y}(\lambda)$, $\bar{z}(\lambda)$ can be determined experimentally by the following procedure: On a trichromatic matching colorimeter, observers reproduce the individual spectral colors, each with an intensity of $m = 1$, from the primaries

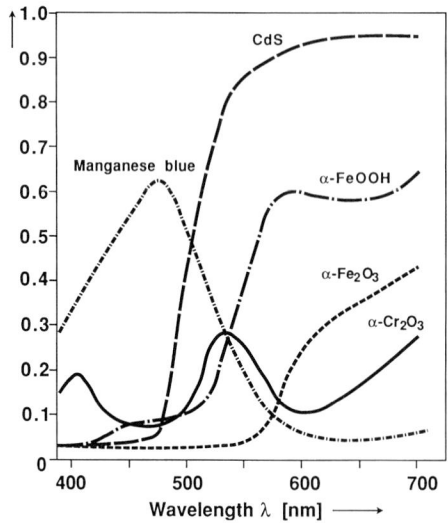

Fig. 2-3. Reflection spectra of some pigments in lacquer binders.

---

* Or of the transmittance $\tau(\lambda)$ or the transmission factor $T(\lambda)$, in the case of light passing through a medium.

X, Y, Z by mixing. The relative mixing proportions are then determined for every λ.

It should not be forgotten that this Young-Helmholtz theory of trichromatic color matching has been brilliantly confirmed by physiological research. Three and only three different visual receptors with photosensitive pigments are found in the human eye, their chemical constitutions have been determined (rhodopsins), and their spectral sensitivities have been measured. For color vision there are three types of receptors (cones), each of which contains one and only one such visual pigment.

Thus a color stimulus can be defined by three tristimulus values, which are of the nature of integrals. Two color stimuli are then alike if the respective values of their integrals are equal. But the integrals change if either of the factors $\varrho(\lambda)$ or $S(\lambda)$ is replaced by a different function. In general, this occurs if the same sample is illuminated with a different type of light: The two colored surfaces are then no longer alike but have a different appearance. If two colors appear alike only under certain conditions (e. g., under a certain type of illumination), they are, logically, said to be "looking alike" or, in Ostwald's term, "metameric"; the phenomenon is called "metamerism."

2.2.3 CIE System

The colorimetric system currently in use all over the world is suggested and introduced by the CIE (Commission *I*nternationale de l'*E*clairage, International Commission on Illumination). The system includes fixed definitions of the reference stimuli **X, Y, Z** and the color matching functions $\bar{x}(\lambda)$, $\bar{y}(\lambda)$, $\bar{z}(\lambda)$. In order that all actual colors can be described, color stimuli lying outside of the spectral locus were introduced ("virtual stimuli"); these three primary stimuli are obtained from three spectral colors by an affine transformation. The CIE principles are reflected in ISO Standard 7724*. The quantities X, Y, Z are called "CIE tristimulus values"; x, y, z, "CIE chromaticity coordinates"; and $\bar{x}(\lambda)$, $\bar{y}(\lambda)$, $\bar{z}(\lambda)$, "CIE color matching functions." There are two sets of color matching functions, for 2° and 10° fields of view, which have been found experimentally to differ somewhat (Fig. 2-4). The CIE tristimulus values** are calculated with equations (2.3):

$$X = \int \bar{x}(\lambda)\varrho(\lambda)\,d\lambda$$

$$Y = \int \bar{y}(\lambda)\varrho(\lambda)\,d\lambda$$

$$Z = \int \bar{z}(\lambda)\varrho(\lambda)\,d\lambda$$

---

* For further standards see "Colorimetry" in Appendix 1.
** Because no limits of integration other than 400 and 700 nm are used, these are omitted from here on.

2 How Colors Depend on Spectra (Colorimetry)

CIE color matching functions

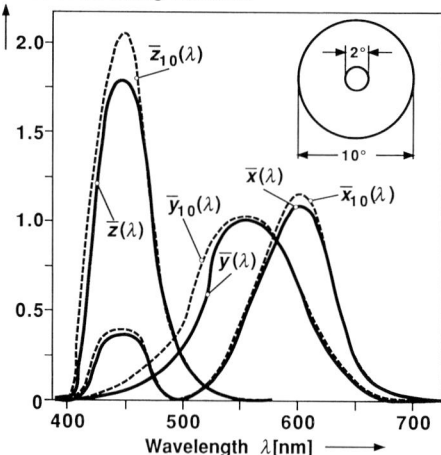

Fig. 2-4. CIE color matching functions for 2° and 10° fields of view. (The circles at top right, viewed from a distance of 10 cm, show the size ratio of the two fields of view.)

If light other than pure white light is used for illumination, $\varrho(\lambda)$ must be replaced by the product $\varrho(\lambda) \cdot S(\lambda)$, as shown previously. A number of "CIE standard illuminants" $S(\lambda)$ have been defined (see Section 2.1.2 and standards*). Table 2-2 presents some values.

From the tristimulus values, the CIE chromaticity coordinates are calculated as

$$x = \frac{X}{X+Y+Z}; \quad y = \frac{Y}{X+Y+Z} \qquad (2.4)$$

These are treated as coordinates in the color plane and define the CIE chromaticity diagram.

Table 2-2. CIE tristimulus values and CIE chromaticity coordinates of some illuminants. ($Y_n = 100.00$ for all standard illuminants and observers.)

CIE illuminant	2°-standard observer				10°-standard observer			
	X_n	Z_n	x_n	y_n	X_n	Z_n	x_n	y_n
A	109.85	35.58	0.4476	0.4074	111.14	35.20	0.4512	0.4059
C	98.07	118.22	0.3001	0.3162	97.28	116.14	0.3104	0.3191
D 65	95.05	108.90	0.3127	0.3290	94.81	107.34	0.3138	0.3310

* For standards see "Colorimetry" in Appendix 1.

2.2 CIE Standard Colorimetric System

The system exhibits the following properties (see Fig. 2-5):
- In the x, y chromaticity diagram, the additive mixture of two color stimuli lies on the straight line connecting the components, at a position determined by the center of gravity principle.
- The shoe-sole-shaped locus of spectrum colors ("spectral locus") in the x, y plane is cut off below by the purple line. Colors lying on this line do not occur in the spectrum but are obtained as mixtures of red and violet.
- In the case of equal-energy illumination*, the "equi-energy stimulus" or white locus lies at $x_w = y_w = z_w = 1/3$.

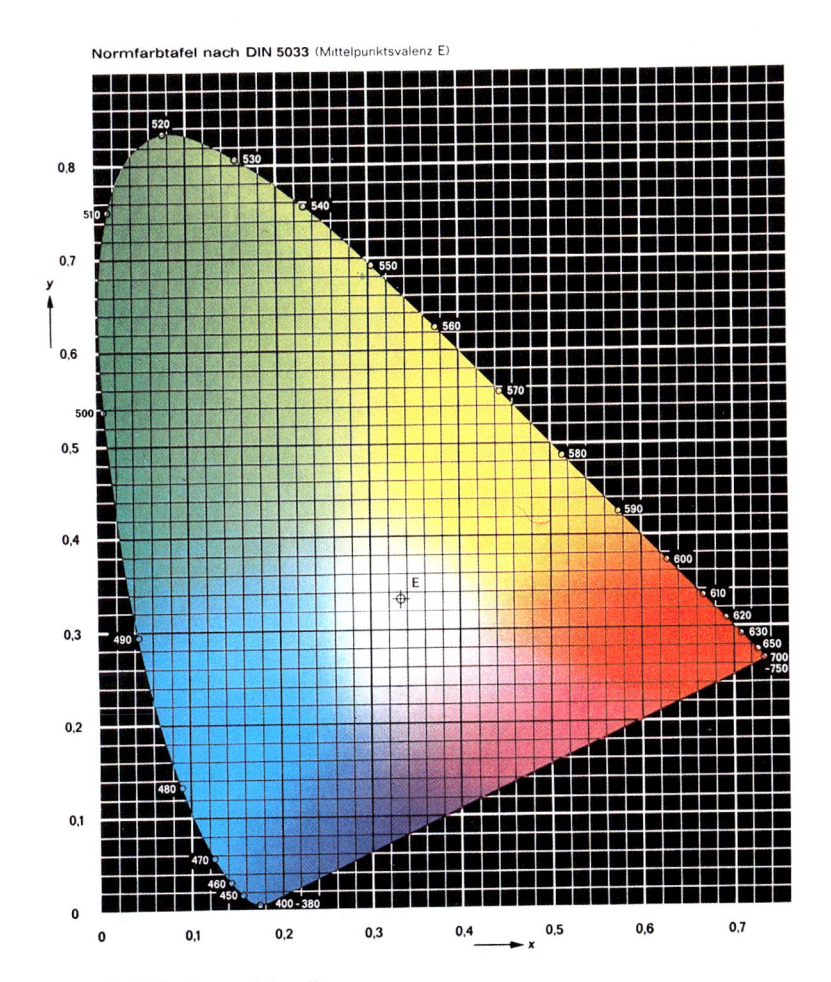

Fig. 2-5. CIE chromaticity diagram.

---

* Otherwise, the equi-energy stimulus is the color locus of the illuminant (standard illuminant); see Table 2-1.

26 2 How Colors Depend on Spectra (Colorimetry)

- The CIE color matching function $\bar{y}(\lambda)$ coincides with the spectral luminosity curve $V(\lambda)$ of the human eye. In place of m, the luminance $Y = m \cdot y$ is cited as a third color coordinate (along with x and y).

Another interesting fact of physiology should also be mentioned here. From the discussion it follows that there may be a very great number of spectral distributions $\varrho(\lambda)$ that belong to one and the same color stimulus **F** (metameric colors). Because the human eye can distinguish of the order of 100 spectral colors and a roughly similar number of brightness levels, it should be possible to prepare the respectable number of 100^{100} distinguishable spectral distributions by mixing. In practice, however, by virtue of the trichromatic principle, the eye can only distinguish 100^3 colors. The eye and the organs responding to it in the perceptual system thus perform a tremendous reduction in the quantity of information offered. In cybernetic terms, the huge amount of information contained in a single color stimulus (600–700 bits), which would be nearly impossible to process in a reasonably short time, is reduced in a consistent way to some 20 bits, a quantity of information that the consciousness is well able to process in one second, as we know from a comparison with other sensory organs.

2.3 Sensation-Based Systems

2.3.1 Lightness, Hue, Saturation

The CIE system offers a worldwide understanding of colors. Using the three numerical values x, y, and Y (and some few additional pieces of information*), the recipient of a communication can localize the color locus under consideration without having seen the color. Since this system came into existence, it has proved suitable for many different technical applications of color, including color television. In order to assess colored systems and pigments, however, there are additional demands. The two most important shortcomings are that it is hard to visualize information in the CIE system and that it is not suitable for direct determination of physiologically equidistant color differences.

  A few explanatory words follow:
- Difficulty of visualization: The values of x, y, and Y make it possible to put colors in a systematic order. It is hard, however, to visualize the nature of a color with only the coordinate values. As a way of coping with this problem, colors are classified and ordered according to another three properties so that one may come to an understanding with other parties. These are *brightness* (one can rank colors as "brighter" and "darker"), *hue* (corresponding, in

---

* For example, information on the geometry of the colorimeter, the standard illuminant used, and exclusion of surface reflection (all of which can be established once and for all).

a way yet to be defined, to the spectral colors), and *saturation* (every color is regarded as a mixture of the "saturated" hue and white; the color is more saturated the smaller the content of white).

– Physiologically equidistant color differences: The introduction of the concepts of brightness, hue, and saturation has made it easier to visualize information in the system. The second difficulty, that of determining color differences, still remains. A glance at Figure 2-5 shows that the CIE chromaticity diagram cannot represent physiologically equidistant differences: The greens (from blue-green to yellow-green) occupy more than half the area enclosed by the spectral locus, while all other colors must make do with what remains. The determination of color differences is so important in practice that it must be deemed one of the principal tasks of colorimetry.

One can also understand directly that the CIE diagram cannot correctly reproduce all color differences at once. The chromaticity diagram is based on the *equality* of color stimuli, since the mixing experiment (Section 2.2.2), in which the color matching functions are determined from the primary stimuli on the trichromatic matching colorimeter, involves using the eye solely as a kind of "null indicator" for the equality of the reference and matched colors. The chromaticity diagram represents only a suitable (but arbitrary) metrical arrangement of equality judgments. In color difference measurements, however, the *distinction* between two color stimuli is of importance. Distinctions can be determined only through new color matching tests performed in a different way. The next section will go into more detail on this point.

2.3.2 Physiologically Equidistant Systems

To determine color differences objectively, one needs to start with a notion that plays an idealized role in the physiology of colors. This is the concept of "absolute color space," which can be thought of as a three-dimensional arrangement of color stimuli such that the distance between two color stimuli in an arbitrary spatial direction corresponds to the perceived difference. A color space with these properties cannot, however, be derived on a purely theoretical basis; its structure must be determined experimentally through color matching tests on samples having small color differences.

As discussed in the previous section, such a space can be constructed with the color qualities of brightness, hue, and saturation. A configuration like that shown in Figure 2-6 results. The material needed for this purpose comprise a large number of color samples arranged by steps of hue, saturation, and finally brightness, and having physiologically equidistant differences. The derived solid is an onion-shaped body with an irregular boundary (for example, more hues can be distinguished in the dark blue than in the light red). The vertical axis is the gray scale. A horizontal cut yields a color plane like that shown in Figure 2-6; the "rings" of the onion are loci of equal saturation. There are a number of systems built up in this way which vary in the type of color samples employed,

28   2 How Colors Depend on Spectra (Colorimetry)

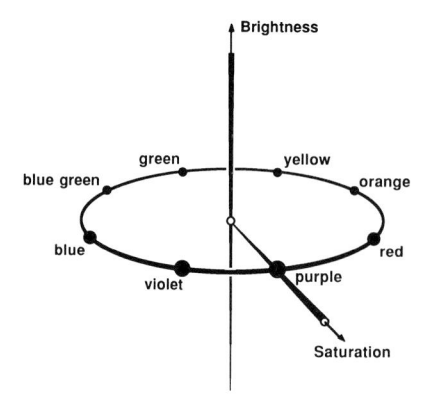

Fig. 2-6. Color order based on qualities: hue, saturation, and brightness.

the illumination, etc. Probably the most widely used system of this type is the Munsell system, which is available in the form of an atlas. Brightness, hue, and saturation are termed value, hue, and chroma respectively in the Munsell system.

When *calculating* color differences, one looks for conversion relations between the CIE system (in which color measurement must be done) and the sensation-based color system. Once these relations are known, the color differences can be calculated. Many color-difference systems have been developed, largely as a result of industrial color examination requirements. The acceptance of the CIELAB system, however, has pushed these into the background, and they need not be discussed further.

2.3.3 CIELAB System

The CIELAB system has been adopted worldwide for the examination of pigmented coatings. This system, derived from the Munsell system, uses the three spatial coordinates a^* (red-green axis), b^* (yellow-blue axis), and L^* (lightness axis) for brightness. In Figure 2-6, the a^* axis is approximately the line connecting red and blue-green, while the b^* axis is approximately the line connecting yellow and violet; the L^* axis is the vertical lightness axis. The name "CIE-LAB" is a contraction from CIE and $L^*a^*b^*$. In order to calculate the CIELAB coordinates, X, Y, and Z are first converted to the "value functions" X^*, Y^*, and Z^*. The functional relation used is similar to the logarithm function and preserves physiologically equidistant (uniform) brightness steps (in a similar way to the Weber-Fechner law):

$$X^* = \sqrt[3]{\frac{X}{X_n}}; \quad Y^* = \sqrt[3]{\frac{Y}{Y_n}}; \quad Z^* = \sqrt[3]{\frac{Z}{Z_n}} \qquad (2.5)$$

(here X_n, Y_n, and Z_n are the CIE tristimulus values for the illuminant used, with $Y_n = 100$ for all standard illuminants). For X/X_n, $Y/100$, and Z/Z_n ≤ 0.008856, however,

$$X^* = 7.787 \frac{X}{X_n} + 0.138$$

$$Y^* = 7.787 \frac{Y}{100} + 0.138$$

$$Z^* = 7.787 \frac{Z}{Z_n} + 0.138 \tag{2.6}$$

With these X^*, Y^*, and Z^* values, the a^*, b^*, and L^* are then obtained as follows:

$$a^* = 500 (X^* - Y^*)$$

$$b^* = 200 (Y^* - Z^*)$$

$$L^* = 116 \, Y^* - 16 \tag{2.7}$$

The color locus in the color plane (a*,b* plane) can also be described in polar coordinates:

$$C_{ab}^* = \sqrt{a^{*2} + b^{*2}} \tag{2.8}$$

(chroma, as radius vector);

$$h_{ab} = \text{Arctan} \frac{b^*}{a^*} \tag{2.9}$$

(hue angle, as polar angle).

The expression for L^* in equation (2.7) calls for some explanation. Why was the simpler form $L^* = 100 \, (Y/100)^{1/3}$, which is also similar to the logarithm function, not chosen? The answer is that the equation was based on color matching tests with gray color samples differing in brightness, which the matching colorists had to arrange on an equidistant gray scale from dark to light. The points in the diagram (Fig. 2-7) show the result for which approximation was made. It turned out that the simpler curve was too steep in the middle

2 How Colors Depend on Spectra (Colorimetry)

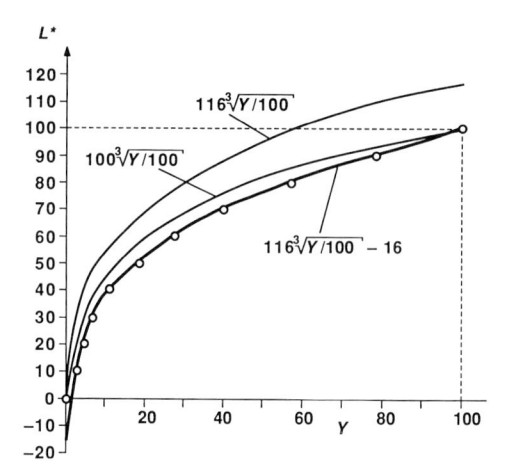

Fig. 2-7. Derivation of the L^* curve (see text for explanation).

range, so it was "stretched" with a factor of 116 (instead of 100), then shifted downward by subtracting 16 so that it passed through the data points, and the condition $L^*(100) = 100$ was restored. After this shifting, however, the curve no longer cut the Y axis at the origin but rather at a $Y > 0$, so that $L^* < 0$ below this value. Negative lightness values are meaningless, so the convention was adopted that this portion of the curve be replaced by the tangent to the curve that also passes through the origin* (Fig. 2-8).

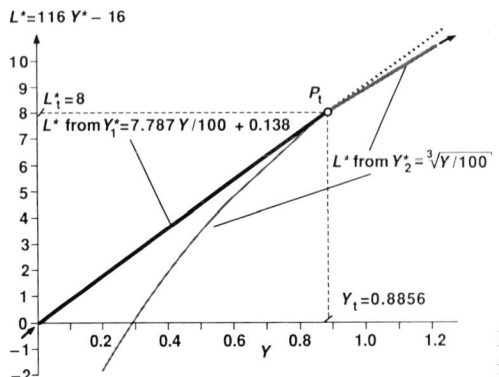

Fig. 2-8. Derivation of the L^* curve (see text for explanation).

---

* Unfortunately, a logarithmic formulation was not immediately chosen when the CIELAB system was established; for example, $L^* = 100\ Y^* = 100 \cdot \log(1 + 13\ Y/100)/\log 14$ or a similar formula.

2.3 Sensation-Based Systems

For the point of tangency (Y_t, L_t^*), we have $(dL^*/dY)_{Y_t} = L_t^*/Y_t$, so that according to equations (2.5) and (2.7)

$$\frac{L_t^*}{Y_t} = \frac{1}{Y_t}\left(116\sqrt[3]{\frac{Y_t}{100}} - 16\right) \text{ and }$$

$$\left.\frac{dL^*}{Y_t}\right|_{Y_t} = \frac{116}{3 \cdot 100 \left(\frac{Y_t}{100}\right)^{2/3}}$$

Setting the right-hand sides equal gives solutions for the coordinates of the point of tangency:

$$\frac{Y_t}{100} = \left(\frac{3 \cdot 16}{2 \cdot 116}\right)^3 = 0.008856,$$

$$L_t^* = 16\left(\frac{3}{2} - 1\right) = 8$$

so that the equation of the tangent line becomes $L^*(Y) = (L_t^*/Y_t) \cdot Y$. Using equation (2.7), Y^* is then given by

$$Y^* = \frac{L^*(Y) + 16}{116} = \frac{100\, L_t^*}{116\, Y_t} \cdot \frac{Y}{100} + \frac{16}{116} = 7.787\frac{Y}{100} + 0.138,$$

which is just equation (2.6) within the range from $Y_0 = 0$ to $Y_t = 0.8856$. Note that $Y^*(0) > 0$; for $X^*(X)$ and $Z^*(Z)$, equality of this equation must nonetheless be preserved so that the additive term 0.138 disappears when differences are taken for a^* and b^* in equation (2.7).

Although the formation of differences for a^* and b^* in equations (2.7) seems strange at first, it has a physiological justification. In the retina, the primary excitation of the cones is processed in six types of nerve cells, four of which function in a "spectrally opponent" antagonistic manner while the other two act in a "spectrally nonopponent" alinear manner. The spectrally nonopponent processors form the black/white system, the spectrally opponent ones the color system.

If one imagines two colors measured on pigmented coatings as radius vectors in the CIELAB color space, namely $\mathbf{V}_T = \{a_T^*, b_T^*, L_T^*\}$ for the test sample (T) and $\mathbf{V}_R = \{a_R^*, b_R^*, L_R^*\}$ for the reference sample (R), then $\Delta E_{ab}^* = |\mathbf{V}_T - \mathbf{V}_R|$ is the color difference in the CIELAB space, measured in "CIELAB units":

32   2 How Colors Depend on Spectra (Colorimetry)

$$\Delta E^*_{ab} = \sqrt{\Delta a^{*2} + \Delta b^{*2} + \Delta L^{*2}} \qquad (2.10)$$

with the components

$$\Delta a^* = a^*_T - a^*_R$$
$$\Delta b^* = b^*_T - b^*_R$$
$$\Delta L^* = L^*_T - L^*_R \qquad (2.11)$$

In the following presentation, all differences notated as Δ will be defined in this sense, namely as test sample minus reference sample.

A significant advantage of the CIELAB system is that the color difference can be broken down into components in another way: into brightness, "chroma" (corresponding to saturation), and "hue" (corresponding to the hue component used earlier), in analogy to the way the color space was constructed (Fig. 2-9):

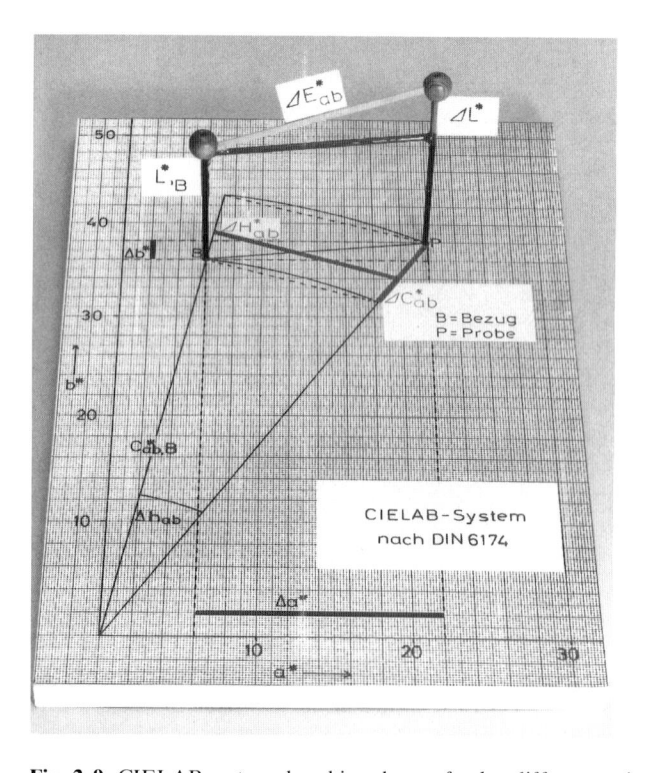

Fig. 2-9. CIELAB system: breaking down of color differences into components.

2.3 Sensation-Based Systems 33

$$\Delta L^* = L_T^* - L_R^* \text{ (lightness difference as above)}$$

$$\Delta C_{ab}^* = C_{ab,T}^* - C_{ab,R}^* = (a_T^{*2} + b_T^{*2})^{1/2} - (a_R^{*2} + b_R^{*2})^{1/2} \quad (2.12)$$

(chroma difference)

$$\Delta H_{ab}^* = (\Delta E_{ab}^{*2} - \Delta L^{*2} - \Delta C_{ab}^{*2})^{1/2} \quad (2.13)$$

(hue difference)

ΔH_{ab}^* is positive if $a_R^* b_T^* - a_T^* b_R^* > 0$ and negative if not. It represents the L^* component of the vector product $\mathbf{V}_R \times \mathbf{V}_T$ and characterizes the sense of rotation in the associated vector plane. If the coincidence of the vector $\mathbf{V}_R$ of the reference sample with the vector $\mathbf{V}_T$ of the test sample involves a counterclockwise rotation of $\mathbf{V}_R$, then ΔH_{ab}^* is positive, and vice versa. The hue angle component is finally given by

$$\Delta h_{ab} = h_{ab,T} - h_{ab,R} = \text{Arctan} (b_T^*/a_T^*) - \text{Arctan} (b_R^*/a_R^*). \quad (2.14)$$

The sign convention for ΔH_{ab}^* ensures that ΔH_{ab}^* and Δh_{ab} will have the same sign. When these are compared with the other differences, it must not be neglected that Δh_{ab} is measured in degrees rather than CIELAB units, and thus has a different metric from these other differences.

A note on the geometric meaning of ΔH_{ab}^*: From Figure 2-10 and equation

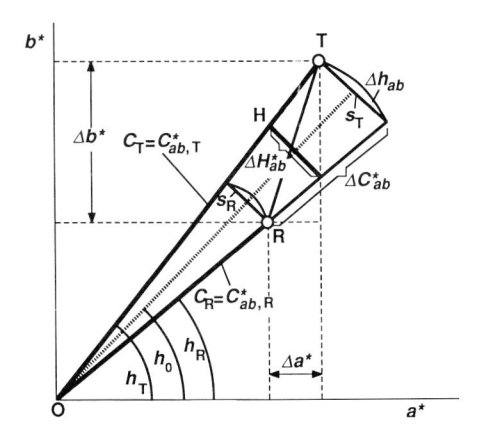

Fig. 2-10. Geometric meaning of ΔH_{ab}^* (see text for explanation).

(2.10), it follows that the line segment $\overline{RT}$ is the projection of ΔE_{ab}^* on the a^*, b^* color plane and has a length given by

$$\overline{RT}^2 = \Delta a^{*2} + \Delta b^{*2} = \Delta E_{ab}^{*2} - \Delta L^{*2}$$

while, on the other hand, it can be calculated from the chromas of the test sample and reference sample and the angle between them ($C_R = C^*_{ab,R}$, $C_T = C^*_{ab,T}$, $\Delta C = \Delta C^*_{ab}$, $\Delta h = \Delta h_{ab}$):

$$\overline{RT}^2 = C_R^2 + C_T^2 - 2C_R C_T \cos \Delta h$$

From equations (2.13) and (2.12), it follows that ΔH^{*2}_{ab} is then:

$$\Delta H^{*2}_{ab} = \Delta E^{*2}_{ab} - \Delta L^{*2} - \Delta C^2 = \overline{RT}^2 - (C_T - C_R)^2 = 2C_R C_T (1 - \cos \Delta h)$$

In terms of the half-angle,

$$\Delta H^{*2}_{ab} = 2C_R C_T [1 - \cos^2(\Delta h/2) + \sin^2(\Delta h/2)] = 4C_R C_T \sin^2(\Delta h/2)$$

The right-hand side is simply the product of the two chords (Fig. 2-10) associated with C_R and C_T respectively, $s_R = 2C_R \sin(\Delta h/2)$ and $s_T = 2C_T \sin(\Delta h/2)$, so that the hue difference

$$\Delta H^*_{ab} = \sqrt{s_R s_T} \qquad (2.15)$$

is the geometric mean of the two chords associated with the hues of the reference sample and the test sample.

For nearly achromatic, (i.e., nearly white, black, or gray) samples, it is usually advantageous to break ΔE^*_{ab} down in a different way, because the usual breakdown as in equations (2.11) to (2.14) can lead to poor agreement with the observed color qualities. For example, suppose the color locus of the reference sample lies at or close to the zero point of the CIELAB system ($a^*_R \approx b^*_R \approx 0$) while the color locus of the test sample lies at an arbitrary distance from the zero point ($a^*_T \neq 0$, $b^*_T \neq 0$). Since $\Delta a^* = a^*_T$ and $\Delta b^* = b^*_T$, the hue difference is then given by equation (2.13) as

$$\Delta H^{*2}_{ab} = \Delta E^{*2}_{ab} - \Delta L^{*2} - \Delta C^{*2}_{ab} = \Delta a^{*2} + \Delta b^{*2} - (a^{*2}_T + b^{*2}_T) = 0.$$

This, however, does not agree with the shade actually observed in the direction of the test sample.

For this reason it is better to calculate this direction and the reference sample-to-test sample distance in the a^*, b^* color plane, as corresponding to segment $\overline{RT}$ in Figure 2-10. Its length Δs can be found for example, using equation (2.10) with $\Delta L^* = 0$; the direction angle of the shade $\Delta \varphi$ is obtained as the tangent of the fraction $\Delta b^* / \Delta a^*$:

$$\Delta s = \sqrt{\Delta a^{*2} + \Delta b^{*2}} \;; \qquad \Delta \varphi = \mathrm{Arctan} \frac{\Delta b^*}{\Delta a^*} \qquad (2.16)$$

2.3 Sensation-Based Systems

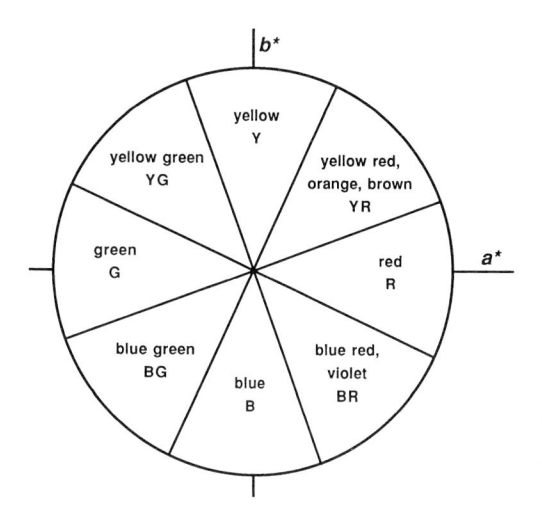

Fig. 2-11. Subdivision of the CIELAB color plane for calculation of the shade of achromatic samples.

Table 2-3. Color names.

Sign of Δb^*	Sign of Δa^*		$|\Delta b^*/\Delta a^*|$
	−	+	
	yellow (Y)		>2.5
+	yellow-green (YG)	yellow-red (orange) (YR)	0.4 to 2.5
	green (G)	red (R)	< 0.4
−	blue-green (BG)	blue-red (violet) (BR)	0.4 to 2.5
	blue (B)		> 2.5

To simplify the evaluation, the color plane is divided into octants beginning at $h = 22.5°$, and a color name is assigned to each of the resulting sectors (see Fig. 2-11). Table 2-3 then offers a very simple method of evaluation (with $\tan 22.5° \approx 0.4$ and $\tan 67.5° \approx 2.5$). For the example worked out above ($a_R^* \approx 0$, $b_R^* \approx 0$), the formula

$$\Delta s = \sqrt{\Delta a^{*2} + \Delta b^{*2}} = \sqrt{a_T^{*2} + b_T^{*2}} \neq 0$$

correctly gives the strength while the nature of the shade follows from $\Delta b^*/\Delta a^* = b_T^*/a_T^*$.

36   *2 How Colors Depend on Spectra (Colorimetry)*

The definitions and equations of the CIELAB system* are set forth in standards**. Accompanying the system when it was introduced was a recommendation for its preferential use and testing. This proved to be justified, for the system has now gained widespread acceptance. A model of the CIELAB color space appears in Figure 2-12. The crossed cords indicate where the a^* and b^* axes cross; the cord running from the left foreground to the right background lies in the plane of the a^* axis (left foreground is red, $a^* > 0$), while the cord running from right to left lies in the plane of the b^* axis (right is yellow, $b^* > 0$). The circles marked on the plates are spaced 10 CIELAB units apart. As in the Munsell system, it is significant that there are a large number of distinguishable dark violet and light yellow hues. The formalism of the CIELAB system can easily be programmed into a desktop computer, which can also be linked on-line with a colorimeter.

Despite the initial success of the CIELAB system, it should be noted that there is still room for improvement. For instance, the differences obtained by breaking down color differences are not always correctly reproduced when the

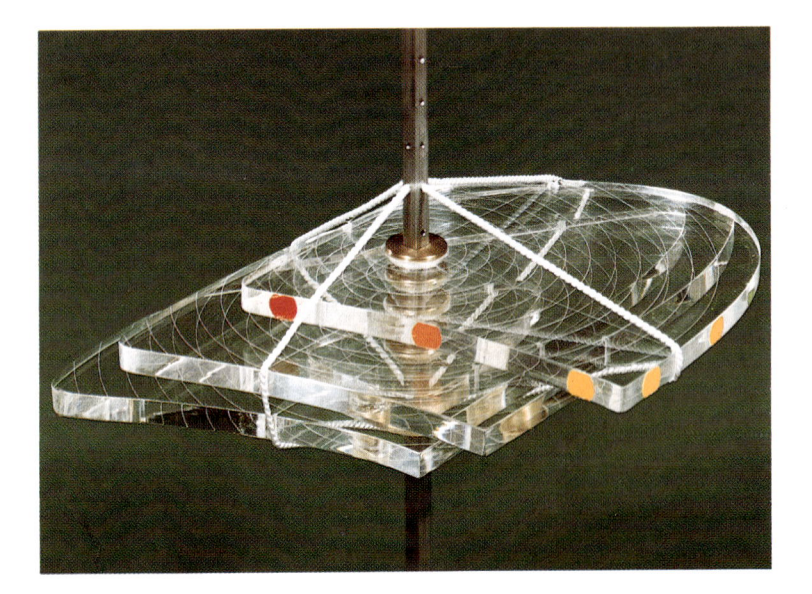

Fig. 2-12. Model of the CIELAB color space.

---

* On the notation: the * was selected to avoid confusion with the similarly named quantities in the earlier Adams-Nickerson system. The subscript "*ab*" is necessary because there is also a CIE color system with coordinates L^*, u^*, and v^* (the "CIELUV" system). Because CIELUV is not used for pigmented coatings, it will not be discussed here.

** For standards see "Color Differences" in Appendix 1.

system is applied to pigmented coatings. The following discrepancies are regularly established:
- Perceived brightness differences are undervalued by the CIELAB lightness difference ΔL^*.
- Perceived saturation differences are overvalued by the CIELAB chroma difference ΔC_{ab}^*.

Details on these points will be presented further on, in the practical section. The only remedies at this point comprise changes in the system which would eliminate these problems while preserving the differential-geometric properties.

For these reasons, the CIE has recently proposed a new color difference formula based on CIELAB. The new "CIELAB 92" formula is:

$$\Delta E_{CH}^* = \sqrt{\left(\frac{\Delta L^*}{k_l S_l}\right)^2 + \left(\frac{\Delta C_{ab}^*}{k_c S_c}\right)^2 + \left(\frac{\Delta H_{ab}^*}{k_h S_h}\right)^2}$$

where $S_l = 1$, $S_c = 1 + 0.047 C_{ab,R}^*$, $S_h = 1 + 0.014 C_{ab,R}^*$, and $C_{ab,R}^*$ is the chroma of the reference. Here the k_i are the parameters of externally imposed conditions, such as conditions specific to a certain application or conditions differing from the standard conditions. The standard conditions are as follows: Illuminant D65, ca. 1000 lx, surroundings gray $L^* = 50$, ca. 100 cd/m², sample size over 4°, sample pair without gap between, homogeneous texture, $0 < \Delta E_{CH}^* < 5$. Under these conditions, $k_l = k_c = k_h = 1$. CIELAB 92 is to be tested as quickly as possible in various industries in order to promote further discussion of the formula and suitable parameters k_i.

The formula is based on the CMC (l:c) formula but the correction terms are restricted to the chroma effect only. This is a drawback, since it must result in uncorrected color differences in the unsaturated and achromatic region. Another disadvantage is that color differences are no longer Euclidean, and spanning of a color space is therefore not possible.

2.3.4 Color Order Systems

Color order systems arose from the practical requirement that manufacturer and purchaser should have a common vocabulary for the manufacture of colored objects. The origins of such systems go back a long way, and from the very beginning there has been an effort to verify all orderings visually on the basis of a suitable collection of color samples. This means, however, that the difficulty inherent in any color order system was programmed into it from the start, i.e., color samples are subject to the restrictions of whether the coloring materials (pigments or dyes) are available. The controversy therefore exists already: of whether available coloring materials (or their mixtures) should be systematized, or whether a new system should be developed for color appearance. This dilemma has not yet been resolved by any color order system; it is apparent, and a problem, more or less in all systems to be discussed.

What distinguishes such color order systems from the physiologically equidistant systems discussed earlier? This justifiable question has no easy answer, for if both systems were perfect from the physiological standpoint then they would have to be identical. In other words, the "absolute color space" (to the extent that it is Euclidean for color differences as well) would offer the ideal metric for both requirements. This fact provides the best starting point for describing the distinction. A physiologically equidistant system imposes more stringent requirements: In general, it must satisfy the postulate that the geometry of the color space is Euclidean, since otherwise it would be difficult to give an unambiguous definition of color differences (i.e., distances between color loci). Color order systems do not require this postulate, since they generally involve merely some ordering (often a naive one) of color samples. Although the samples possess some systematic features, these are not always immediately obvious. Most of these systems employ the qualities of brightness, hue, and saturation for ordering.

Color order systems are still important for the classification of pigments and dyes, but play no role at all in the colorimetric inspection of colorants. This book could thus meet its objectives without taking these systems into account; the best-known representatives will, however, be discussed briefly, because a description of them may lead to a deeper understanding of the problems encountered in colorimetry.

We have chosen the following classification in the knowledge that it is not free of arbitrariness:
– Empirical color order systems
– Colorimetry-based color order systems

1) Empirical Systems

Of all the empirical systems, the *Munsell system* is the most important; we introduced this system in Section 2.3.2. The system approximately satisfies the postulates of a physiologically equidistant system; otherwise it would not have been such a simple matter to make it the basis for the CIELAB system. New editions of the color samples and color atlases for the Munsell system are available.

The *Natural Color System (NCS)* was developed in Sweden and is in use there. It takes the form of a color atlas. The three variables are blackness s, chroma c, and hue ϕ. Certain limitations became apparent after the introduction of the system: it does not span the full color space occupied by coloring materials, and it does not always give reproducible color identification codes for glossy samples. Like any empirical system, NCS is not tied to colorimetry, and so it cannot be directly converted to other color systems. A set of tristimulus values is, however, defined for one set of measuring conditions.

"*RAL-Farbregister 840 HR*" *(RAL Color Register)* is not a color order system in the narrow sense. It imposes a numbered order, with color names, on industrially manufactured coloring materials required by public authorities (railways, postal service, etc.) and by makers of consumer goods. The ordering is not highly systematic and features only a rough orientation to color relationships (e. g., 3 = red). Colors can be obtained as samples (paint chips); one example is RAL 3013 "Tomato Red". The RAL Register has recently been linked to the DIN Color System through colorimetry (see below).

The *Color Index* is also not strictly a color order system, since it lacks consistency. It includes names based on origin ("generic names") and numbers based on chemical makeup ("constitution numbers") for individual pigments. Adequate differentiation is possible only for organic pigments.

2) Colorimetry-Based Systems

Among schemes based on colorimetry, *the DIN Color System* (DIN 6164) should be mentioned first because it features the most complete systematization and offers many advantages. It employs the measures blackness value D, hue T, and saturation S; these are linked to the CIE system. The available color samples include only matte colors. The division of the color circle into 24 parts makes the system somewhat unclear. Because it is not widely used except in Germany, this system is not likely to become an international standard.

The Optical Society of America Uniform Color Scales System (OSA-UCS) should also be mentioned. It offers a set of formulae (link to CIE) and a collection of color samples. This system is virtually restricted to the U.S.A.

The *Ostwald system* is really only of historical interest, especially as no sample collections are available.

All the color order systems mentioned up to now share the problem that they do not permit direct determination of color differences between the test sample and the reference sample or between two specimens. To make such a determination, it is necessary to use a color difference system, most commonly the CIELAB system. This in turn is possible only for the colorimetry-based order systems. The best way to avoid having to work with two fundamentally different coordinate systems (the selected color order system and the CIELAB system) would be to employ CIELAB (which spans a color space) as an order system. This is the approach taken in the new *RAL Design System (RAL-DS)* (a successor to the Eurocolor system, which is no longer available). The notation is directly derived from the CIELAB coordinates: hue $H = h_{ab}$, lightness $L = L^*$, and chroma $C = C^*_{ab}$. Thus 090.80.70 would denote a yellow color ($h_{ab} = 90°$) with lightness 80 and chroma 70. The RAL-DS color atlas serves as the collection of color samples.

Given the stated reservations about the CIELAB system, this color order system unites all the advantages that such a system can at present. First, it offers a simple connection to colorimetry; second, it permits direct color difference determination in the same system; and since CIELAB has become accepted for color difference measurement anyway, RAL-DS should lead to an international consensus. Along with national reservations, it appears that the main obstacle to agreement on a binding color order system is conservativism.

2.4 Mathematical Statistics of Color Coordinates

2.4.1 Normal Distribution in Three Dimensions

The values obtained from a color measurement are generally the coordinates of a color locus in color space; that is, they are components of three-dimensional vectors. This is why it is more complicated to determine, for example, a measurement error in three dimensions than in one. The components of three-dimensional vectors can no longer be treated as independent variables; it must be taken into account that correlations exist between the components, and statistical analysis must allow for these correlations. For example, a^* and b^* are formed from the same tristimulus value Y; furthermore, the color matching functions $\bar{x}(\lambda)$, $\bar{y}(\lambda)$, and $\bar{z}(\lambda)$, which lead to the tristimulus values X, Y, and Z used in calculating a^*, b^*, and L^*, have significant overlaps (see Fig. 2-4). Mathematical treatment must therefore be oriented to extending the normal distribution from one to three dimensions.

For one-dimensional variables, the probability that a measured value will lie between x and $x + dx$ is

$$w\, dx = A e^{-\frac{1}{2}\delta s^2}\, dx \tag{2.17}$$

where

$$A = \frac{1}{\sigma_x \sqrt{2\pi}} \tag{2.18}$$

$$\boxed{\delta s = \pm \frac{x - \bar{x}}{\sigma_x}} \tag{2.19}$$

$$\bar{x} = \frac{1}{n} \sum_{i=1}^{n} x_i \quad \text{(mean)} \tag{2.20}$$

$$\sigma_x = \sqrt{V} \quad \text{(standard deviation)}$$

and

$$V = \frac{1}{n-1} \sum_{i=1}^{n} (x_i - \bar{x})^2 \quad \text{(variance)} \tag{2.21}$$

If $\delta s^2 = 1$, then from equation (2.19) the limiting value x_e is

$$x_e = \bar{x} \pm \sigma_x \tag{2.22}$$

2.4 Mathematical Statistics of Color Coordinates

and, from equation (2.17), the frequency is

$$p_D = \frac{w}{w_{max}} = e^{-\frac{1}{2}} \approx 0.6 \tag{2.23}$$

$$(w_{max} = A)$$

This means that the probability of encountering a measured value at $x_e = \bar{x} \pm \sigma_x$ is 60 % as great as that of encountering the mean value.

In the case of two-dimensional vectors $\mathbf{r} = \{x, y\}$, the probability that the measured value lies between $\mathbf{r}$ and $\mathbf{r} + \mathbf{dr}$ is given by

$$w \, dx \, dy = B e^{-\frac{1}{2} \delta s^2} dx \, dy \tag{2.24}$$

where

$$B = \frac{\sqrt{\det \mathbb{G}}}{2\pi} \tag{2.25}$$

$$\mathbb{G} = \begin{pmatrix} g_{11} & g_{12} \\ g_{12} & g_{22} \end{pmatrix} \quad \text{and}$$

$$\det \mathbb{G} = g_{11} g_{22} - g_{12}^2 \quad \text{(determinant)} \tag{2.26}$$

$$\boxed{\delta s^2 = g_{11} \, \delta x^2 + 2 g_{12} \, \delta x \, \delta y + g_{22} \, \delta y^2} \tag{2.27}$$

$$\delta \mathbf{r} = \{\delta x, \delta y\} = \mathbf{r} - \bar{\mathbf{r}} = \{x - \bar{x}, x - \bar{y}\} \tag{2.28}$$

$$\bar{x} = \frac{1}{n} \Sigma x; \quad \bar{y} = \frac{1}{n} \Sigma y \quad \text{(means)} \tag{2.29}$$

The expression δs^2 is thus replaced by the quadratic expression in equation (2.27). If δs^2 is set equal to a constant, the function $\delta y(\delta x)$ describes an ellipse in the $\delta x, \delta y$ plane. For $\delta s^2 = 1$, this ellipse is called a "standard deviation ellipse". It represents the two-dimensional analog of the standard deviation σ_x which is the measure of deviations in the one-dimensional case. $\mathbb{G}$ is the matrix formed from the coefficients of the ellipse, and $\det \mathbb{G}$ is its determinant.

The step to the third dimension now presents no difficulty. Because the quantities whose deviation is of interest here are always color differences, the remainder of the development can be stated in terms of the CIELAB color space. The formula for the probability that a triple of measured values lies between $\mathbf{V} = \{a^*, b^*, L^*\}$ and $\mathbf{V} + d\mathbf{V} = \{a^* + da^*, b^* + db^*, L^* + dL^*\}$ is now

2 How Colors Depend on Spectra (Colorimetry)

$$w \, da^* \, db^* \, dL^* = C e^{-\frac{1}{2} \delta s^2} \, da^* \, db^* \, dL^* \tag{2.30}$$

where

$$C = \frac{\sqrt{\det \mathbb{G}}}{(2\pi)^{3/2}} \tag{2.31}$$

$$\mathbb{G} = \begin{pmatrix} g_{11} & g_{12} & g_{13} \\ g_{12} & g_{22} & g_{23} \\ g_{13} & g_{23} & g_{33} \end{pmatrix} \tag{2.32}$$

(symmetric matrix of ellipsoid coefficients)

$$\det \mathbb{G} = g_{11} g_{22} g_{33} + 2 g_{12} g_{23} g_{13} - g_{11} g_{23}^2 - g_{22} g_{13}^2 - g_{33} g_{12}^2 \tag{2.33}$$

(determinant of $\mathbb{G}$)

$$\delta s^2 = g_{11} \, \delta a^{*2} + g_{22} \, \delta b^{*2} + g_{33} \, \delta L^{*2} + 2 g_{12} \, \delta a^* \, \delta b^* + 2 g_{23} \, \delta b^* \, \delta L^* + 2 g_{13} \, \delta a^* \, \delta L^* \tag{2.34}$$

$$\delta \mathbf{V} = \{\delta a^*, \delta b^*, \delta L^*\} = \mathbf{V} - \bar{\mathbf{V}} = \{a^* - \bar{a}^*, b^* - \bar{b}^*, L^* - \bar{L}^*\} \tag{2.35}$$

$$\bar{a}^* = \frac{1}{n} \Sigma a^*; \quad \bar{b}^* = \frac{1}{n} \Sigma b^*; \quad \bar{L}^* = \frac{1}{n} \Sigma * \quad \text{(means)} \tag{2.36}$$

The exponent in the normal distribution is dominated by a quadratic form, which for $\delta s^2 = $ const describes an ellipsoid in the color space. The normal form $\delta s^2 = 1$ now represents the "standard deviation ellipsoid". The probability of encountering a measured value on the ellipsoid is, once again, ca. 60% of the probability of the mean, since equation (2.23) holds here as well.

The following statistic applies to frequency information based on the *probability density*:

$$p_{\text{den}} = \exp[-(1/2) \delta s_{\text{den}}^2] \quad \text{and} \quad \delta s_{\text{den}}^2 = 2 \ln(1/p_{\text{den}}) \tag{2.37}$$

Here, as we already know from equation (2.34), δs_{den}^2 signifies a certain ellipsoid. The collection of concentric ellipsoids with parameter p_{den}, for example with p_{den} in 10% steps as listed in Table 2-4, forms an image of the three-dimensional normal distribution $w(a^*, b^*, L^*)$.

To be distinguished from this is statistical information based on the *cumulative probability*. The cumulative probability is found by summing the probability densities between certain limits of integration, for example from zero (i.e., the mean itself) up to an ellipsoid δs_{cum}^2 specified as the limit of integration. The values answer the following question: How many measured values are located

2.4 Mathematical Statistics of Color Coordinates

Table 2-4. δs^2 values for the three-dimensional normal distribution versus frequency p, for the probability density (subscript den) and the cumulative probability (subscript cum).

p	δs^2_{den}	δs^2_{cum}
0 %	∞	0
10 %	4.61	0.584
20 %	3.219	1.005
30 %	2.408	1.424
40 %	1.833	1.869
50 %	1.386	2.366
60 %	1.022	2.946
70 %	0.713	3.665
75 %	0.575	4.11
80 %	0.446	4.64
90 %	0.2016	6.25
95 %	0.1026	7.81
99 %	0.02010	11.35
99.5 %	0.01025	12.84
100 %	0	∞

inside the ellipsoid δs^2_{cum}? If one refers to the set of all measured values, the cumulative frequency p_{cum} is obtained; this again yields a family of concentric ellipsoids, but now these image the cumulative distribution. This family of bounding ellipsoids obeys another well-known statistic, the chi-square distribution with three degrees of freedom:

$$p_{cum} = \frac{1}{(2\pi)^{1/2}} \int_0^{\delta s^2_{cum}} x^{1/2} e^{-x/2} \, dx \tag{2.38}$$

which can also be tabulated. Table 2-4 includes recently re-calculated values of the cumulative frequency of the three-dimensional normal distribution for p_{cum} at 10 % intervals, plus some further values in the range beyond 90 %.

In the one-dimensional case, probability density and cumulative probability can also be represented in terms of simple graphs (Gaussian bell-shaped curve or S-shaped cumulative curve). In the three-dimensional case, distributions can be visualized only through families of ellipsoids, as shown here; furthermore, this is a representation suitable to the problem under consideration. The next section will discuss how the spatial position of ellipsoids can be visualized by projection into the three principal planes of the color space.

The determination of the typical coefficients g_{ik} for the standard deviation ellipsoid is presented in Sections 2.4.4 and 2.4.5.

2.4.2 Standard Deviation Ellipsoid

The equation of an ellipsoid (2.34) is a special case of the equation for the general surface of order two. The necessary and sufficient conditions for the coefficients g_{ik} to describe an ellipsoid are as follows:

$$g_{ii} > 0 \ (i = 1, 2, 3), \text{ i.e., } tr(\mathbb{G}) = \sum_{i=1}^{3} g_{ii} > 0$$

$$\det \hat{\mathbb{G}}_i > 0 \ (i = 1, 2, 3), \text{ i.e., } \sum_{i=1}^{3} \det \hat{\mathbb{G}}_i > 0$$

$$\det \mathbb{G} > 0 \tag{2.39}$$

Here $tr(\mathbb{G})$ is the trace and $\det \mathbb{G}$ the determinant of matrix $\mathbb{G}$ (eq. 2.32); the $\hat{\mathbb{G}}_i$ are the cofactors (the 2 × 2 matrices left after the i-th row and the i-th column are deleted). Explicitly,

$$tr(\mathbb{G}) = g_{11} + g_{22} + g_{33}$$

$$\sum_{i=1}^{3} \det \hat{\mathbb{G}}_i = \det \hat{\mathbb{G}}_1 + \det \hat{\mathbb{G}}_2 + \det \hat{\mathbb{G}}_3 = (g_{11} g_{22} - g_{12}^2) + (g_{22} g_{33} - g_{23}^2) +$$

$$+ (g_{11} g_{33} - g_{13}^2)$$

$$\det \mathbb{G} = g_{11} g_{22} g_{33} + 2 g_{12} g_{23} g_{13} - g_{11} g_{23}^2 - g_{22} g_{13}^2 - g_{33} g_{12}^2 \tag{2.40}$$

In order to judge the spatial orientation of the ellipsoid, it is important to know the positions of its principal axes. The partial derivatives of normal form in equation (2.34) are calculated for this purpose; the three partial derivatives are simultaneously the components of the gradient on the surface $E = \delta s^2$:

$$\text{grad } E = \left\{ \frac{\partial E}{\partial a^*}, \frac{\partial E}{\partial b^*}, \frac{\partial E}{\partial L^*} \right\} = 2 \begin{Bmatrix} g_{11} a^* + g_{12} b^* + g_{13} L^*, \\ g_{12} a^* + g_{22} b^* + g_{23} L^*, \\ g_{13} a^* + g_{23} b^* + g_{33} L^*, \end{Bmatrix} = 2 \mathbb{G} \cdot \mathbf{V} \tag{2.41}$$

The gradient is normal to the ellipsoid surface (Fig. 2-13). The axes of the ellipsoid (i.e., the vectors **a**, **b**, and **c**) have the property that the gradient is parallel to each axis; mathematically:

$$\frac{1}{2} \text{grad } E = \mathbb{G} \cdot \mathbf{V} = \{\lambda a^*, \lambda b^*, \lambda L^*\} = \lambda \cdot \mathbf{V}$$

$$(\lambda = \text{scalar factor}) \tag{2.42}$$

The second and last expressions lead to three equations for the three unknown coordinates, each for one principal axis of the ellipsoid.

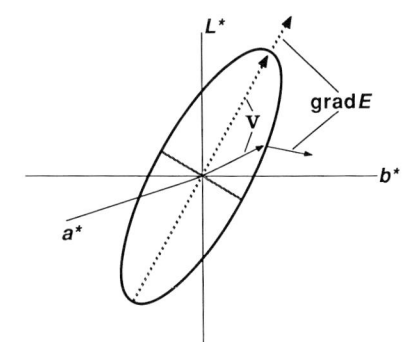

2.4 Mathematical Statistics of Color Coordinates

Fig. 2-13. Gradients on the ellipsoid surface (see text for explanation).

For the system of equations to have solutions for a^*, b^*, and L^*, the functional determinant ("characteristic polynomial" of matrix $\mathbb{G}$) must vanish. This requirement gives the "characteristic equation" of $\mathbb{G}$, also called the "secular equation":

$$\det(\mathbb{G} - \lambda \mathbb{E}) = \det \begin{pmatrix} (g_{11} - \lambda) & g_{12} & g_{13} \\ g_{12} & (g_{22} - \lambda) & g_{23} \\ g_{13} & g_{23} & (g_{33} - \lambda) \end{pmatrix} = \lambda^3 - a_2 \lambda^2 + a_1 \lambda - a_0 = 0 \quad (2.43)$$

where

$$a_0 = \det \mathbb{G}; \quad a_1 = \sum_{i=1}^{3} \det \mathbb{\hat{G}}_i; \quad a_2 = tr(\mathbb{G})$$

(see eq. 2.40) ($\mathbb{E}$ = identity matrix)

This is a third-degree equation in λ and can be solved by well-known methods. Its "characteristic" roots or eigenvalues λ_a, λ_b, and λ_c are all real; each is associated with one of the ellipsoid axes **a**, **b**, **c**. The coordinate system spanned by the three ellipsoid axes is called the "principal axis system."

If the gradients in the three directions **a**, **b**, **c** are formed with equation (2.42), each gradient, expressed in the principal axis system, has just one component in one of the three axis directions. In this way, $\mathbb{G}$ from equation (2.32) becomes a diagonal matrix in which only the three diagonal components are nonzero; these positions are occupied by λ_a, λ_b, and λ_c. From equation (2.34), this corresponds to a normal form reduced to the coefficients of the quadratic terms g_{ii}, with $g_{11} = \lambda_a$, $g_{22} = \lambda_b$, and $g_{33} = \lambda_c$. From the principal axis equation of the ellipsoid

$$\frac{a^{*2}}{a^2} + \frac{b^{*2}}{b^2} + \frac{L^{*2}}{c^2} = 1 \quad (2.44)$$

2 How Colors Depend on Spectra (Colorimetry)

it follows that the lengths of the principal axes are

$$a = |\mathbf{a}| = \frac{1}{\sqrt{\lambda_a}}$$

$$b = |\mathbf{b}| = \frac{1}{\sqrt{\lambda_b}}$$

$$c = |\mathbf{c}| = \frac{1}{\sqrt{\lambda_c}} \tag{2.45}$$

If we have one root λ, two of equations (2.42) give the ratios $u = a^*/L^*$ and $v = b^*/L^*$, that is, the directions of the principal axes:

$$u = \frac{a^*}{L^*} = \frac{g_{12}(g_{33} - \lambda) - g_{13} g_{23}}{g_{23}(g_{11} - \lambda) - g_{12} g_{13}}$$

$$v = \frac{b^*}{L^*} = \frac{g_{12}(g_{33} - \lambda) - g_{13} g_{23}}{g_{13}(g_{22} - \lambda) - g_{12} g_{23}} \tag{2.46}$$

Then, for the principal axis **a**, the system of equations (2.45) and (2.46) yields

$$a_{a^*}^2 + a_{b^*}^2 + a_{L^*}^2 = \frac{1}{\lambda_a}$$

$$\frac{a_{a^*}}{a_{L^*}} = u_a, \qquad \frac{a_{b^*}}{a_{L^*}} = v_a \tag{2.47}$$

and with $\mathbf{a} = \{a_{a^*}, a_{b^*}, a_{L^*}\}$

$$\boxed{\begin{aligned} a_{L^*} &= \frac{1}{\sqrt{\lambda_a (1 + u_a^2 + v_a^2)}} \\ a_{a^*} &= u_a\, a_{L^*} \\ a_{b^*} &= v_a\, a_{L^*} \end{aligned}} \tag{2.48}$$

and similar solutions for **b** and **c**.

The axis directions can also be characterized in terms of the polar coordinate angles φ and ϑ (Fig. 2-14):

2.4 Mathematical Statistics of Color Coordinates 47

$$\tan \varphi_a = \frac{a_{b*}}{a_{a*}} = \frac{v_a}{u_a},$$

$$\tan \vartheta_a = \frac{\sqrt{a_{a*}^2 + a_{b*}^2}}{a_{L*}} = \sqrt{u_a^2 + v_a^2} \quad (2.49)$$

and similarly for φ_b, ϑ_b and φ_c, ϑ_c.

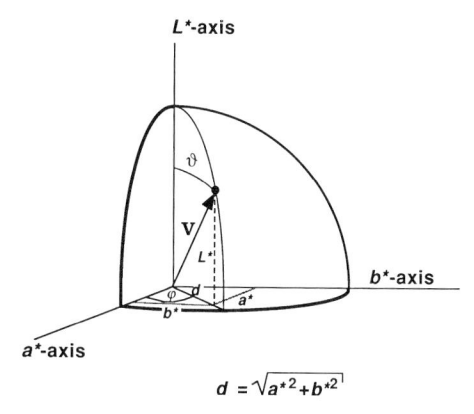

Fig. 2-14. Three-dimensional polar coordinates.

$d = \sqrt{a^{*2} + b^{*2}}$

To arrive at a three-dimensional picture of the position and orientation of the ellipsoid in color space, it is useful to calculate the projections of the ellipsoid on the coordinate planes and plot these, using a plotter connected to the computer if possible. The projection on the a^*, b^* plane is defined by the condition that the gradient has no L^* component (Fig. 2-15):

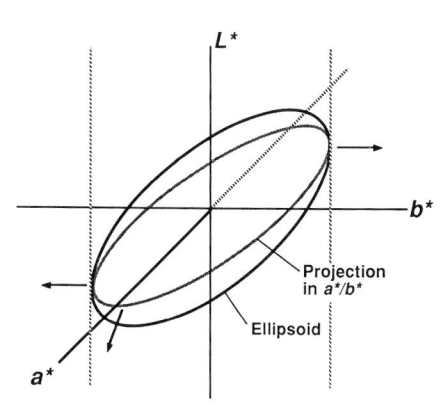

Fig. 2-15. Projections of the ellipsoid (see text for explanation).

$$\frac{1}{2}\frac{\partial E}{\partial L^*} = g_{13}\, a^* + g_{23}\, b^* + g_{33}\, L^* = 0 \tag{2.50}$$

from which it follows that

$$L^* = -\frac{1}{g_{33}}(g_{13}\, a^* + g_{23}\, b^*) \tag{2.51}$$

Now the normal form of the ellipsoid equation (2.34) becomes

$$E_{a^*b^*} = g'_{11}\, a^{*2} + g'_{22}\, b^{*2} + 2\, g'_{12}\, a^*\, b^* = 1 \tag{2.52}$$

where

$$g'_{11} = g_{11} - \frac{g_{13}^2}{g_{33}}$$

$$g'_{22} = g_{22} - \frac{g_{23}^2}{g_{33}}$$

$$g'_{12} = g_{12} - \frac{g_{13}\, g_{23}}{g_{33}}$$

Equation (2.52) is the desired ellipse in the a^*, b^* plane (horizontal projection). The two axes of this ellipse are obtained from the "small" secular equation

$$\det \begin{pmatrix} (g'_{11} - \lambda') & g'_{12} \\ g'_{12} & (g'_{22} - \lambda') \end{pmatrix} = 0 \tag{2.53}$$

which is quadratic. By analogy to equation (2.46), its two solutions λ'_a, λ'_b imply two linear equations for the axes:

$$b^* = \frac{1}{g'_{12}}(\lambda' - g'_{11})\, a^* \tag{2.54}$$

The three projections of the ellipsoid can so be calculated and then drawn as a horizontal and two vertical projections. To do this, it is desirable to connect pairs of exterior points by "section lines". These are determined as follows: The gradient of equation (2.52) also has no b^* component:

$$\frac{1}{2}\frac{\partial E_{a^*b^*}}{\partial b^*} = g'_{12}\, a^* + g'_{22}\, b^* = 0 \tag{2.55}$$

2.4 Mathematical Statistics of Color Coordinates

Then

$$b^* = -\frac{g'_{12}}{g'_{22}} \cdot a^* \qquad (2.56)$$

From equation (2.52) it now follows that the lines are tangent to the ellipse at $\pm a^*_{sec}$:

$$a^*_{sec} = \pm [g'_{22}/(g'_{11}g'_{22} - g'^2_{12})]^{1/2} \qquad (2.57)$$

($g'_{11}g'_{22} > g'^2_{12}$ by virtue of the ellipse conditions).

The projections, axes, and section lines for the other two coordinates are calculated in a similar fashion, using appropriatly modified equations.

2.4.3 Standard Deviations

We refer to the discussion in Section 2.4.1; given the coefficients of the standard deviation ellipsoid, we now wish to calculate the standard deviations of the color measurements. In color space, the standard deviation is a vector **s**, defined as the point on the ellipsoid to which **s** points. Because the components of a vector can be expressed as a direction and a magnitude, the vectorial standard deviation of direction **n** is described by

$$\mathbf{s_n} = \{\alpha\sigma_n, \beta\sigma_n, \gamma\sigma_n\} \qquad (2.58)$$

where

$$\sigma_n = |\mathbf{s_n}|$$

Here $\mathbf{n} = \{\alpha, \beta, \gamma\}$ is a unit vector, which accordingly has to characterize only the direction of $\mathbf{s_n}$. Its components are the direction cosines, that is, the cosines of the angles between $\mathbf{s_n}$ and the three coordinate axes. Substitution in the normal form of the standard deviation ellipsoid, equation (2.34), yields

$$\delta s^2 = (g_{11}\alpha^2 + g_{22}\beta^2 + g_{33}\gamma^2 + 2g_{12}\alpha\beta + 2g_{23}\beta\gamma + 2g_{13}\alpha\gamma) \cdot \sigma_n^2 = 1$$

so that for σ_n

$$\boxed{\sigma_n = \frac{1}{\sqrt{g_{11}\alpha^2 + g_{22}\beta^2 + g_{33}\gamma^2 + 2g_{12}\alpha\beta + 2g_{23}\beta\gamma + 2g_{13}\alpha\gamma}}} \qquad (2.59)$$

Here σ_n gives the length of the ellipsoid radius in spatial direction **n**. Thus, if a specified spatial direction is substituted in Equation (2.59), the radius for this direction is obtained. For a color coordinate or a color difference (or a difference obtained by breaking down the color difference) having the direction of **n**, the radius found is the desired standard deviation σ_n. The task thus consists of determining only the directions of the color coordinates or color difference and its differences obtained by breaking down the color difference, then substituting these in equation (2.59).

We begin with the *color coordinates**. The calculation is particularly simple for the standard deviations of a^*, b^*, and L^*. For a^*, $\alpha_a = 1$ and $\beta_a = \gamma_a = 0$. Substitution in equation (2.59) gives the following for σ_a (similarly for σ_b and σ_L):

$$\sigma_a = \frac{1}{\sqrt{g_{11}}}; \quad \sigma_b = \frac{1}{\sqrt{g_{22}}}; \quad \sigma_L = \frac{1}{\sqrt{g_{33}}} \tag{2.60}$$

The standard deviations of C^*_{ab} and h_{ab} also follow from the coefficients of the ellipsoid. For C^*_{ab},

$$\alpha_C = \frac{a}{C} = \cos h; \quad \beta_C = \frac{b}{C} = \sin h; \quad \gamma_C = 0$$

Hence it follows that

$$\sigma_C = \frac{1}{\sqrt{g_{11} \cos^2 h + 2 g_{12} \sin h \cdot \cos h + g_{22} \sin^2 h}}$$

$$= \frac{C}{\sqrt{g_{11} a^2 + 2 g_{12} a b + g_{22} b^2}} \tag{2.61}$$

The direction in which h_{ab} can vary is given by the chord perpendicular to C^*_{ab}, $s = 2\sigma_H$. The magnitude is $\pm (s/2) = \pm \sigma_H$ (Fig. 2-16):

$$\alpha_H = -\sin h; \quad \beta_H = \cos h; \quad \gamma_H = 0$$

$$\sigma_H = \frac{1}{\sqrt{g_{11} \sin^2 h - 2 g_{12} \sin h \cdot \cos h + g_{22} \cos^2 h}} \tag{2.62}$$

---

* To simplify the notation, the subscript ab and the * are dropped from the color coordinates and color differences; thus $C = C^*_{ab}$, $\Delta E = \Delta E^*_{ab}$, etc.

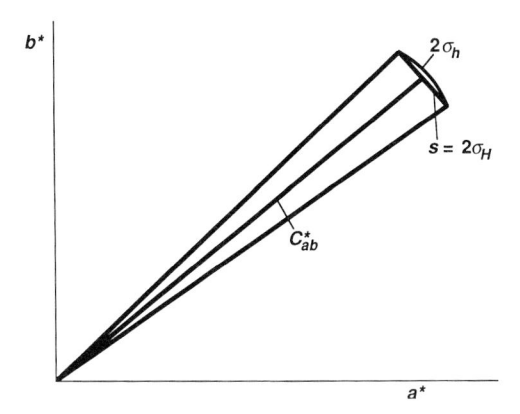

2.4 Mathematical Statistics of Color Coordinates

Fig. 2-16. Calculation of σ_h (see text for explanation).

To calculate σ_h, half the chord must be divided by $C = C_{ab}^*$ and converted back to an angle by taking the tangent:

$$\sigma_h = \text{Arctan} \frac{1}{C \sqrt{g_{11} \sin^2 h - 2 g_{12} \sin h \cdot \cos h + g_{22} \cos^2 h}}$$

$$= \text{Arctan} \frac{1}{\sqrt{g_{11} b^2 - 2 g_{12} a b + g_{22} a^2}} \qquad (2.63)$$

Now on to the standard deviations of the *color differences*. Here the standard deviation ellipsoid $\mathbb{G}_\Delta$ of the color differences must be known. It is defined as in equation (2.32) except that the g_{ik} are now replaced by g_{ik}^Δ; δa^*, δb^*, and δL^* by the differences $\delta \Delta a^* = \Delta a^* - \overline{\Delta a^*}$, $\delta \Delta b^* = \Delta b^* - \overline{\Delta b^*}$, and $\delta \Delta L^* = \Delta L^* - \overline{\Delta L^*}$, where $\overline{\Delta a^*}$, $\overline{\Delta b^*}$, $\overline{\Delta L^*}$ are the arithmetic means. We use arguments presented in the next section (Section 2.4.4), where it is shown that the difference ellipsoid $\mathbb{G}_\Delta$ is linked to the coordinate ellipsoid $\mathbb{G}_R$ of the reference by the relation

$$\mathbb{G}_\Delta = g_{ik}^\Delta = (1/2) g_{ik}^R = (1/2) \mathbb{G}_R. \qquad (2.64)$$

Thus, if $\mathbb{G}_\Delta$ has not been determined separately, i.e., if only the reference ellipsoid $\mathbb{G}_R$ is available, then the coefficients of $\mathbb{G}_\Delta$ can be calculated with equation (2.64). If, on the other hand, only $\mathbb{G}_\Delta$ is known, then $\mathbb{G}_R$ can be obtained from equation (2.64).

Proceeding in the same way as before, we substitute the direction cosines of the respective color differences in equation (2.59), thus obtaining the following relationships.

For the differences Δa^*, Δb^*, ΔL^*:

$$\sigma_{\Delta a} = \frac{1}{\sqrt{d_{11}^\Lambda}}, \quad \sigma_{\Delta b} = \frac{1}{\sqrt{d_{22}^\Lambda}}, \quad \sigma_{\Delta L} = \frac{1}{\sqrt{g_{33}^\Lambda}} \tag{2.65}$$

For the *hue difference* ΔH_{ab}^*:

The hue difference is perpendicular (see Fig. 2-10) to the bisector of the angle $h_0 = (1/2)(h_R + h_T)$.
Therefore,

$$\alpha_{\Delta H} = -\sin h_0; \quad \beta_{\Delta H} = \cos h_0; \quad \gamma_{\Delta H} = 0$$

so that by equations (2.59) and (2.62)

$$\sigma_{\Delta H} = \sigma_H \; (g_{ik}^\Lambda, h_0) \tag{2.66}$$

For the *hue angle* Δh_{ab}:

Its standard deviation is the angle associated with $\sigma_{\Delta H}$, that is, $\tan \sigma_{\Delta h} = \sigma_{\Delta H}/\overline{OW}$ (Fig. 2-10). The length $\overline{OW}$ can be expressed in either of two ways, in terms of ΔH or $\overline{OH}$, or in terms of both. From Figure 2-10 and equation (2.15), it is also easy to show that $\overline{OH}^2 = C_R C_T$. Accordingly,

$$\begin{aligned}\sigma_{\Delta H} &= \operatorname{Arctan}\,[(2\sigma_{\Delta h}/\Delta H)\tan(\Delta h/2)] \\ &= \operatorname{Arctan}\,\{\sigma_{\Delta h}/[(C_R C_T)^{1/2}\cos(\Delta h/2)]\} \\ &= \operatorname{Arctan}\,\{\sigma_{\Delta h}/[C_R C_T - (\Delta H/2)^2]^{1/2}\}\end{aligned} \tag{2.67}$$

For the *chroma difference* ΔC_{ab}^*:

Because, in general, the chromas of reference sample and test sample differ in direction (Fig. 2-10), their standard deviations $\sigma_{C.R}$ and $\sigma_{C.T}$ must be calculated separately using equation (2.61). This must be done with the ellipsoid $\mathbb{G}_R$ of the reference sample; if it is not available, equation (2.64) is employed to calculate it from $\mathbb{G}_\Lambda$. Then, by the additive rule for variances,

$$\sigma_{\Delta C} = (\sigma_{C.R}^2 + \sigma_{C.T}^2)^{1/2} \tag{2.68}$$

2.4 Mathematical Statistics of Color Coordinates

For the *color difference* ΔE_{ab}^*:

The direction cosines are

$$\alpha_{\Delta E} = \frac{\Delta a}{\Delta E}, \quad \beta_{\Delta E} = \frac{\Delta b}{\Delta E}, \quad \gamma_{\Delta E} = \frac{\Delta L}{\Delta E}$$

The magnitude of the standard deviation is then

$$\boxed{\sigma_{\Delta E} = \frac{\Delta E}{\sqrt{g_{11}^\Delta \Delta a^2 + g_{22}^\Delta \Delta b^2 + g_{33}^\Delta \Delta L^2 + 2 g_{12}^\Delta \Delta a\, \Delta b + 2 g_{23}^\Delta \Delta b\, \Delta L + 2 g_{13}^\Delta \Delta a\, \Delta L}}}$$

(2.69)

These equations fulfill the objective: The standard deviations of the three color coordinates, the chroma, and the hue angle, as well as the standard deviations of the color difference and of all its composant differences, can now be calculated. All that is necessary to know are the six coefficients g_{ik}^Δ of the standard deviation ellipsoid of the differences (or g_{ik}^R of the reference) for a specified threshold confidence level.

2.4.4 Color Measurement Errors and Significance

The ellipsoid coefficients g_{ik} needed for all the calculations in the preceding section can be determined by known methods. Firstly, from the number of measurements n, the arithmetic means are formed in accordance with equation (2.36), then the variances of a^*, b^*, and L^* are formed:

$$\text{Var}(a^*) = \frac{1}{n-1} \sum_i^n (a_i^* - \overline{a^*})^2,$$

$$\text{Var}(b^*) = \frac{1}{n-1} \sum_i^n (b_i^* - \overline{b^*})^2,$$

$$\text{Var}(L^*) = \frac{1}{n-1} \sum_i^n (L_i^* - \overline{L^*})^2, \qquad (2.70)$$

In the case of vector quantities, the variances alone are not enough to characterize the deviations in space. Correlations between the individual components must also be taken into account and can strongly influence their standard deviations: If, for example, large values of a^* are always associated with large b^* and L^* values, then the standard deviation of a^* will be smaller than if a^*, b^*, and L^* varied in random sequence with no correlations. The strengths of the correlations between the three pairs of components are expressed in terms of the covariances:

2 How Colors Depend on Spectra (Colorimetry)

$$\text{Cov}(a^*, b^*) = \frac{1}{n-1} \sum_i^n (a_i^* - \overline{a^*})(b_i^* - \overline{b^*})$$

$$\text{Cov}(b^*, L^*) = \frac{1}{n-1} \sum_i^n (b_i^* - \overline{b^*})(L_i^* - \overline{L^*})$$

$$\text{Cov}(a^*, L^*) = \frac{1}{n-1} \sum_i^n (a_i^* - \overline{a^*})(L_i^* - \overline{L^*}) \qquad (2.71)$$

Now the variances and covariances are entered in a matrix $\mathbb{V}$:

$$\mathbb{V} = v_{ik} = \begin{pmatrix} \text{Var}(a^*) & \text{Cov}(a^*, b^*) & \text{Cov}(a^*, L^*) \\ \text{Cov}(a^*, b^*) & \text{Var}(b^*) & \text{Cov}(b^*, L^*) \\ \text{Cov}(a^*, L^*) & \text{Cov}(b^*, L^*) & \text{Var}(L^*) \end{pmatrix} = \begin{pmatrix} v_{11} & v_{12} & v_{13} \\ v_{12} & v_{22} & v_{23} \\ v_{13} & v_{23} & v_{33} \end{pmatrix} \qquad (2.72)$$

This matrix is symmetric, because $v_{ik} = v_{ki}$. It is the three-dimensional analog of the variance σ_x^2 of the one-dimensional normal distribution.

In the one-dimensional case, the precision can be specified in terms of the confidence interval, the interval within which a given percentage of the measurements are expected to fall. Similarly, the precision in the three-dimensional case can be specified by the ellipsoid, which has the point $\{\overline{a^*}, \overline{b^*}, \overline{L^*}\}$ as its center (reference). The ellipsoid coefficients g_{ik} are the elements of the inverse matrix of $\mathbb{V}$; thus*

$$\boxed{g_{ik} = \mathbb{G} = \mathbb{V}^{-1}} \qquad (2.73)$$

$\mathbb{G}$ is also symmetric, since it was obtained from a symmetric matrix by inversion; $g_{ik} = g_{ki}$ is also one of the necessary ellipsoid conditions.

From this we see the following: Only by using the covariances can the standard deviation ellipsoid be obtained; in general, the ellipsoid is not oriented to the principal axes, because the measures of deviations must be invariant to the coordinate system chosen.

If no correlation exists between the measured values of the reference and the test sample, then the additive rule for variances and covariances leads to the following relation for the reference, test sample, and color difference matrices:

$$\mathbb{V}_\Delta = \mathbb{V}_T + \mathbb{V}_R \qquad (2.74)$$

or, after substitution of the respective standard deviation matrices,

$$\mathbb{G}_\Delta^{-1} = \mathbb{G}_T^{-1} + \mathbb{G}_R^{-1}$$

If $\mathbb{G}_T$ and $\mathbb{G}_R$ are equal ($\mathbb{G}_T = \mathbb{G}_R$), then

$$\mathbb{G}_\Delta^{-1} = 2\mathbb{G}_R^{-1}$$

---

* For an example of matrix inversion, see Chapter 6.

or, after inverting,

$$\mathbb{G}_\Delta = (1/2)\,\mathbb{G}_R$$

This relation was used in Section 2.4.3 (eq. 2.64).

The preceding arguments used to determine g_{ik} thus close the circle: All the standard deviations of color measurement values can be calculated using equations (2.60) to (2.69). The formulae were obtained from the normal form of the ellipsoid $\mathbb{G}$, that is, with $\delta s_{cum}^2 = 1$. A look at Table 2-4, however, shows that according to the χ^2 distribution this ellipsoid contains only 20 % of all the measured values; we thus say that the "confidence interval" $1 - \alpha$ is just 20 %. It would be desirable to have a confidence interval of 70 % (which would roughly correspond to the range of $\pm\sigma_x$ customarily used with the one-dimensional normal distribution). Again from Table 2-4, this confidence interval is obtained with a value $\delta s_{70\%}^2 = 3.665$ of the χ^2 distribution. The following procedure results:

- If the ellipsoid is to be plotted as in Section 2.4.2, the g_{ik} should be divided by 3.665.
- If the standard deviations are to be calculated as in Section 2.4.3, the σ_x values should be multiplied by $3.665^{1/2} = 1.914$.

If a different confidence interval is required, the corresponding value for δs_{cum}^2 from Table 2-4 should be used instead.

The statistical concept of *significance* offers quantitative information as to whether a measured value – here, the color difference and its composant differences – is "true" or whether it is falsified solely by measurement errors. This risk is always present when the color differences are very small, of the order of the error deviations. In the language of mathematical statistics, this means that it is necessary to determine whether the color difference found is larger or smaller than a "critical value" t (also called the "significance threshold"). This value is formed in the same way as illustrated for the measurement error, with one additional step: division by the square root of the number of measurements $n^{1/2}$. This extra step takes account of the fact that the mean becomes more and more accurate as the number of measurements increases, so that the deviations of the measurements about the mean are increasingly attributed to their real value. The value of t thus is given by the number of measurements n, the standard deviation σ_x, and the desired confidence interval $1-\alpha$:

$$\boxed{t_{1-\alpha} = \sigma_x(\delta s_{1-\alpha}^2/n)^{1/2}\;(\delta s_{cum}^2 \text{ from Table 2-4})} \qquad (2.75)$$

The measured color difference ΔF is significant if $|\Delta F| > |t_{\Delta F}|$, and not significant if $|\Delta F| \leq |t_{\Delta F}|$.

The principles and methods for determining the significance of color differences are presented in DIN 55600. The standard purposely avoids calculating the standard deviations of the CIELAB coordinates, since the nonlinearity of the relations used for the transformation from the CIE system means that these values do not necessarily have a normal distribution. Even so, there is no objection to applying this system to the CIELAB coordinates themselves. The arithmetic mean of the coordinates of the reference also corresponds better to the true value than does the color locus of a selected individual preparation.

The application to color differences alone is justified because color differences can be represented to a good approximation as differentials. As in the theory of errors, the total differential can be used; higher-order derivatives need not be considered. The total differential in fact represents a linear transformation in a small neighborhood, and since its coefficients in the neighborhood of a color locus are largely constant, it follows that da^*, db^*, and dL^* have just as normal a distribution as the dX, dY, and dZ, or as the $d\varrho$ of the reflectances, to which dX, dY, and dZ are related in a linear function.

2.4.5 Acceptability

The manufacturer of pigments or pigmented coatings is expected to make products that match specified samples ("references"). The instrument for determining the matches is the human eye; products that pass this examination by the colorist are "accepted" and are said to be "conforming" (to a standard). In general, to say that the test sample and the reference match is not to say that the measurements are absolutely identical; instead, it means that the sample values lie in a specified tolerance range.

This initially unknown tolerance range cannot be derived from the general rules of colorimetry but must be determined experimentally in each individual case. The technique involves the coloristic matching of color samples close to the reference. The justification for this procedure was stated in Section 2.3.1; furthermore, none of the existing color difference systems is perfect in exactly reproducing the perceived color differences (not even the CIELAB system, Section 2.3.3, which has become the standard chiefly because of its consistent use). For this reason, a systematic series of graded color samples are placed around the reference in each instance, and this set is presented to the colorist for evaluation. As a rule, the submitted samples are divided into two sets ("accepted" and "not accepted") in accordance with the colorist's judgment. (An intermediate range identified as "conditionally conforming" can also be established.)

The goal of statistical analysis of the colorist judgments is to derive the coefficients of the "acceptability ellipsoid," since this supplies the computer with the needed selection criteria in a computer-controlled production control system. The most important procedure for deriving usable estimation functions for the parameters of a distribution is the "maximum likelihood method". As applied to our three-dimensional case, the steps are as follows: In accordance with equations (2.30) and (2.34), the probability that a test sample i will be judged as conforming to a reference is expressed in terms of Δa^*, Δb^*, and ΔL^*, the color differences of the i-th sample. If the samples match the reference, the color differences vanish. Obviously, from equation (2.34), this implies that

2.4 Mathematical Statistics of Color Coordinates 57

$\delta s_i^2 = 0$ and $w = 1$; that is, the sample is in principle always judged to conform. In practice, a psychophysical experiment such as the one just described is always assumed to have a certain error rate p (false alarm rate). For example, a sample colorimetrically identical to the reference may not be judged as conforming. Thus, the probability that the colorist will adjudge such a sample as conforming to the reference is equal to $q = 1 - p$, and the probability that any sample i will be judged as conforming is $q \exp(-\delta s_i^2/2)$. The complementary probability, that is, the probability that the sample will be judged as nonconforming, is then $1 - q \exp(-\delta s_i^2/2)$.

If the colorist making these judgments is permitted only the choices "conforming" and "nonconforming," the probabilities of these two answers can be linked through the binomial distribution and a likelihood function can be constructed:

$$M = \prod_{i=1}^{n} \binom{N}{j_i} \left(q\, e^{-\frac{1}{2}\delta s_i^2}\right)^{j_i} \left(1 - q\, e^{-\frac{1}{2}\delta s_i^2}\right)^{N-j_i} \tag{2.76}$$

In this formula, n is the number of color samples; j_i is the number of "conforming" judgments; and N is the number of colorists making the judgments. The maximum likelihood method approximates the unknown parameters q and g_{11}, g_{22}, g_{33}, g_{12}, g_{23}, g_{13} by taking values which maximize M. A necessary condition for M, as a differentiable function of q and the g_{ik}, to have a maximum (except at the boundary) is that the partial derivatives with respect to q and the g_{ik} vanish:

$$\frac{\partial M}{\partial q} = \frac{\partial M}{\partial g_{11}} = \frac{\partial M}{\partial g_{22}} = \frac{\partial M}{\partial g_{33}} = \frac{\partial M}{\partial g_{12}} = \frac{\partial M}{\partial g_{23}} = \frac{\partial M}{\partial g_{13}} = 0 \tag{2.77}$$

Because the factors in equation (2.76) are nonnegative, M is generally positive at a maximum. The natural logarithm ln M, which is a monotonically increasing function, therefore has a maximum precisely where M has a maximum. This result suggests replacing M with the function ln M, so that instead of equation (2.76) we have

$$\ln M = \sum_{i=1}^{n} \left(-\frac{1}{2} j_i\, \delta s_i^2 + j_i \ln q\right) +$$

$$+ \sum_{i=1}^{n} (N - j_i) \ln\left(1 - q\, e^{-\frac{1}{2}\delta s_i^2}\right) + \text{const} \tag{2.78}$$

and instead of equation (2.77)

$$\frac{\partial \ln M}{\partial q} = \frac{\partial \ln M}{\partial g_{ik}} = 0 \tag{2.79}$$

Now, instead of the laborious differentiation of the products, only the sums have to be differentiated. To determine the desired values of g_{ik} and p, iterative procedures for zero finding can be applied; these need not be discussed further here.

By following the procedure described, we obtain ellipsoids of the probability density in a different way from the method based on the covariance matrix. Thus, the shape and position of such an ellipsoid in color space represents the probability with which a color locus lying on the ellipsoid will be deemed as conforming.

This method is somewhat complicated, both experimentally and mathematically, but it is the best of all the methods tested; the mathematical difficulties are taken care of by the development of a functioning program. It must be pointed out that there are simpler methods of estimating the parameters. These are not discussed here because either they presuppose some simplified version of the psychophysical experiment or else they are not efficient enough.

In the equations written above, δs_i^2 is the quadratic expression δs^2 of the i-th sample:

$$\delta s^2 = g_{11} \Delta a^{*2} + g_{22} \Delta b^{*2} + g_{33} \Delta L^{*2} +$$
$$+ 2g_{12} \Delta a^* \Delta b^* + 2g_{23} \Delta b^* \Delta L^* + 2g_{13} \Delta a^* \Delta L^* \qquad (2.80)$$

with ellipsoid $\mathbb{G} = g_{ik}$.

It is often desirable to relate the quadratic expression not to the $L^*a^*b^*$ coordinates but, instead, to the coordinate system spanned by the three orthogonal difference vectors ΔC_{ab}^*, ΔH_{ab}^*, ΔL^*. It will now be shown that this is possible; the new ellipsoid $\mathbb{G}' = g'_{ik}$ is obtained by transformation to the orthogonal ΔC_{ab}^*, ΔH_{ab}^*, ΔL^* system. First, the a^* and b^* of the reference are expressed in terms of its polar coordinates C_{ab}^* and h_{ab}:

$$a^* = C_{ab}^* \cos h_{ab} \qquad b^* = C_{ab}^* \sin h_{ab}$$

Then the color differences (with $\Delta H_{ab}^* = 2[C_{ab}^*(C_{ab}^* + \Delta C_{ab}^*)]^{1/2} \sin(\Delta h_{ab}/2) \approx C_{ab}^* \Delta h_{ab}$) are given by the following formulae (see eq. 2.15 and Fig. 2-10):

$$\Delta a^* \approx da^* = \cos h_{ab}\, dC_{ab}^* - C_{ab}^* \sin h_{ab}\, dh_{ab} \approx \frac{1}{C_{ab}^*}(a^* \Delta C_{ab}^* - b^* \Delta H_{ab}^*)$$

$$\Delta b^* \approx db^* = \sin h_{ab}\, dC_{ab}^* + C_{ab}^* \cos h_{ab}\, dh_{ab} \approx \frac{1}{C_{ab}^*}(b^* \Delta C_{ab}^* + a^* \Delta H_{ab}^*)$$

After substituting these expressions into equation (2.80) and comparing the coefficients

$$\delta s^2 = g'_{11} \Delta C^{*2}_{ab} + g'_{22} \Delta H^{*2}_{ab} + g'_{33} \Delta L^{*2} +$$
$$+ 2g'_{12} \Delta C^*_{ab} \Delta H^*_{ab} + 2g'_{23} \Delta H^*_{ab} \Delta L^* + 2g'_{13} \Delta C^*_{ab} \Delta L^*$$

we obtain the new ellipsoid

$$g'_{11} = \frac{1}{C^{*2}_{ab}} (g_{11} a^{*2} + 2g_{12} a^* b^* + g_{22} b^{*2})$$

$$g'_{22} = \frac{1}{C^{*2}_{ab}} (g_{11} b^{*2} - 2g_{12} a^* b^* + g_{22} a^{*2})$$

$$g'_{33} = g_{33}$$

$$g'_{12} = \frac{1}{C^{*2}_{ab}} [g_{12} (a^{*2} - b^{*2}) + (g_{22} - g_{11}) a^* b^*]$$

$$g'_{23} = \frac{1}{C^*_{ab}} (g_{23} a^* - g_{13} b^*)$$

$$g'_{23} = \frac{1}{C^*_{ab}} (g_{23} b^* + g_{13} a^*) \tag{2.81}$$

2.5 List of Symbols Used in Formulas

a, b, c	magnitudes of **a**, **b**, **c**
ab	(as subscript) denotes CIELAB
a^*, b^*, L^*	CIELAB coordinates
a, b, c	vectors of the ellipsoid axes
α	significance level; $1 - \alpha$, confidence interval
α, β, γ	direction cosines of **s**; components of **n**
A	normalization factor for one-dimensional normal distribution
B	normalization factor for two-dimensional normal distribution
cum	(as subscript) denotes cumulative probability
C	normalization factor for three-dimensional normal distribution
C, C^*_{ab}	chroma (CIELAB)
Cov	covariance
den	(as subscript) denotes probability density
δ	difference (measured value minus its mean)
δs^2	quadratic form of normal distributions
Δ	color difference expressed as difference between CIELAB coordinates (test sample minus reference)

$\Delta E^*_{ab}, \Delta E$	color difference (CIELAB)
ΔE^*_{CH}	color difference (CIELAB 92)
$\Delta h_{ab}, \Delta h$	hue angle difference
$\Delta H^*_{ab}, \Delta H$	hue difference (CIELAB)
Δs	magnitude of shade
e	(as subscript) denotes limiting value
E	area of ellipsoid
E_{a^*,b^*}	circumference of ellipse
F	color stimulus in CIE system; CIE color locus
g_{ik}	matrix element in i-th row and k-th column of $\mathbb{G}$
$\mathbb{G}$	matrix of ellipsoid coefficients
h_{ab}, h	hue angle (CIELAB) [in degrees]
i	number of sample evaluated
j	number of "conforming" judgments
λ	wavelength
λ	scalar factor in secular equation
L^*	(see a^*)
m	physical measure of intensity
max	(as subscript) denotes maximum
M	likelihood function
n	number of measurements
n	(as subscript) denotes standard illuminant
n	unit vector in direction of **s**
N	number of colorists
p	frequency (statistics)
$\varphi(\lambda)$	spectral distribution
φ	meridian angle (spherical coordinates)
q	individual probability
r	radius vector
ϱ	reflectance
R	(as subscript) denotes reference
s_R, s_T	circle chords in CIELAB system
sec	(as subscript) denotes section line
$\mathbf{s_n}$	vector of standard deviation in direction of **n**
σ_x	standard deviation of quantity x
t	significance threshold
t	(as subscript) denotes point of tangency
T	(as subscript) denotes test pigment
ϑ	polar angle (spherical coordinates)
u, v	auxiliary variables for principal axis directions
v_{ik}	matrix element in i-th row and k-th column of $\mathbb{V}$
V	CIELAB color locus
$\mathbb{V}$	covariance matrix
Var, V	variance
w	probability function
$x, \bar{x}$	measured value and its arithmetic mean

x, y, z	(CIE) chromaticity coordinates
$\bar{x}_\lambda, \bar{y}_\lambda, \bar{z}_\lambda$	tristimulus values of spectral colors; CIE color matching functions
X, Y, Z	(CIE) reference stimuli
X, Y, Z	(CIE) tristimulus values
X^*, Y^*, Z^*	CIELAB value functions
$y, \bar{y}$	measured value and its arithmetic mean

2.6 Summary

The basic principle of colorimetry is that, using just three selected lights, all other color stimuli can be simulated fully by additive mixing. On the other hand, a color stimulus can also be prepared from the spectral colors by mixing; thus it has a spectral distribution, which in the case of object colors has been given the name "spectral reflectance." The trichromatic principle makes it possible to construct a three-dimensional color space once three reference stimuli (primaries) have been defined. In this color space, the coordinates are to be understood as components of a vector. This trichromatic theory has recently gained physiological confirmation. The CIE color notation system and the CIE standard observer were created on this basis in 1931; the principles are stated in standards. The CIE system imitates the human eye: The huge number of possible optical stimuli is reduced in a meaningful way to a much smaller set that can be comprehended by the mind.

The CIE color notation system has, since its creation, proved its unrestricted suitability for use in all technical applications of colorimetry. It permits worldwide understanding as to colors. As to pigment evaluation, however, there are additional demands. The system is not easy to visualize, and physiologically equidistant color differences cannot be determined with it. For ease of understanding, Helmholtz proposed a different three-variable description: brightness, hue, and saturation. In the determination of color differences, one of the principal tasks of colorimetry, there was no major progress until recently. For this purpose, a color space must be constructed through matching experiments on samples with small color differences; the best-known system of this type is the Munsell system. After a CIE recommendation, the CIELAB system (based on the Adams-Nickerson system) has been used in pigment testing for some years. In CIELAB, quantities evaluated physiologically can first be calculated from the CIE tristimulus values, and by taking differences in pairs between the results the CIELAB coordinates are then obtained. This process of forming differences finds its physiological justification in the structure of the nervous system that processes information from the eye. The determination of actual color differences is then accomplished by calculating the vectorial distance between the color loci of the reference and test pigments and its breakdown into the three components lightness, chroma, and hue. Thus color measurement is performed in the CIE system and the results are then transformed into the CIELAB system (preferably with the aid of a computer, since CIELAB lends itself to programming).

Despite the success of CIELAB, there are needs for improvement (ΔL^* undervalues the perceived brightness, ΔC_{ab}^* overvalues the perceived saturation). For that, a new formula "CIELAB 92" has been created.

This chapter dealt briefly with color order systems, which were classified as empirical and colorimetry-based systems. The second group included the DIN Color System and the OSA-UCS system. Special attention was focused on the new RAL Design System with its color atlas, which provides an ideal link between color order and CIELAB.

In pigment testing, determining the precision of color measurement is the key to assessing the significance of color differences. The precision serves as a figure of merit both for the colorimetric measurements in full shade comparisons and for the colorimeter being employed (with the help of a collection of samples). The treatment begins with known relations describing the statistics of colorimetric data. The first step is to calculate the variances of the CIELAB coordinates. These are not, however, sufficient to describe the deviations of the measurements; if one colorimetric value shows large deviations, it is important to know whether the other two also vary widely or whether they are stochastically distributed. For this reason, the covariances are also necessary for a complete description. The three variances and the three covariances are entered in a symmetric matrix, the inversion of which yields the coefficients of the standard deviation ellipsoid. The equation of the ellipsoid always contains a term on the right-hand side for which the tabulated χ^2 distribution value for a given confidence interval must be substituted. A relatively new procedure is the use of the resulting ellipsoid for the calculation of errors in the measured CIELAB coordinates, in the color differences, and in the composant color differences. These are equal to the radii of the ellipsoid in the spatial direction corresponding to the coordinate or color difference in question. The formulas for calculating these standard deviations are stated in this chapter, along with formulas useful for visualizing the spatial position of the error ellipsoid.

In the manufacture of pigments and pigmented coatings, the requirement of conformity (to a standard or reference) is leading to a more rational procedure for acceptability testing. Rapid development in the computer field has opened up the possibility of letting a computer make the decision as to conformity, using specified criteria. But such criteria cannot be derived from purely colorimetric considerations; they must be obtained by having the responsible colorists perform matching tests on appropriate samples. In order to calculate the ellipsoid coefficients, it is desirable to employ the maximum likelihood method; while this is an old technique, it has recently been applied to this problem in the U. S. A. It has to be kept in mind that misjudgments also occur and must be smoothed out statistically. The likelihood function can be maximized with the help of a formula based on a binomial distribution of judgments. The mathematical operations can be substantial and a computer is needed. The calculation yields ellipsoids of probability density.

2.7 Historical and Bibliographical Notes

The crucial step in color theory occurred when Newton [1] laid the foundations for a scientific understanding of color phenomena by breaking sunlight down into its spectral components and correctly interpreting this process. The invention of the spectrograph by Young [2] also made it possible to measure spectral distributions. Young further discovered empirically the physiological relationships that were later formulated by Grassmann [3]. For the physiological principles of the Young-Helmholtz theory of color vision see, e. g., [4]; Helmholtz himself reported on the trichromatic theory [5]. Nonetheless, it was not until 1931 that a definition of the primary stimuli was agreed on. In that year the CIE, on the basis of the first usable experimental studies of the color matching functions by Wright [6] and Guild [7], decided to propose such primary stimuli (primaries, reference stimuli) as worldwide standards. The curves evaluated by various observers were averaged and the result was defined as the "CIE standard observer".

The sensory organs as information-processing systems are described in, e.g., [8]. A number of excellent publications on the theoretical principles of colorimetry have appeared both in English [13], [14], [14a], [14b] and in German [9], [10], [11], [12]; further citations can also be found in these. Colorimetry with special application to pigments is described by Gall [15].

For the construction of a physiologically equidistant color space see, e. g., [16]; [17] is the Munsell atlas. The DIN Color System was introduced by Richter [18]; the RAL Design System, by Gall [18a]. The Adams-Nickerson (AN) system for testing pigments [19], [20], derived from the Munsell system, had been well known for decades when, in 1973, the German Standards Committee (DIN) and later the CIE proposed its testing in the form of the CIELAB system. CIELAB 92 has appeared as CIE Publication No. 101.

The physiology of color vision is described in, e. g., [21].

A comprehensive treatment of the mathematical statistics of color coordinates can be found in [22]. The theoretical basis for characterizing test errors in the colorimetric evaluation of color differences was set forth earlier and fully developed, in a different context, by Brown and MacAdam [23]. The problem solved by these authors was to determine experimentally the perceptibility threshold for color differences. Their analysis led to "tolerance ellipsoids" localized in color space. The reference was located at the center of each such ellipsoid; all colors which were not distinguishable from the reference by the observer are located within the ellipsoid, while outside the ellipsoid are the colors that the observer can distinguish from the reference. It had already been demonstrated that the figure must be an ellipsoid for normally distributed color samples [24]. Later, Robertson also proposed the Brown-MacAdam principle for test errors [25]. The justification is that the two problems have identical mathematical structure: As has been shown, both cases involve the extension of the Gaussian normal distribution from one to three dimensions; see, e. g., [26]. Tables giving the values of statistical functions such as the χ^2 distribution can also be found in statistics textbooks and collections of mathematical tables, e.g., [27].

The maximum likelihood method goes back to GAUSS (1880) and was later generalized by FISHER in 1912 [28]. The principles of the technique used here were set forth in a publication by BILLMEYER and co-workers [29] and later recommended by ROBERTSON [30] on behalf of the CIE. Another calculation method is described by BROCKES and co-workers [31].

It must not be ignored that cybernetics, more than any other field of knowledge, has contributed enormously to the development of colorimetric test methods. Cybernetics has figured prominently in at least three areas and will help to determine the course of future development:
- Through development of computers for use specifically in color examination.
- Through the application of cybernetic conceptual models to measurement procedures and instruments.
- Through the application of cybernetic conceptual and analytical methods to the phenomenon of vision [32].

The last of these points is important because an improved knowledge of vision is an essential condition, e.g., for the construction of better colorimeters.

On the use of computers in the laboratory see, e.g., [33].

The history of colorimeters and color measuring has been dealt with elsewhere in detail [11], and it is therefore not taken up here.

References

[1] I. Newton: *Optics. Innys,* London 1704–1730 (reissued by McGraw-Hill, New York 1931).
[2] T. Young, *Phil. Trans. Roy. Soc.* London **92** (1802) 12.
[3] H. Grassmann, *Ann. Phys.* (Leipzig) **89** (1853) 69.
[4] G. Wald, *Angew. Chemie. Internat. Edit.* **80** (1968) 857.
[5] H. v. Helmholtz: *Handbuch der physiologischen Optik.* Voss, Leipzig 1911.
[6] W. D. Wright, *Trans. Opt. Soc.* London **30** (1928/29) 141.
[7] H. Guild, *Phil. Trans. Roy. Soc.* London A **230** (1931/32) 149.
[8] K. Steinbuch: *Automat und Mensch,* Springer, New York 1971, 4. edit., 200 ff.
[9] E. Schrödinger. *Ann. Phys.* (Leipzig) **63** (1920) 397, 427.
[10] M. Born, Naturwissenschaften **50** (1963) No. 2, 3.
[11] M. Richter, *Einführung in die Farbmetrik,* W. de Gruyter, Coll. Göschen, Berlin-Heidelberg 1976.
[12] A. Brockes, D. Strocka, A. Berger-Schunn, *Bayer*-Farben-Revue, Sonderheft 3/2 D, 1986.
[13] W. S. Stiles, G. Wyszecki, *Color Science,* Wiley & Sons, New York, 2. edit., 1983.
[14] D. L. MacAdam, *Color Measurement,* Springer, New York, 2. edit., 1985.
[14a] F. W. Billmeyer, jr. Kirk-Othmer, *Encyclopedia of Chemical Technology,* 3. edit., J. Wiley & Sons, New York 1979, Vol. 6, 523.
[14b] R. W. G. Hunt, *Measuring Color,* J. Wiley & Sons, New York, 1987.
[15] L. Gall, *The Color Science of Pigments,* BASF Ludwigshafen, 1971.
[16] G. Wyszecki: *Farbsysteme,* Musterschmidt, Göttingen 1960, 81.
[17] A. H. Munsell: *Atlas of the Munsell Color System,* Wadsworth-Howland, Boston 1915.
[18] M. Richter, *Farbe* I (1953) 85.

2.7 Historical and Bibliographical Notes

[18a] L. Gall, *farbe + lack* **98** (1992) 863.
[19] E. Q. Adams, *J. Opt. Soc. Am.* **32** (1942), 168.
[20] D. Nickerson, K. F. Stultz, *J. Opt. Soc. Am.* **34** (1944) 550.
[21] D. H. Hubel, *Eye, Brain and Vision*, Sci. Am. Lib., New York 1988.
[21a] J. Walraven, 7th Aic Congress 1993 Budapest, Congress book.
[22] H. G. Völz, *farbe + lack* **88** (1982), S. 264 und 443.
[23] W. R. J. Brown, D. L. MacAdam, *J. opt. Soc. Am.* **39** (1949) 808.
[24] L. Silberstein, D. L. MacAdam, *J. opt. Soc. Am.* **35** (1945) 32.
[25] A. R. Robertson, The specification of the performance characteristics of colorimetric instruments, CIE TC-1.3, June 1973.
[26] K. Stange, *Angewandte Statistik*, Bd. I (1970) u. Bd. 2 (1971), Springer, New York.
[27] W. Abramowitz, I. Stegun, *Handbook of Mathematical Functions*, Nat. Bur. Stand., Appl. Math. Ser. 55, US-Gov. Print Off., Washington, 3. edit. 1965.
[28] E. Kreyszig, *Statistische Methoden und ihre Anwendungen*, Vandenhoeck u. Ruprecht, Göttingen, 3. edit., 3. reprint, 1974, 174 ff.
[29] F. W. Billmeyer, W. G. Howe, R. M. Rich, *J. opt. Soc. Am.* **65** (1975) 956.
[30] A. R. Robertson, *Col. Res. Appl.* **3** (1978) Nr. 3, 149.
[31] D. Strocka, A. Brockes, W. Paffhausen, *Farbe + Design* **1980** Nr. 15/16, 75.
[32] H. G. Völz, XI. *Fatipec*-Kongress 1972 Florence, Edizione Ariminum, Mailand, 43.
[33] J. Hengstenberg, K.-H. Schmitt, B. Sturm, O. Winkler (Ed.) *Messen, Steuern, und Regeln in der chemischen Technik*, Springer New York, 3. edit., 1985.

3 How Spectra Depend on the Scattering and Absorption of Light (Phenomenological Theory)

3.1 Introduction

3.1.1 Phenomenological Theory and Its Significance

The name "phenomenological theory" refers to all theoretical statements about how the radiance of light incident on a layer is affected by absorption and scattering. Such a theory affords the needed physical-theoretical and mathematical tools for describing how the reflection spectrum of a layer can be derived from its absorption and scattering properties as well as its thickness. The model employed is that of light pathways or channels, roughly as known from geometrical optics; the absorption or scattering action of the medium is represented by optical material constants, that is, purely phenomenological quantities. The phenomenological theory treats the medium as a continuum characterized by the physical qualities of absorption and scattering; accordingly, the presence of discrete particles as a necessary (but not sufficient) condition for scattering is of no significance in this theory. Speaking casually, one can say that "the phenomenological theory does not know anything about particles"; in other words, it deals in physical qualities that can be described without reference to pigment particles and their special properties.

To attribute absorption and scattering properties to a single particle, one must resort to the "corpuscular theory" (see Chapter 5, also Figs. 1-2 in Chapter 1). Only that theory makes further statements about the optical material constants from phenomenological theory, indeed attributing them to the refractive index, size, and shape of the individual particle. In a certain sense, descriptions of the concentration dependences of the optical constants provide a bridge between the phenomenological and corpuscular theories, defining coefficients related to the pigment or dye concentration and then treating these coefficients as independent of the concentration.

In this context, the position of the phenomenological theory of turbid media clearly implies the fundamental importance of the theory for the pigmenting of lacquers, paints, plastics, and so forth, and thus for all matters relating to the application and economics of coloring materials. In this chapter, we will briefly present the many-flux theory as a complete, general phenomenological theory, with a detailed treatment of the four-flux theory, and derive the most important formulas. It will be seen that the theory of Ryde is a special case of the four-flux theory; the well-known Kubelka-Munk theory, a special case of the two-flux theory; and the Lambert (Lambert-Bouguer) law, a special case of the one-flux theory. Lambert's law and the Kubelka-Munk theory long ago became accepted throughout the pigment and dye world; the four-flux theory, in contrast, has been a presence chiefly in the literature. Nonetheless, it seems important to us to deal with this theory in more depth, since it contains possibilities that have as yet been utilized little or not at all. It includes, for example, the cases in which radiation (either directional or diffuse) is incident on both sides of the layer. Such configurations are encountered everywhere and are also of technical interest (e.g., lamp enclosures that are viewed in both reflected and transmitted light). A further advantage is that the four-flux theory makes it exceedingly easy to handle special cases, not just the two named above (Kubelka-Munk theory, Lambert's law) but also the extreme cases of "absorption only" and "scattering only," as well as the limiting cases of a black background and of an opaque ("optically infinitely thick") layer.

In the four-flux theory, scattering and absorption are controlled through a few material constants, the scattering and absorption coefficients. After the derivation of the theory, it will be shown how these coefficients can be exactly determined by reflection or transmission measurements.

3.1.2 Many-Flux Theory

The many-flux theory is the most generalized version of the phenomenological theory. It is presented briefly here because it can aid in the better understanding of absorption and scattering phenomena, thus making for an easier route to the important special cases that will be treated later.

The development begins with a coordinate system having position coordinate z and using specified, fixed polar angles θ to describe the angular distribution of the radiation. By virtue of the spherical symmetry of the problem, two coordinates will be enough.

This coordinate system is illustrated in Figure 3-1. Space is subdivided into conical segments, each labeled with the polar angle θ; these bound regions of space shaped like ice-cream cones, each called a "channel." In this way, all light traveling in direction θ_1 (in the positive z direction perpendicular to the plane bounding the layer) is in channel 1; light between θ_1 and θ_2 is in channel 2; and so forth. It is useful to select an even number n of channels so that the channels in the positive and negative z directions will be symmetrical.

The radiative transfer problem is described by a differential equation that takes away from each channel the radiances that it emits into all other channels and adds to it the radiances that it receives from all other channels. If there are n channels,

$$dL_i = s_{i1} L_1 \, dz + s_{i2} L_2 \, dz \ldots + s_{in} L_n \, dz = \sum_{j=1}^{n} s_{ij} L_j \, dz \qquad (3.1)$$

3.1 Introduction

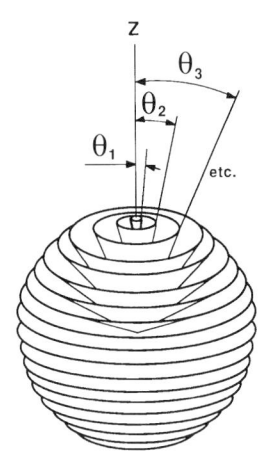

Fig. 3-1. Many-Flux theory: Subdivision of directions in space into n = 22 channels.

where L_i is the radiance in channel i, s_{ij} is the scattering coefficient from channel j into channel i, and dz is the path element in which the change described by equation (3.1) takes place. If $j = i$, the coefficient s_{jj} is the total scattering from channel j into all other channels, including the absorption contribution k_j in channel j:

$$s_{jj} = -k_j - \sum_{i=1}^{n} (1 - \delta_{ij}) s_{ij} \qquad (i = 1, 2, \dots n) \qquad (3.2)$$

where δ_{ij} is the Kronecker delta ($\delta = 1$ for $i = j$, $\delta = 0$ otherwise). It also has to be considered that the contribution s_{ij} is negative if $i > n/2$, since the radiation is traveling opposite to the positive z direction. This situation can be handled by introducing a factor $\eta_i = (n - 2i + 1)/|(n - 2i + 1)|$, which becomes –1 for this case but is 1 otherwise. Now equation (3.2) can be rewritten as

$$\frac{dL_i}{dz} = \sum_{j=1}^{n} \eta_i \, s_{ij} \, L_j \qquad (i = 1, 2, \dots n) \qquad (3.3)$$

The coefficients $\eta_i s_{ij} = S_{ij}$ form a matrix $\mathbb{S} = S_{ij}$, while the radiances L_i make up a vector $\mathbf{L} = \{L_1, L_2, \dots, L_n\}$, so that the fundamental vector differential equation can be written as

$$\frac{d\mathbf{L}}{dz} = \mathbb{S}\mathbf{L} \qquad (3.4)$$

The general solution for $\mathbf{L}$ can be found in mathematics textbooks:

$$L_i = \sum_{j=1}^{n} A_{ij} \, C_j \, e^{\lambda_j z} \qquad (i = 1, 2, \dots n) \qquad (3.5)$$

where the λ_j are the eigenvalues of the matrix $\mathbb{S}$, the A_{ij} are the eigenvectors $\mathbf{A}_j = \{A_{1j}, A_{2j},..., A_{nj}\}$ corresponding to the λ_j, and the C_j are constants of integration determined by the boundary conditions.

The eigenvalues λ_j can also be obtained in a well-known way by solving the "characteristic equation" (see also Section 2.4.2):

$$\sum_{j=1}^{n} S_{ij} A_{jk} - \lambda_k A_{ik} = 0 \qquad (i = 1, 2, ... n) \qquad (3.6)$$

This equation can be solved only with powerful computers, since simultaneous differential equations with i = 1, 2,..., n have to be integrated. So that the L_i can be stated as functions of z, finally, and thus the reflectance ϱ or the transmittance τ obtained, the constants of integration C_j must still be eliminated. These may be governed by a great variety of boundary conditions. Conditions of special practical interest are those of the substrates beneath the layer, which have their own reflectance ϱ_0, or the change in radiance at the boundary of the layer, which is caused by the discontinuity in refractive index at the surface. These cases will be treated in depth further on, in connection with special phenomenological theories.

It remains now to determine the scattering and absorption coefficients s_{ij} in this system. Equation (3.6) shows that they are included in the eigenvalues and eigenvectors and can be calculated from these with equation (3.5). If there are n channels, there are n^2 coefficients s_{ij}; by the symmetry of the channels in the half-spaces, $s_{n-i+1, n-j+1} = s_{ij}$ for all i, j, leaving $n^2/2$ coefficients.* With eight channels, for example, 64/2 = 32 coefficients are required, and this large number makes the many-flux theory unsuitable for practical use, because just as many measurements would have to be made on coatings in order to determine these coefficients.

Another way to determine the coefficients is to specify the radiance distribution of the incident light in terms of appropriate functions of the polar angle. A method has been described in which the radiation distribution is represented by Legendre polynomials, but it remains limited to citations in the literature.

Software is available to aid in calculations under the many-flux theory by numerically integrating differential equations ("implicit" solution). One difficulty arises in that initial values must be provided for the scattering and absorption coefficients. Of course, the coefficients from Mie calculations (see Chapter 5) could be used, although obtaining them is an intricate and laborious process, but convenient approximate formulas do exist.

In summary, the theoretical value of the many-flux theory is great but its practical significance is virtually nil. We employ it here as a basis for understanding some portions of the phenomenological theory that are used in practice and thus serve as principles underlying the results obtained in this book.

---

* These include n/2 distinct s_{jj}, each of which contains one absorption coefficient.

In particular, we will treat special four-flux, two-flux, and one-flux theories derived directly from the model or else through mathematical integration of the differential equations ("explicit" solution). We begin with surface phenomena.

3.1.3 Surface Phenomena

In order to determine the absorption and scattering coefficients from the phenomenological theory, it is necessary to use the reflectances prevailing in the interior of the layer under study. They can differ greatly from the reflectances measured at the surface, because of the discontinuity in refractive index that is usually present between air and medium.* When light enters the layer a part of it is reflected. This refelected portion (ca. 4%) cannot take any further part in scattering phenomena occurring inside the layer, but just the same it can enter the measuring aperture of the colorimeter, thus contributing to the reflectance.

An even stronger effect takes place inside the layer: much of the light scattered back from the interior is reflected into the interior again before exiting the layer, thus becoming involved in new scattering processes. The light reflected into the interior of the layer consists in part of totally reflected light, so that this fraction is large (ca. 60%). This interaction is repeated, and in the end the net radiation leaving the layer governs the reflectance measured at the surface.

In view of these relationships, measured reflectances must be "surface-corrected" before these values are put into the equations of the phenomenological theory. In other words, the external measured reflectance or reflection factor is converted to the corresponding internal quantities, which can then be subjected to phenomenological analysis. We then speak of "corrected" reflectances or reflection factors; the formulas used have come to be known as the Saunderson correction. One of these formulas will now be derived. The formulas contain, along with the already known reflection coefficient r_0 at the interface, two other coefficients.

Accordingly, three constants figure in the derivation:

r_0: reflection coefficient at the surface for directional light incident perpendicularly from outside
r_1: reflection coefficient for light incident diffusely from outside
r_2: reflection coefficient for light incident on the surface diffusely from inside the specimen

These three coefficients can be calculated with Fresnel's equations (Table 3-1), provided the diffuse light is ideally diffuse, that is, obeys the Lambert cosine distribution.

---

* For media in practical use (lacquers, paints, plastics), n ≈ 1.5 is a good approximation to the refractive index.

3 How Spectra Depend on the Scattering and Absorption of Light

Table 3-1. Fresnel reflection coefficients for light of directional and diffuse incidence.

Refractive index n	Directional incidence: r_0	Ideally diffuse incidence External reflection: r_1	Internal reflection: r_2
1.40	0.02778	0.0769	0.528
1.41	0.02894	0.0784	0.536
1.42	0.03012	0.0800	0.543
1.43	0.03131	0.0815	0.550
1.44	0.03252	0.0830	0.557
1.45	0.03373	0.0845	0.564
1.46	0.03497	0.0860	0.571
1.47	0.03621	0.0875	0.577
1.48	0.03746	0.0890	0.584
1.49	0.03873	0.0904	0.590
→ 1.50	0.04000	0.0919	0.596
1.51	0.04129	0.0934	0.602
1.52	0.04258	0.0948	0.608
1.53	0.04389	0.0963	0.614
1.54	0.04520	0.0977	0.619
1.55	0.04652	0.0992	0.624
1.56	0.04785	0.1006	0.630
1.57	0.04919	0.1020	0.635
1.58	0.05054	0.1035	0.640
1.59	0.05189	0.1049	0.645
1.60	0.05325	0.1063	0.650

We impose the following condition in order to derive the formula: The layer must be opaque, "infinitely thick" as specialists say; furthermore, the light must be ideally diffuse before exiting from the surface, that is, its radiance must not depend on the angle. Naturally, all quantities are generally functions of wavelength.

We begin by considering a high-gloss specimen and a 0/d measuring geometry (0° incidence angle, diffuse observation) of the photometer. The radiance of the incident light is denoted by L_0^+, the internal reflectance by ϱ^*. Then from Figure 3-2 we have immediately:

3.1 Introduction

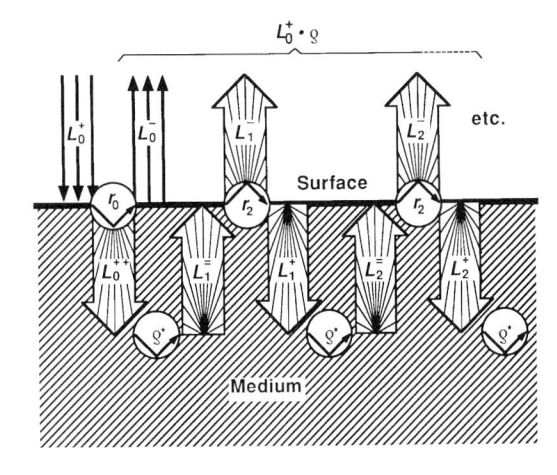

Fig. 3-2. Surface phenomena: The available radiances outside and inside the layer.

$$L_0^{++} = L_0^+ (1-r_0) \qquad\qquad L_0^- = L_0^+ \cdot r_0$$

$$L_1^= = L_0^{++} \cdot \varrho^* = L_0^+ (1-r_0)\varrho^*$$

$$L_1^+ = L_1^= \cdot r_2 = L_0^+ (1-r_0)\varrho^* r_2 \qquad L_1^- = L_1^= (1-r_2) = L_0^+ (1-r_0) \cdot (1-r_2)\varrho^*$$

$$L_2^= = L_1^+ \cdot \varrho^* = L_0^+ (1-r_0)\varrho^* r_2 \varrho^*$$

$$L_2^+ = L_2^= \cdot r_2 = L_0^+ (1-r_0)\varrho^* r_2^2 \varrho^* \qquad L_2^- = L_0^+ (1-r_0)(1-r_2)\varrho^* (r_2\varrho^*)$$

$$\vdots$$

$$L_n^- = L_0^+ (1-r_0)(1-r_2)\varrho^* (r_2\varrho^*)^{n-1}$$

$$\sum_{i=0}^{\infty} L_i^- = L_0^+ \left[r_0 + \frac{(1-r_0)(1-r_2)\varrho^*}{1-r_2\varrho^*} \right]$$

The radiance received by the instrument is the sum of all outward radiances L_i^- (written at right). The summation is done with the formula for an infinite geometric progression with first term $(1-r_0)(1-r_2)\varrho^*$ and $r_2\varrho^*$ as common ratio.*

---

* More precisely, the reflectance ϱ_g^* for $L_1^=$ would have to be for directional incidence while the reflectances ϱ_{dif}^* for all subsequent L_i would have to be for diffuse incidence. The slight difference between the two values is definitely of higher order in the context of this correction formula, so that for present purposes we can set $\varrho_g^* = \varrho_{dif}^* = \varrho^*$.

3 How Spectra Depend on the Scattering and Absorption of Light

The externally measured reflectance ϱ is the ratio between the radiation exiting and the radiation incident, that is, $\Sigma(L_i^-/L_0^+)$:

$$\varrho = r_0 + \frac{(1-r_0)(1-r_2)\varrho^*}{1-r_2\varrho^*} = \frac{(1-r_0-r_2)\varrho^* + r_0}{1-r_2\varrho^*} \tag{3.7}$$

For practical application, this equation is mostly written for ϱ^*:

$$\varrho^* = \frac{\varrho - r_0}{(1-r_0)(1-r_2) + r_2(\varrho - r_0)} = \frac{\varrho - r_0}{1 - r_0 - r_2(1-\varrho)} \tag{3.8}$$

The expressions on the right-hand sides are alternative forms.

For the ideal white standard with a totally matte surface, $\varrho^* = 1$, and so equation (3.7) implies that ϱ is also equal to 1. This means that the correction formulas stated above apply in the same form to matte and glossy specimens.

Now let us consider the measuring geometry d/0 (diffuse incidence, 0° observation). By similar arguments, the reflection factor R can be written as

$$R = a\,r_0 + \frac{(1-r_0)(1-r_1)\varrho^*}{n^2(1-r_2\varrho^*)} \tag{3.9}$$

The fraction on the right-hand side differs from the one in equation (3.7) by a factor of $n^2(1-r_2)/(1-r_1)$, which is equal to 1 for all refractive indices, as can easily be seen by examining Table 3-1. The term ar_0 means that it is possible to design the instrument in such a way that the surface-reflection contribution is included ($a = 1$) or suppressed ($a = 0$). In the first case, equation (3.9) becomes equation (3.7); in other words, equations (3.7) and (3.8) also hold for the 0/d and d/0 measuring geometries.

It can also be shown that the corresponding formulas for the 45/0 geometry are not much different from the ones already stated. This means that the correction formulas written out can be applied universally. For refractive index $n = 1.5$, Table 3-1 gives $r_0 = 0.04$ and $r_2 = 0.6$. Then the correction formulas become

$$\varrho = 0.04 + \frac{0.384\,\varrho^*}{1-0.6\,\varrho^*} = \frac{0.36\,\varrho^* + 0.04}{1-0.6\,\varrho^*} \tag{3.10}$$

$$\varrho^* = \frac{\varrho - 0.04}{0.384 + 0.6\,(\varrho - 0.04)} = \frac{\varrho - 0.04}{0.96 - 0.6\,(1-\varrho)} = \frac{\varrho - 0.04}{0.36 + 0.6\,\varrho} \tag{3.11}$$

The ϱ in equation (3.11) must be replaced by $\varrho + 0.04$ when the measurement is performed with gloss trap (specular exclusion).

These formulas have been used in practice for many years.

3.2 Four-Flux Theory

3.2.1 The Differential Equations and Their Integration

The "explicit" solution of the four-flux theory to be discussed now is based on the model of Figure 3-3. An absorbing and scattering layer of thickness h is traversed by four radiances,* identified by arrow symbols and corresponding to the four channels:
- Directional incident light: l^+
- Directional light opposite to direction of incidence: l^-
- Diffuse incident light: L^+
- Diffuse light opposite to direction of incidence: L^-

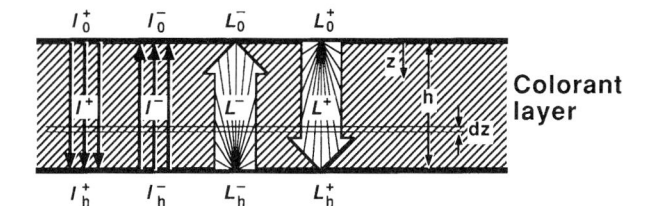

Fig. 3-3. Four-Flux theory: The radiances, directional (l^+, l^-) and diffuse (L^+, L^-), together with the boundary conditions.

Let z be the position coordinate in the direction of propagation. We now want to know the amount by which the individual radiances decrease or increase as the light traverses the short distance dz. As in the differential statement of Lambert's law, $-k'l^+ dz$ is taken away from l^+ by absorption (here k' is the absorption coefficient for directional light). If scattering is present, l^+ also loses $-s^+l^+ dz$ scattered forward and $-s^-l^+ dz$ scattered backward (s^+ and s^- are the forward and backward scattering coefficients for directional light). Thus we can write

$$dl^+ = -(k' + s^+ + s^-)\, l^+ \, dz \qquad (3.12)$$

In distinction to the absorbed fraction, however, the two scattered components do not disappear without trace; they must later be added into the respective radiances for diffuse light. Similarly, in the case of directional light propagating in the opposite direction, we can write for l^-

$$-dl^- = -(k' + s^+ + s^-)\, l^- \, dz \qquad (3.13)$$

---

*The terminology and notation used in this chapter are based on those of DIN 1349; departures have been made whenever there is a possibility of confusion with quantities already defined.

3 How Spectra Depend on the Scattering and Absorption of Light

The minus sign on dl^- takes care of the fact that l^- is in the direction opposite to the positive z direction.

We now consider the radiance L^+ of diffuse light. First, similarly to l^+, it loses the absorbed component $-KL^+dz$ and the scattered component $-SL^+dz$ (where K is the absorption coefficient and S the scattering coefficient for diffuse light). Further, however, the light scattered from the l^+ and l^- beams must now be added to L^+; these are s^+l^+dz and s^-l^-dz respectively. Finally, the scattering SL^-dz from the opposite diffuse light must be added as well, so that

$$dL^+ = s^+ l^+ dz + s^- l^- dz - (K + S) L^+ dz + SL^- dz \tag{3.14}$$

and similarly

$$-dL^- = s^- l^+ dz + s^+ l^- dz - (K + S) L^- dz + SL^+ dz \tag{3.15}$$

The coefficients and their definitions are written out once again:

$$k' = \frac{1}{l}\frac{dl}{dz}\bigg|_{abs} \; ; \quad s^+ = \frac{1}{l}\frac{dl}{dz}\bigg|_{sca \, forw} \; ; \quad s^- = \frac{1}{l}\frac{dl}{dz}\bigg|_{sca \, back} \tag{3.16}$$

$$K = \frac{1}{L}\frac{dL}{dz}\bigg|_{abs} \; ; \quad S = \frac{1}{L}\frac{dL}{dz}\bigg|_{sca} \tag{3.17}$$

By the rule stated at the end of Section 3.1.2, the four-flux theory would require $4^2/2 = 8$ coefficients, of which $4/2 = 2$ contain absorption coefficients. The reduction of the remaining six scattering coefficients to three comes about because, in three cases, there is no radiative transfer by scattering:

from l^+ to l^-
from L^+ to l^-
from L^- to l^+.

The associated scattering coefficients are therefore equal to zero.

The following abbreviations are now introduced into the equations:

$$\boxed{a = 1 + \frac{K}{S} \quad \text{(auxiliary variable)}} \tag{3.18}$$

$$\boxed{\mu = k' + s^+ + s^- \quad \text{(attenuation coefficient)}} \tag{3.19}$$

The system of differential equations then becomes

$$\frac{dl^+}{dz} = -\mu l^+ \tag{3.20}$$

$$\frac{dl^-}{dz} = \mu l^- \tag{3.21}$$

$$\frac{dL^+}{dz} = s^+ l^+ + s^- l^- - aSL^+ + SL^- \tag{3.22}$$

$$\frac{dL^-}{dz} = -s^- l^+ - s^+ l^- + aSL^- - SL^+ \tag{3.23}$$

This coupled system of four simultaneous differential equations for the functions $l^+(z)$, $l^-(z)$, $L^+(z)$, and $L^-(z)$ must now be decoupled and integrated. It will be sufficient to integrate out to the boundary conditions. Thus the integrals of the first two equations can be written immediately. With

$$l^+(0) = l_0^+; \qquad l^-(h) = l_h^- \tag{3.24}$$

we get

$$l^+ = l_0^+ \, e^{-\mu z}; \qquad l^- = l_h^- \, e^{-\mu(h-z)} \tag{3.25}$$

The film thickness h then becomes

$$l^+(h) = l_h^+ = l_0^+ \, e^{-\mu h}; \qquad l^-(0) = l_0^- = l_h^- \, e^{-\mu h} \tag{3.26}$$

The first two differential equations have now been decoupled and integrated by themselves. These solutions are needed for the integration of the last two equations (3.22) and (3.23). In order to decouple the remaining simultaneous system, one of the two equations must be differentiated; we choose equation (3.22). The derivatives of the directional radiances l^+ and l^- are replaced by expressions (3.20) and (3.21); dL^+/dz and dL^-/dz, by expressions (3.22) and (3.23). Using the partial integrals in equation (3.26) along with

$$\boxed{b = \sqrt{a^2 - 1} \ \ \text{(auxiliary variable)}} \tag{3.27}$$

we obtain finally

$$\frac{d^2 L^+}{dz^2} = b^2 S^2 L^+ - l_0^+ [Ss^- + (\mu + aS) s^+] e^{-\mu z} -$$
$$- l_h^- [Ss^+ - (\mu - aS) s^-] e^{-\mu(h-z)} \qquad (3.28)$$

The homogeneous equation associated with this second-order differential equation has the well-known exponential solution with two constants of integration. For the inhomogeneous equation, two terms with the exponents $-\mu z$ and $-\mu(h-z)$ must be considered, so that the solution becomes

$$L^+ = B_1 e^{bSz} + B_2 e^{-bSz} - l_0^+ p e^{-\mu z} - l_h^- q e^{-\mu(h-z)} \qquad (3.29)$$

Here B_1 and B_2 are the two constants of integration of the second-order differential equation; the boundary conditions will be applied later in order to eliminate these. On the other hand, p and q are fixed in value and can therefore be found immediately by comparing coefficients with differential equation (3.28). For this purpose, expression (3.29) is substituted in (3.28), an the other hand, equation (3.29) is differentiated twice. The comparison of coefficients then gives

$$p = \frac{Ss^- + (\mu + a S) s^+}{\mu^2 - b^2 S^2} = \frac{(K + k') s^+ + (s^+ + s^-) (S + s^+)}{(k' + s^+ + s^-)^2 - K (K + 2 S)} \qquad (3.30)$$

$$q = \frac{Ss^+ - (\mu - a S) s^-}{\mu^2 - b^2 S^2} = \frac{(K - k') s^- + (s^+ + s^-) (S - s^-)}{(k' + s^+ + s^-)^2 - K (K + 2 S)} \qquad (3.31)$$

Accordingly, equation (3.29) is already the general integral.

The next step is to impose the boundary conditions on the general integral and its first derivative. The boundary conditions themselves are obtained from the cases $z = 0$ and $z = h$ for L^+ and dL^+/dz; the starting differential equation (3.22) is employed for the latter. We then have

$$L^+(0) = B_1 + B_2 - l_0^+ p - l_h^- q e^{-\mu h} = L_0^+ \qquad (3.32)$$

$$L^+(h) = B_1 e^{bSh} + B_2 e^{-bSh} - l_0^+ p e^{-\mu h} - l_h^- q = L_h^+ \qquad (3.33)$$

$$\left.\frac{dL^+}{dz}\right|_{z=0} = B_1 b S - B_2 b S + l_0^+ \mu p - l_h^- \mu q e^{-\mu h}$$
$$= s^+ l_0^+ + s^- l_0^- - a S L_0^+ + S L_0^- \qquad (3.34)$$

$$\left.\frac{dL^+}{dz}\right|_{z=h} = B_1 b S\, e^{bSh} - B_2 b S\, e^{-bSh} + l_0^+ \mu p\, e^{-\mu h} - l_h^- \mu q$$
$$= s^+ l_h^+ + s^- l_h^- - a S L_h^+ + S L_h^- \tag{3.35}$$

We recall that l_0^+, l_h^-, L_0^+, and L_h^- are given (as the incident radiances; see Fig. 3-3). The equation system (3.32)–(3.35) thus contains four further unknowns, B_1, B_2, L_h^+, and L_0^-, which can now be determined from the four equations. Only L_h^+ and L_0^-, the quantities sought, are of interest here. The solutions are

$$L_h^+ = l_0^+ P + l_h^- Q + L_0^+ \frac{b}{A} + L_h^- \frac{\sinh x}{A} \tag{3.36}$$

$$L_0^- = l_0^+ Q + l_h^- P + L_0^+ \frac{\sinh x}{A} + L_h^- \frac{b}{A} \tag{3.37}$$

where we have introduced the notation*

$$x = bSh; \qquad A = a \sinh x + b \cosh x \tag{3.38}$$

$$\boxed{P = \frac{1}{A}(bp + q\, e^{-\mu h} \sinh x) - p\, e^{-\mu h}} \tag{3.39}$$

$$\boxed{Q = \frac{1}{A}(p \sinh x + bq\, e^{-\mu h}) - q} \tag{3.40}$$

Equations (3.36) and (3.37) together with (3.26) give the radiances for a layer irradiated from both sides. Interchanging l_0^+ with l_h^- and L_0^+ with L_h^- takes each equation into the other. This is a necessary condition, because the experimental radiances L_h^+ and L_0^- as well as l_h^+ and l_0^- must be invariant to reversal of the sides of the layer.

3.2.2 Transmittance and Transmission Factor

The most general configuration that can be imagined for the measurement of transmission is described by the following model: "Mixed" irradiation (i.e., directional and diffuse radiation) on the incidence side, observation of total radiation (i.e., directional and diffuse once again) on the observation side. The

---

* P and Q, unlike p and q, are not constants of the material, since they depend on the film thickness h.

3 How Spectra Depend on the Scattering and Absorption of Light

corresponding formula for the transmittance in the interior of the layer* is as follows (see Fig. 3-3):

$$\tau^* = \frac{l_h^+ + L_h^+}{l_0^+ + L_0^+} \tag{3.41}$$

If diffuse light is not recorded on the measurement side, L_h^+ vanishes and τ is replaced by the transmission factor T. In the *most general model*, the following equation is obtained by merging (3.26) and (3.36) and the fact that no light is incident on the specimen from the observation side (i.e., $l_h^- = L_h^- = 0$):

$$\tau^* = (1 - \gamma)(P + e^{-\mu h}) + \gamma \frac{b}{A} \tag{3.42}$$

where $\gamma = L_0^+/(l_0^+ + L_0^+)$ is the diffuse fraction and $1 - \gamma = l_0^+/(l_0^+ + L_0^+)$ the directional fraction of the net incident light.

This equation immediately yields the case of irradiation with *directional (quasiparallel)* light if γ is set equal to 0:

$$\boxed{\tau_g^* = P + e^{-\mu h}} \tag{3.43}$$

This is identical to the equation stated by Ryde if we set $k' = 0$, for Ryde did not introduce a second absorption coefficient.

In the same way, for irradiation with *diffuse* light ($\gamma = 1$),

$$\boxed{\tau_{\text{dif}}^* = \frac{b}{A} = \frac{b}{a \sinh x + b \cosh x}} \tag{3.44}$$

This is the transmittance in the Kubelka-Munk theory. The same equation is also obtained, incidentally, if the substitutions $s^+ = 0$, $s^- = S$, and $k' = K$ are made in equation (3.43). This would mean that, as far as scattering and absorption in the interior of the medium are concerned, there is no difference between directional and diffuse radiation: The substitutions lead to $p = 1$ and $q = 0$ from equations (3.30) and (3.31), so that equation (3.43) actually does become (3.44). In reality, however, s^- cannot be identical to S nor k' with K

---

* In general, $\tau^* = \tau/\tau_0$ (where τ_0 is the blank value). If there is much scattering in the layer, however, it may become necessary to apply the Saunderson corrections, equation (3.11), writing τ in place of ϱ.

since the mean path length through the layer for diffuse light is longer than the film thickness. As a consequence, more light is scattered and absorbed than from a directional beam perpendicular to the layer, and this fact must affect the numerical values of the scattering and absorption coefficients. The mean path length of ideally diffuse radiation through a layer is, according to Kubelka, twice the film thickness (corresponding to an incidence angle of 60°); hence S and K are about twice as large as s^- and k' respectively.

Two other configurations of measuring geometries are possible and must be considered. On the irradiation side, suppose that incidence is solely directional and perpendicular to the layer surface, while on the observation side only the quasi-perpendicular directional radiation component is measured. Then what is measured is the *transmission factor* T_r^*, which is given directly by equation (3.26):

$$T_r^* = \frac{l_h^+}{l_0^+} = e^{-\mu h} \tag{3.45}$$

The second possibility is the transmittance $\bar{\tau}_g^*$, which is obtained if irradiation is *directional* and the directly transmitted beam l_h^+ is excluded, so that only the remaining diffuse light is measured. The formula for this case is

$$\bar{\tau}_g^* = \tau_g^* - T_r^* = P \tag{3.46}$$

All the important transmittance equations for the four-flux theory have now been derived. We need only add the equations obtained by examining the two important extreme cases

- Absorption only (no scattering)
- Scattering only (no absorption)

For the *"absorption only"* case, we have k', $K \gg s^+$, s^-, S; $a \to b \to K/S \to \infty$; $\mu \to k'$; and $p \to q \to 0$. The *general* transmittance is then

$$\tau_{abs}^* = (1 - \gamma)\, e^{-k'h} + \gamma\, e^{-Kh} \tag{3.47}$$

while the transmittance and transmission factor for *directional* irradiation are given by

$$\boxed{\tau_{abs.g}^* = T_{abs.r}^* = e^{-k'h}} \tag{3.48}$$

and the transmittance for *diffuse* irradiation by

$$\boxed{\tau_{abs.dif}^* = e^{-Kh}} \tag{3.49}$$

3 How Spectra Depend on the Scattering and Absorption of Light

Since there is no incident diffuse light in the case of directional irradiation, the transmittance after exclusion of directional radiation on the measurement side is, as expected, $\bar{\tau}^*_{abs,g} = \tau^*_{abs,g} - T^*_{abs,r} = 0$.

Equation (3.48) is the well-known transmission factor from Lambert's law.

In the second extreme case, "*scattering only,*" s^+, s^-, $S \gg k'$, K; therefore $a \to 1$, $b \to 0$, $\mu \to u = s^+ + s^-$. For p and q, we obtain $p \to v = (S + s^+)/(s^+ + s^-)$ and $q \to w = (S - s^-)/(s^+ + s^-)$, with $v - w = 1$. Taking the limit gives the following for the *general transmittance*:

$$\tau^*_{sca} = (1-\gamma)\frac{1 + w(1 - e^{-uh})}{1 + Sh} + \gamma \frac{1}{1 + Sh} \tag{3.50}$$

whence, for *directional* irradiation,

$$\boxed{\tau^*_{sca,g} = \frac{1 + w(1 - e^{-uh})}{1 + Sh}} \tag{3.51}$$

for *diffuse* irradiation,

$$\boxed{\tau^*_{sca,dif} = \frac{1}{1 + Sh}} \tag{3.52}$$

while the *transmission factor* is given by

$$T^*_{sca,r} = e^{-uh} \tag{3.53}$$

and the *transmittance after exclusion of directional radiation* by

$$\bar{\tau}^*_{sca,g} = \tau^*_{sca,g} = T^*_{sca,r} = \frac{1 + w(1 - e^{-uh})}{1 + Sh} - e^{-uh} \tag{3.54}$$

The following auxiliary variables have been used here:

$$u = s^+ + s^-; \qquad w = \frac{S - s^-}{s^+ + s^-} \tag{3.55}$$

All the equations derived in Section 3.2.2 are summarized and related to one another by the transmission function pedigree of Figure 3-4.

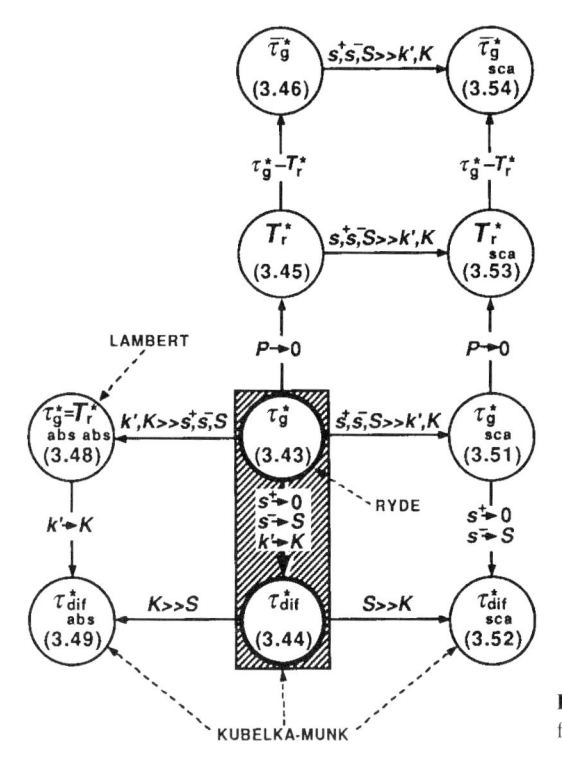

Fig. 3-4. Pedigree of the transmission function.

3.2.3 Reflectance and Reflection Factor

As Figure 3-3 shows, the reflectance in the interior of the layer for irradiation with both types of light is given by

$$\varrho^* = \frac{l_0^- + L_0^-}{l_0^+ + L_0^+} \tag{3.56}$$

In the model that serves as basis for the present discussion, the absorbing and scattering layer is applied directly on a substrate. This kind of model is of greatest importance in practice, since it describes lacquer and paint coatings. Because the substrate generally has its own reflection properties, these must be taken into account when the equations of Section 3.2.1 are employed. The intrinsic reflection of the substrate will, in general, give rise to diffusely reflected light, although specularly reflecting substrates are increasingly significant (e.g., for transparent lacquers). For the specular case, the radiance reflected from the substrate is (see Fig. 3-5)

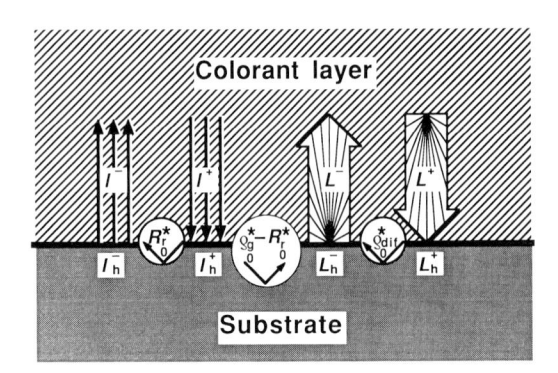

84   3 How Spectra Depend on the Scattering and Absorption of Light

Fig. 3-5. Reflection with substrate.

$$l_h^- = R_{r,0}^* \cdot l_h^+ = R_{r,0}^* \, l_0^+ \, e^{-\mu h} \tag{3.57}$$

Here $R_{r,0}^*$ is the reflection factor of the specular substrate; the subscript r identifies the 0/0 geometry (i.e., the angles of incidence and reflection are both 0°), while the * identifies the reflection factor as an internal one. If there is no specular reflection, however, only diffuse radiation comes from the substrate; it consists of two components, the first $\varrho_{g,0}^* \cdot l_h^+$ originating in directional irradiation and the second $\varrho_{\mathrm{dif},0}^* \cdot L_h^+$ in diffuse irradiation (here $\varrho_{g,0}^*$ and $\varrho_{\mathrm{dif},0}^*$ are the reflectances of the substrate for directional and diffuse irradiation, respectively; see Fig. 3-5). The definition of reflectance in equation (3.56) requires that all the radiation – directional and diffuse – must be received on the observation side. The same holds, of course, for the reflectance of the substrate. Accordingly, if specular reflection is present along with diffuse reflection, it must be recalled that the specular reflection has already been taken care of by equation (3.57), so that $R_{r,0}^* \cdot l_h^+$ must be subtracted, else it will be counted twice. The radiance diffusely reflected by the substrate is then

$$L_h^- = (\varrho_{g,0}^* - R_{r,0}^*) \, l_h^+ + \varrho_{\mathrm{dif},0}^* \, L_h^+ \tag{3.58}$$

From this with equations (3.26) and (3.36), we obtain an expression containing L_h^- twice; solution for L_h^- yields

$$L_h^- = \frac{A}{A - \varrho_{\mathrm{dif},0}^* \sinh x} \left\{ l_0^+ \left[(\varrho_{g,0}^* - R_{r,0}^*) \, e^{-\mu h} + \right. \right.$$
$$\left. \left. + \varrho_{\mathrm{dif},0}^* (P + R_{r,0}^* \, Q \, e^{-\mu h}) \right] + L_0^+ \, \varrho_{\mathrm{dif},0}^* \, \frac{b}{A} \right\} \tag{3.59}$$

Now, with the aid of equation (3.56) and the use of equations (3.26), (3.37), and (3.59), we immediately obtain the reflectance for the *general model* (parallel and diffuse irradiation):

3.2 Four-Flux Theory

$$\varrho^* = \frac{1-\gamma}{A - \varrho^*_{\text{dif},0} \sinh x} (A Q + \overline{R} + \overline{\varrho} + \overline{\overline{R}}) +$$

$$+ \gamma \left[\frac{\sinh x}{A} + \varrho^*_{\text{dif},0} \frac{b^2}{A (A - \varrho^*_{\text{dif},0} \sinh x)} \right] \quad (3.60)$$

In this expression we have used the notation

$$\overline{R} = R^*_{r,0} e^{-\mu h} [A (P + e^{-\mu h}) - b] \quad (3.61)$$

$$\overline{\varrho} = \varrho^*_{g,0} b e^{-\mu h} + \varrho^*_{\text{dif},0} (b P - Q \sinh x) \quad (3.62)$$

$$\overline{\overline{R}} = R^*_{r,0} \varrho^*_{\text{dif},0} e^{-\mu h} [b Q - (P + e^{-\mu h}) \sinh x] \quad (3.63)$$

Hence, with $\gamma = 0$, we have the following for irradiation with *directional* light:

$$\boxed{\varrho^*_g = \frac{A Q + \overline{R} + \overline{\varrho} + \overline{\overline{R}}}{A - \varrho^*_{\text{dif},0} \sinh x}} \quad (3.64)$$

Taking $\gamma = 1$ leads to the reflectance for irradiation with *diffuse* light; this equation can be transformed with the easily derived relation $\sinh^2 x - b^2 = A(a \sinh x - b \cosh x)$. Since $\cosh x/\sinh x = \coth x$,

$$\boxed{\varrho^*_{\text{dif}} = \frac{1 - \varrho^*_{\text{dif},0} (a - b \coth x)}{a - \varrho^*_{\text{dif},0} + b \coth x}} \quad (3.65)$$

This is the reflectance in the Kubelka-Munk theory, which we will use frequently.

Wholly similarly to the transmission factor T_r, a *reflection factor* R_r applicable to directionally reflected light can be defined. This is nonzero only when $R^*_{r,0} \neq 0$, that is, when there is a specularly reflecting substrate:

$$R^*_r = \frac{I_0^-}{I_0^+} = R^*_{r,0} e^{-2\mu h} \quad (3.66)$$

With a measuring setup that can *exclude* directional reflected light, a reflectance $\overline{\varrho}^*_g = \varrho^*_g - R^*_r$ can be obtained similarly to the transmittance:

$$\overline{\varrho}^*_g = \frac{1}{A - \varrho^*_{\text{dif},0} \sinh x} (A Q + \overline{R} - R^*_{r,0} A e^{-2\mu h} + \overline{\varrho} + \overline{\overline{R}} +$$
$$+ R^*_{r,0} \varrho^*_{\text{dif},0} e^{-2\mu h} \sinh x) \quad (3.67)$$

3 How Spectra Depend on the Scattering and Absorption of Light

As in Section 3.2.2, there are two extreme cases. Let us first consider *"absorption only."* Taking the limit yields the following for the *general* reflectance:

$$\varrho^*_{abs} = (1 - \gamma) [R^*_{r,0} e^{-2k'h} + (\varrho^*_{g,0} - R^*_{r,0}) e^{-(k' + K)h}] + \gamma \varrho^*_{dif,0} e^{-2Kh} \quad (3.68)$$

for *directional* irradiation,

$$\varrho^*_{abs,g} = R^*_{r,0} e^{-2k'h} + (\varrho^*_{g,0} - R^*_{r,0}) e^{-(k' + K)h} \quad (3.69)$$

for *diffuse* irradiation

$$\varrho^*_{abs,dif} = \varrho^*_{dif,0} e^{-2Kh} \quad (3.70)$$

With an appropriate measuring setup, we can again split into the *reflection factor*

$$R^*_{abs,r} = R^*_{r,0} e^{-2k'h} \quad (3.71)$$

and the reflectance *after exclusion* of directional radiation

$$\overline{\varrho}^*_{abs,g} = (\varrho^*_{g,0} - R^*_{r,0}) e^{-(2k' + K)h} \quad (3.72)$$

Taking the limit for the second extreme case, *"scattering only,"* leads to the general reflectance

$$\varrho^*_{sca} = \frac{1 - \gamma}{(1 - \varrho^*_{dif,0}) Sh + 1} [Q_{sca}(1 + Sh) + \overline{R}_{sca} + \overline{\varrho}_{sca} + \overline{\overline{R}}_{sca}] + $$
$$+ \gamma \frac{(1 - \varrho^*_{dif,0}) Sh + \varrho^*_{dif,0}}{(1 - \varrho^*_{dif,0}) Sh + 1} \quad (3.73)$$

where the barred expressions are somewhat simplified:

$$\overline{R}_{sca} = R^*_{r,0} e^{-uh} [(1 + Sh)(P_{sca} + e^{-uh}) - 1] \quad (3.74)$$

$$\overline{\varrho}_{sca} = \varrho^*_{g,0} e^{-uh} + \varrho^*_{dif,0} (P_{sca} - Q_{sca} Sh) \quad (3.75)$$

$$\overline{\overline{R}}_{sca} = R^*_{r,0} \varrho^*_{dif,0} e^{-uh} [Q_{sca} - (P_{sca} + e^{-uh}) Sh] \quad (3.76)$$

and

$$P_{sca} = \frac{1}{1 + Sh} [(1 + w)(1 - e^{-uh}) - Sh\, e^{-uh}] \quad (3.77)$$

$$Q_{sca} = \frac{1}{1 + Sh} [Sh - w(1 - e^{-uh})] \quad (3.78)$$

The symbols u and w have the same meanings as in equation (3.55).

From the above, we immediately get the expressions for *directional* light

$$\varrho^*_{\text{sca,g}} = \frac{1}{(1 - \varrho^*_{\text{dif,0}}) Sh + 1} \cdot [Q_{\text{sca}} (1 + Sh) + \overline{R}_{\text{sca}} + \overline{\varrho}_{\text{sca}} + \overline{\overline{R}}_{\text{sca}}] \quad (3.79)$$

and for *diffuse* light

$$\varrho^*_{\text{sca,dif}} = \frac{(1 - \varrho^*_{\text{dif,0}}) Sh + \varrho^*_{\text{dif,0}}}{(1 - \varrho^*_{\text{dif,0}}) Sh + 1} \quad (3.80)$$

as well as for the *reflection factor*

$$R^*_{\text{sca,r}} = R^*_{\text{r,0}} e^{-2uh} \quad (3.81)$$

and the reflectance *after exclusion* of the directional component:

$$\overline{\varrho}^*_{\text{sca,g}} = \frac{1}{(1 - \varrho^*_{\text{dif,0}}) Sh + 1} \cdot \{Q_{\text{sca}} (1 + Sh) + \overline{R}_{\text{sca}} + \overline{\varrho}_{\text{sca}} + \overline{\overline{R}}_{\text{sca}} -$$
$$- R^*_{\text{r,0}} e^{-2uh} [1 + (1 - \varrho^*_{\text{dif,0}}) Sh]\} \quad (3.82)$$

3.2.4 Limiting Cases of Reflection

The following limiting cases are of practical interest:
- The substrate exhibits specular reflection only
- The substrate exhibits diffuse reflection only
- The substrate is black
- The layer is opaque ("infinitely thick")

1) Substrate with exclusively specular reflection

The equations for this limiting case are obtained by setting the reflectances and reflection factors of the substrate in equation (3.64) equal (i.e., $\varrho^*_{\text{g,0}} = \varrho^*_{\text{dif,0}} = R^*_{\text{r,0}}$); the equation for *directional irradiation* is then

$$\boxed{\varrho^*_{\text{spr,g}} = \frac{1}{A - R^*_{\text{r,0}} \cdot \sinh x} \cdot \langle A Q + R^*_{\text{r,0}} \{A e^{-\mu h} (P + e^{-\mu h}) +$$
$$+ b P - Q \sinh x + R^*_{\text{r,0}} e^{-\mu h} [bQ - (P + e^{-\mu h}) \sinh x]\}\rangle} \quad (3.83)$$

For *diffuse irradiation*, equation (3.65) remains unchanged except that $R^*_{\text{r,0}}$ appears in place of $\varrho^*_{\text{dif,0}}$. The reflection factor of equation (3.66) does not change either. For the reflectance *after exclusion* of directional reflected light, we then have

88   3 How Spectra Depend on the Scattering and Absorption of Light

$$\overline{\varrho}^*_{spr,g} = \frac{1}{A - R^*_{r,0} \cdot \sinh x} \cdot \{AQ + R^*_{r,0}[P(Ae^{-\mu h} + b) - Q\sinh x + R^*_{r,0} e^{-\mu h}(bQ - P\sinh x)]\} \tag{3.84}$$

The equations for the two extreme cases, "absorption only" and "scattering only," will not be written out here, because they can be obtained effortlessly from equations (3.68)–(3.82) by setting $\varrho^*_{g,0} = \varrho^*_{dif,0} = R^*_{r,0}$.

2) Substrate with exclusively diffuse reflection

These equations are derived by setting $R^*_{r,0}$ equal to zero in the respective formulas. First, again, the case of *directional irradiation*:

$$\varrho^*_{dfr,g} = \frac{1}{A - \varrho^*_{dif,0} \cdot \sinh x} \cdot [AQ + \varrho^*_{g,0} b\, e^{-\mu h} + \varrho^*_{dif,0}(bP - Q\sinh x)]$$

$$= Q + b\, \frac{\varrho^*_{g,0} e^{-\mu h} + \varrho^*_{dif,0} P}{A - \varrho^*_{dif,0} \sinh x} \tag{3.85}$$

Here too, the case of *diffuse irradiation* contains nothing not considered in equation (3.65), since that formula does not show $R^*_{r,0}$ at all. Because no directional light can be *excluded*, the reflectance has the trivial solution $R^*_{dfr,r} = 0$, and $\overline{\varrho}^*_{dfr,g}$ is as in equation (3.85). Once more, the equations for the extreme cases are omitted, being easily derivable from the respective formulas with $R^*_{r,0} = 0$.

3) Black substrate

Now all the reflectances and reflection factors of the substrate are zero. From equation (3.64) with $R^*_{r,0} = \varrho^*_{g,0} = \varrho^*_{dif,0} = \overline{\overline{R}} = \overline{\varrho} = \overline{R} = 0$, we have the following: for *directional irradiation*

$$\varrho^*_{g,b} = Q \tag{3.86}$$

(this is the reflectance in the Ryde theory if $k' = 0$), and for *diffuse irradiation*

$$\varrho^*_{dif,b} = \frac{1}{a + b \coth x} \tag{3.87}$$

Again $R^*_{r,b} = 0$, so that the expression for $\overline{\varrho}^*_{g,b}$ is the same as equation (3.86). As in the previous derivations, the *extreme cases* are uninteresting up to "*scattering only.*" For *directional irradiation* we have

$$\varrho^*_{b/sca,g} = Q_{sca} = \frac{1}{1 + Sh}[Sh - w(1 - e^{-uh})] \tag{3.88}$$

and for *diffuse irradiation*

$$\varrho^*_{b/sca,dif} = \frac{Sh}{1 + Sh} \tag{3.89}$$

4) Infinitely thick layer

These important equations are obtained by taking the limit as $h \to \infty$, so that equation (3.86) leads to the following for *directional irradiation*:

$$\varrho^*_{g,\infty} = p(a - b) - q \tag{3.90}$$

and equation (3.87), with $a^2 - b^2 = 1$, gives for *diffuse irradiation*

$$\varrho^*_{dif,\infty} = a - b \tag{3.91}$$

The extreme cases yield trivial solutions.

All the equations derived in Section 3.2.3 are summarized and related to one another by the reflection function pedigree of Figure 3-6.

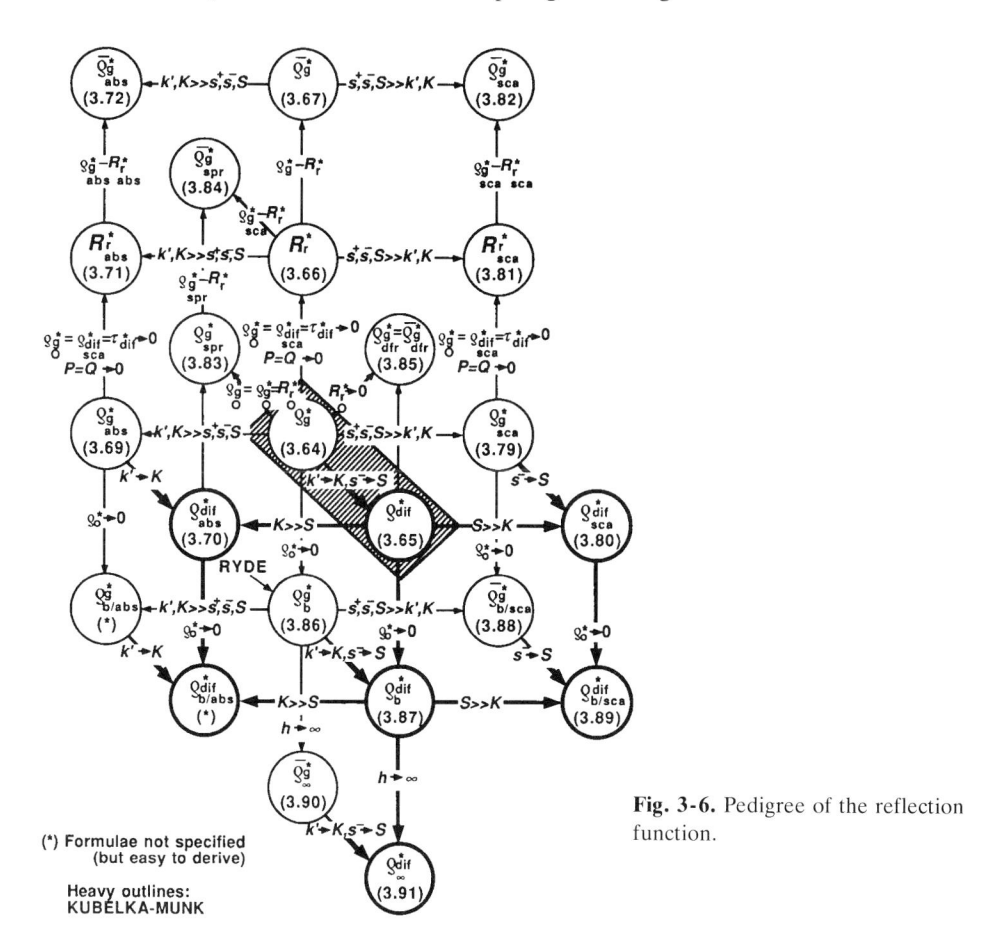

Fig. 3-6. Pedigree of the reflection function.

(*) Formulae not specified (but easy to derive)

Heavy outlines: KUBELKA-MUNK

3.2.5 Determination of the Coefficients

Practical use of the formulas stated in the preceding sections requires that the absorption coefficients k', K and scattering coefficients S, s^+, s^- appearing in them be determinable by reflection or transmission measurements. The problem is easy to solve, because that determination can be carried out in two steps. The measurements labeled with "dif" contain only the coefficients K and S; these are the typical optical parameters of the Kubelka-Munk theory, to be described in the next section along with methods for their determination. Thus K and S must always be determined first, since these coefficients are assumed known when k', s^+, and s^- are determined.

Accordingly, three coefficients remain to be found, so that three equations have to be employed. The following measurements (all for directional irradiation) are available for the purpose:

- R_r^*, the reflection factor for (quasi) 0/0 geometry with a specularly reflecting substrate having reflection factor $R_{r,0}^*$
- $\varrho_{g,\infty}^*$, the reflectance of the infinitely thick layer
- $\varrho_{g,b}^*$, the reflectance over a black substrate
- $\varrho_{g,w}^*$, the reflectance over a white substrate having reflectance $\varrho_{g,ow}^*$
- T_r^*, the transmission factor measured with (quasi) 0/0 geometry
- τ_g^*, the transmittance

Any three of these measurable quantities can be used in determining k', s^+, and s^-. We will distinguish between determinations based on reflection measurements and those based on transmission measurements.

The equations needed are derived from the formulas of Sections 3.2.2 and 3.2.3. All the quantities from the Kubelka-Munk theory can – because they are (abbreviated terms on the right hand side) known – be substituted in these formulas. They are

$$\tau_{\text{dif}}^* = \frac{b}{A} = \tau \tag{3.92}$$

$$\varrho_{\text{dif},\infty}^* = a - b = \varrho_1 \tag{3.93}$$

$$\varrho_{\text{dif},b}^* = \frac{\sinh x}{A} = \varrho_2 \tag{3.94}$$

$$\varrho_{\text{dif,ow}}^* = \varrho_3 \tag{3.95}$$

These values can either be obtained by direct measurement on black and white drawdowns or calculated from K and S (which are assumed to be known) and h. The notation and relations stated previously are used:

$$a = 1 + \frac{K}{S}$$

$$b = \sqrt{a^2 - 1} = \sqrt{\frac{K}{S}\left(\frac{K}{S} + 2\right)}$$

$$x = bSh$$

$$A = a \sinh x + b \cosh x$$

Special value is attached to the measurement of the film thickness h, which must be done with great care because the error in h strongly influences the accuracy of the calculated absorption and scattering coefficients. It is also vital that the layers on the black and white substrates be equally thick; this can be achieved with black and white glass plates in which the white and black portions are joined together without a "step."

The reflection factor R_r^*, on the other hand, can be measured on a drawdown having a different film thickness h_R on the specularly reflecting substrate.

1 Determination of k', s^+, and s^- from reflection measurements

1.1 From R_r^*, $\varrho_{g,\infty}^*$, $\varrho_{g,b}^*$ (and $R_{r,0}^*$)

By equations (3.45) and (3.66),

$$T_r^* = e^{-\mu h} = \left(\frac{R_r^*}{R_{r,0}^*}\right)^{\frac{h}{2h_R}} = T \qquad (3.96)$$

where h_R is the film thickness over the specular reflective substrate; also, by equations (3.86) and (3.90),

$$p = \frac{\varrho_{g,\infty}^* (1 - \tau T) - \varrho_{g,b}^*}{\varrho_1 (1 - \tau T) - \varrho_2}$$

$$q = \varrho_1 p - \varrho_{g,\infty}^* \qquad (3.97)$$

The auxiliary variables p and q are linked to the coefficients by equations (3.30) and (3.31). If these are solved for s^+ and s^-, the following formulas are obtained:

$$s^+ = p(\mu - aS) + qS \qquad (3.98)$$

$$s^- = pS - q(\mu + aS) \qquad (3.99)$$

The quantity μ needed here follows immediately from equation (3.96):

3 How Spectra Depend on the Scattering and Absorption of Light

$$\mu = \frac{1}{h} \ln \frac{1}{T} \qquad (3.100)$$

so that by equation (3.19)

$$k' = \mu - (s^+ + s^-) \qquad (3.101)$$

The three coefficients have thus been determined.

1.2 From R_r^*, $\varrho_{g,b}^*$, $\varrho_{g,w}^*$ (and $\varrho_{g,ow}^*$)

If $\varrho_{g,\infty}^*$ is difficult of access because of problems in achieving a thick layer, a determination based on a layer over black and white is also possible. The equation for $\varrho_{g,b}^*$ is now supplemented by equation (3.85) for $\varrho_{g,w}^*$, specifically the expression following the second equality sign; then, with $Q = \varrho_{g,b}^*$, p and q are obtained as follows from equation (3.86):

$$p = \frac{\varrho_2 \, T \, \varrho_{g,b}^* + C(1 - \tau \, T)}{\varrho_2^2 \, T + (\tau - T)(1 - \tau \, T)}$$

$$q = \frac{p \, \varrho_2 - \varrho_{g,b}^*}{1 - \tau \, T} \qquad (3.102)$$

with T as above. Here

$$C = \frac{1}{\varrho_3} \left[\frac{1}{\tau} (\varrho_{g,w}^* - \varrho_{g,b}^*)(1 - \varrho_2 \varrho_3) - \varrho_{g,ow}^* \, T \right] \qquad (3.103)$$

With μ as defined previously and the formalism of 1.1, we finally obtain k', s^+, and s^-.

1.3 From $\varrho_{g,\infty}^*$, $\varrho_{g,b}^*$, $\varrho_{g,w}^*$ (and $\varrho_{g,ow}^*$)

This variant does not involve a drawdown on a specularly reflecting substrate. The formulas used are again equations (3.86), (3.90), and (3.85), the expression after the first equality sign in eqn. (3.85) being taken. The term $bP - Q \sinh a$ is transformed, and then q is substituted with $\varrho_{g,\infty}^*$ from equation (3.90). Solving for p leads to an expression that has the same denominator as the formula for p in equation (3.97). By setting the two expressions equal, we obtain directly a formula for T containing no quadratic expressions:

$$T = \frac{\varrho_1 \left[\varrho_{g,w}^* - \varrho_{g,b}^* + \varrho_2 \varrho_3 (\varrho_{g,\infty}^* - \varrho_{g,w}^*)\right] - \varrho_3 (\varrho_{g,\infty}^* - \varrho_{g,b}^*)}{\tau (\varrho_1 \varrho_{g,ow}^* - \varrho_3 \varrho_{g,\infty}^*)} \qquad (3.104)$$

The subsequent analysis is most easily done with equations (3.97)–(3.101).

2 Determination of k', s^+, s^- from transmission measurements

Initially, only $T_r^* = T$ from equation (3.45) and τ_g^* from (3.43) are available for the determination. But a third equation is needed if the three coefficients are to be obtained; this can be derived from a further transmission measurement, $\hat{\tau}_g^*$, with a doubled film thickness h. In the case of suspensions, this involves no difficulties; we simply take a cell with twice the thickness. It is possible, however, even for pigmented coatings, provided these are plastic wafers or self-supporting films (films not applied on a substrate – but optical contact is a special concern here, see Section 3.3.3). All quantities relating to the doubled film thickness are identified by $\hat{}$; then

$$\hat{\tau}_{\text{dif}}^* = \frac{b}{\hat{A}} = \hat{\tau} \tag{3.105}$$

$$\hat{\varrho}_{\text{dif,b}}^* = \frac{\sinh 2x}{\hat{A}} = \hat{\varrho}_2 \tag{3.106}$$

where $\hat{A} = a \sinh 2x + b \cosh 2x$.

Now, by equation (3.43) for τ_g^* and $\hat{\tau}_g^*$,

$$p = \frac{\hat{\varrho}_2 \, T \, (\tau_g^* - T) - \varrho_2 \, (\hat{\tau}_g^* - T^2)}{\hat{\varrho}_2 \, T \, (\tau - T) - \varrho_2 \, (\hat{\tau} - T^2)}$$

$$q = \frac{1}{\varrho_2 \, T} \, [\tau_g^* - T - p \, (\tau - T)] \tag{3.107}$$

If p, q, and T are known, we can go on as in 1.1.

We have thus shown how k', s^+, and s^- can be determined without an excessive degree of complication.

3.3 Kubelka-Munk Theory

3.3.1 Significance and Formalism

That two-flux theory is today the most important and successful specialization of the phenomenological theory. It owes its widespread acceptance to three properties:
- It involves only two optical constants (K and S)
- It is easy to work and lends itself to computer programming
- Over a period of decades, it has proved suitable for the description of practical problems.

The first two of these qualities have their source in the fact that the theory is formulated quite simply, on the basis of the following assumptions:
– Ideally diffuse radiation distribution on the irradiation side
– Ideally diffuse radiation distribution in the interior of the layer
– No consideration of surface phenomena resulting from the discontinuity in refractive index

When deviations from the Kubelka-Munk theory are seen in application to practical problems, the reason is nearly always that one or more of these assumptions have been neglected*. The third assumption drops out when the Saunderson corrections are employed, as has become indispensable for most applications today.

We have already encountered the basic equations of Kubelka and Munk in deriving the four-flux theory. Because they are so important, however, they are written out explicitly here, except that the subscript "dif" has been omitted (since there is no radiation other than diffuse).

Transmittance:

General:

$$\tau^* = \frac{b}{a \sinh bSh + b \cosh bSh} \quad \begin{array}{r}(3.44)\\(3.108)\end{array}$$

Absorption only:

$$\tau^*_{abs} = e^{-Kh} \quad \begin{array}{r}(3.49)\\(3.109)\end{array}$$

Scattering only:

$$\tau^*_{sca} = \frac{1}{1 + Sh} \quad \begin{array}{r}(3.52)\\(3.110)\end{array}$$

Reflectance:

General:

$$\varrho^* = \frac{1 - \varrho_0^*(a - b \coth bSh)}{a - \varrho_0^* + b \coth bSh} \quad \begin{array}{r}(3.65)\\(3.111)\end{array}$$

Absorption only:

$$\varrho^*_{abs} = \varrho_0^* e^{-2Kh} \quad \begin{array}{r}(3.70)\\(3.112)\end{array}$$

---

* In such cases, many-flux theories may be helpful; see also the discussion in Section 7.2.3.

Scattering only:

$$\varrho^*_{sca} = \frac{(1 - \varrho^*_0)\, Sh + \varrho^*_0}{(1 - \varrho^*_0)\, Sh + 1} \qquad (3.80)$$
$$\qquad\qquad\qquad\qquad\qquad\qquad (3.113)$$

The symbols here have the following meanings: K and S are the absorption and scattering coefficients defined as in equation (3.17); ϱ^*_0 is the reflectance of the substrate; the auxiliary variables a and b are defined as in equations (3.18) and (3.27), $a = 1 + K/S$ and $b = (a^2 - 1)^{1/2}$; and finally h is the film thickness.

3.3.2 Limiting Cases of Reflection

Because equations (3.111)–(3.113) also hold for substrates with specular reflection only and with diffuse reflection only, there are just two limiting cases of the reflectance to be treated here:
- Reflectance on black substrate
- Reflectance of the opaque ("infinitely thick") layer

To obtain the reflectance on a *black substrate,* we use the equations from Section 3.3.1, setting ϱ^*_0 equal to zero.

General:

$$\varrho^*_b = \frac{1}{a + b \coth bSh} \qquad (3.114)$$

Scattering only:

$$\varrho^*_{sca,b} = \frac{Sh}{1 + Sh} \qquad (3.115)$$

For the "absorption only" case, we get the uninteresting trivial solution $\varrho^*_{abs,b} = 0$.

The easiest way to obtain the reflectance of the *infinitely thick* layer is to take equation (3.144) to the limit as $h \to \infty$. Then $\coth bSh \to 1$, and with the relation $a^2 - b^2 = 1$ implied by equation (3.27),

$$\varrho^*_\infty = a - b = 1 + \frac{K}{S} - \sqrt{\frac{K}{S}\left(\frac{K}{S} + 2\right)} \qquad (3.116)$$

Solving for K/S, we obtain the famous "Kubelka-Munk function" F_{KM}:

96   3 How Spectra Depend on the Scattering and Absorption of Light

$$\boxed{\frac{K}{S} = \frac{(1-\varrho_\infty^*)^2}{2\varrho_\infty^*} \equiv F_{\text{KM}}}$$ (3.117)

(The extreme cases "absorption only" and "scattering only" lead to trivial solutions: $\varrho_{\text{abs},\infty}^* = 0$ and $\varrho_{\text{sca},\infty}^* = 1$.)

For didactic purposes, it may well be desirable to derive the Kubelka-Munk function directly from the differential equations. The typical equations for the Kubelka-Munk theory are obtained from equations (3.22) and (3.23) by setting the directional radiances l^+ and l^- (Fig. 3-3) equal to zero. Dividing the first equation by SL^+ and the second by SL^- and using $L^-/L^+ = \varrho^*$, one gets

$$\frac{1}{S} \cdot \frac{1}{L^+} \cdot \frac{dL^+}{dz} = -\left(1 + \frac{K}{S}\right) + \varrho^*$$

$$\frac{1}{S} \cdot \frac{1}{L^-} \cdot \frac{dL^-}{dz} = \left(1 + \frac{K}{S}\right) - \frac{1}{\varrho^*}$$

The "infinitely thick" layer is characterized by the fact that a further increase in film thickness does not change the reflectance. Thus the percent changes in radiance per unit increase of film thickness in the incidence direction ($1/L^+ \cdot dL^+/dz$) and opposite to the incidence direction ($1/L^- \cdot dL^-/dz$) must be exactly equal. If both right-hand sides are set equal and the equation is solved for K/S, the limit as $\varrho^* \to \varrho_\infty^*$ directly yields the Kubelka-Munk function (3.117).

Equation (3.117) is the most important statement of the Kubelka-Munk theory and the most often used of all formulas in the theory. It links the reflectance of the infinitely thick layer with the absorption and scattering coefficients K and S governing the optical process, making ϱ_∞^* directly dependent on the ratio of the absorption to the scattering coefficient (and not on the value of either). The equation is also available in tabular form; it has found widest use where information about the ratio of absorption to scattering is to be extracted from reflection measurements.

Figure 3-7 is a plot of the Kubelka-Munk function. Typical features are the steep drop from very large values at low reflectance and the branch decreasing slowly toward zero as the reflectance approaches 1. If the variables are interchanged, the reflectance then declines rapidly from unity at low K/S; this fact accounts for the well-known high sensitivity of white specimens to contamination.

If the logarithm of F_{KM} is plotted versus reflectance, the curve of Figure 3-8 results. As can be seen, there is a wide reflectance range in which the curve has a quasilinear shape. The equation for this straight line can be obtained by calculating the point of inflection and the inflectional tangent of the curve. A better approximation can be found by a linear regression calculation in the region of interest:

3.3 Kubelka-Munk Theory

$$\ln F = A - B\varrho_\infty^* \quad \text{or} \quad \log F = A_{10} - B_{10}\varrho_\infty^* \tag{3.118}$$

where $A = 1.68$, $B = 6.182$, $A_{10} = 0.7296$, and $B_{10} = 2.685$ for $0.13 \leq \varrho_\infty^* \leq 0.73$ (i.e., $0.09 \leq \varrho_{\text{in},\infty} \leq 0.54$).*

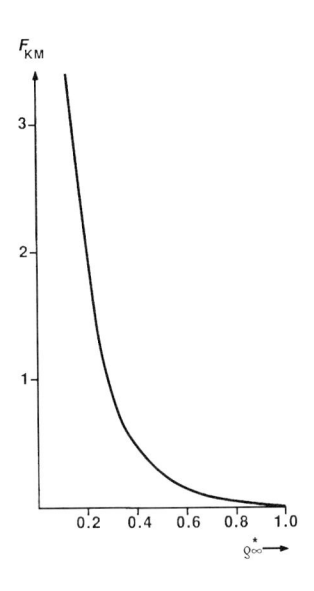

Fig. 3-7. The Kubelka-Munk function $F_{\text{KM}}(\varrho_\infty^*)$.

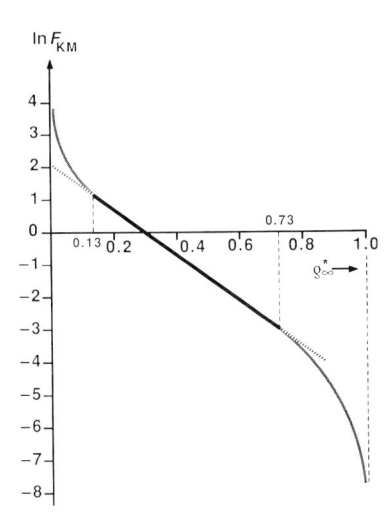

Fig. 3-8. The function $\ln F_{\text{KM}}(\varrho_\infty^*)$.

---

*The subscript "in" refers to measurement including surface reflection (see Section 6.1.2).

3 How Spectra Depend on the Scattering and Absorption of Light

The values of the coefficients are somewhat dependent on higher accuracy in the lower or the upper part of the range. The values stated above represent a good approximation over the entire region. Between $\varrho_\infty^* = 0.20$ and 0.65, the deviation from the true F value is less than 3%, coming to exceed 20% only at the bounds cited above.

What appears to be a bothersome restriction on the use of the approximation formula is actually not much of a hindrance: the requirements on the accuracy of measurement in the determination of F already restrict the field in which the measurements must be done. Let us consider the relative error of determination of F in reflection measurements:

$$\frac{dF}{F} = \frac{\dfrac{d \ln F}{d\varrho^*}}{\dfrac{d\varrho}{d\varrho^*}} d\varrho = -\frac{1+\varrho^*}{1-\varrho^*} \cdot \frac{1-0.6\varrho^*}{\varrho - 0.04} d\varrho \tag{3.119}$$

As Figure 3-9 shows, given a fairly realistic measurement error of $d\varrho = 0.001$, the relative error of F is smaller than 1% only when measurements are performed with $0.2 \leq \varrho_{in,\infty} \leq 0.75$ (i.e., $0.33 \leq \varrho_\infty^* \leq 0.88$). The measurement interval does not expand to $0.1 \leq \varrho_{in,\infty} \leq 0.88$ (i.e., $0.14 \leq \varrho_\infty^* \leq 0.95$) until a 2% error (dF/F) is allowed. Here we have a reflection counterpart to the spectroscopists' old rule, "For extinction measurements, the transmission must always be between 8 and 80 % (or the decadic extinction between 1 and 0.1)."

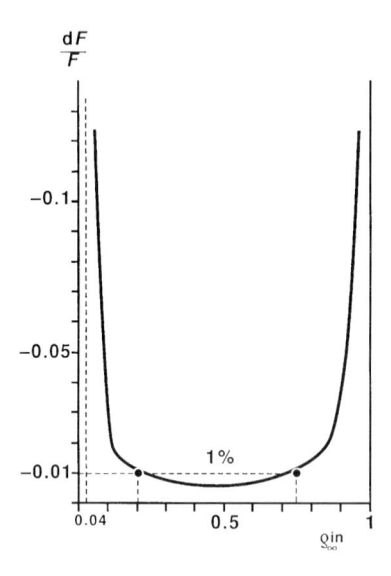

Fig. 3-9. Relative error of the Kubelka-Munk function $F(\varrho_{in,\infty})$ for a measurement error $d\varrho = 0.001$.

3.3.3 Determination of the Absorption and Scattering Coefficients

The *classical method* of determining the coefficients is to measure directly the reflectance ϱ_∞ of the infinitely thick layer and, from this, determine, after Saunderson correction, a and b. Equations (3.116) and (3.27) imply that

$$a = \frac{1}{2}\left(\frac{1}{\varrho_\infty^*} + \varrho_\infty^*\right)$$

$$b = a - \varrho_\infty^* = \frac{1}{2}\left(\frac{1}{\varrho_\infty^*} - \varrho_\infty^*\right) \tag{3.120}$$

K and S can be obtained by solving equation (3.111) for S:

$$S = \frac{1}{bh}\,\text{Arcoth}\,\frac{1 - a\,(\varrho^* + \varrho_0^*) + \varrho^*\,\varrho_0^*}{b\,(\varrho^* - \varrho_0^*)}$$

$$K = S\,(a - 1) \tag{3.121}$$

This procedure requires two coatings: the infinitely thick layer and a coating with thickness h on an arbitrary substrate with reflectance ϱ_0^*. The measured reflectances on these two coatings are ϱ_∞^* and ϱ^* respectively. Accordingly, only three measurements in all (including that of the film thickness h) are needed in order to determine K and S.

A *second method* is to determine the reflectances ϱ_b^* and ϱ_w^* of layers of equal thickness h on black and white substrates having known reflectances ϱ_{ob}^* and ϱ_{ow}^*. Accordingly, coth x can be eliminated from the two equations (3.111), written for ϱ_b^* and ϱ_w^*. The result is an equation for a:

$$a = \frac{(1 + \varrho_w^* \varrho_{ow}^*)\,(\varrho_b^* - \varrho_{ob}^*) + (1 + \varrho_b^* \varrho_{ob}^*)\,(\varrho_{ow}^* - \varrho_w^*)}{2\,(\varrho_b^* \varrho_{ow}^* - \varrho_w^* \varrho_{ob}^*)} \tag{3.122}$$

Once a has been obtained, b is also known from equation (3.27). Then, from equation (3.121), K and S follow (as above) if ϱ^* and ϱ_0^* are replaced by the pair ϱ_w^*, ϱ_{ow}^* or ϱ_b^*, ϱ_{ob}^*.

The classical method is generally more accurate, as an error analysis on white pigments has shown.

Often, the reflectance of the black substrate ϱ_{ob}^* can be set equal to zero, as described in the preceding section. Equations (3.122) and (3.121) then simplify to

3 How Spectra Depend on the Scattering and Absorption of Light

$$a = \frac{1}{2}\left(\varrho_w^* + \frac{\varrho_b^* + \varrho_{ow}^* - \varrho_w^*}{\varrho_b^* \varrho_{ow}^*}\right) \tag{3.123}$$

$$S = \frac{1}{bh}\operatorname{Arcoth}\frac{1 - a\,\varrho_b^*}{b\varrho_b^*} \tag{3.124}$$

Another way to determine K and S is with two *transmission measurements*, provided the film thickness is doubled for the second measurement. With suspensions, this can be done by using a cell with twice the cell thickness; with wafer-shaped plastic specimens, by laying one wafer on another. "Optical contact" is a matter of concern here; it can be insured, for example, by applying an oil film to the surfaces and pressing the two wafers together.

Because of the double film thickness, there are two measured values τ^* and $\hat{\tau}^*$ available to serve as arguments in the respective equations, giving $bSh = x$ in one case and $2bSh = 2x$ in the other. The auxiliary variables a and b are thus not affected by doubling the thickness, since they do not depend on the thickness. The addition theorems for hyperbolic functions make it possible to obtain simple relations for this case:

$$a = \frac{1}{2}\left[\frac{1}{t}\left(\frac{1}{\tau^*} - \tau^*\right) + \tau^*\,t\right] \tag{3.125}$$

Here we have used the abbreviated notation

$$t = \sqrt{\frac{1}{\tau^{*2}} - \frac{1}{\hat{\tau}^*}} \tag{3.126}$$

The following equation can be used to determine S:

$$S = \frac{1}{bh}\operatorname{Arsinh} bt \tag{3.127}$$

3.4 Hiding Power

3.4.1 General

The official definition of the hiding power of a pigmented material is "its ability to hide the color or color differences of the substrate." The reader may be a little perturbed by the fact that this definition* uses the word "hide" to explain "hiding power," but in defense of the writers of the standards it must be said that there is scarcely a better way to express what is meant.

The model underlying the hiding power relates to a black/white contrast substrate on which the coating is to be applied. The determination of the hiding power value involves ascertaining the film thickness h such that the contrast substrate just vanishes for the eye performing the assessment. The thickness h_D (in mm) that satisfies this condition is called the "hiding layer"; its reciprocal $D = 1/h_D$ (in mm^{-1} = m^2/L) is the hiding power value. Like the tinting strength to be discussed further on, it is a measure of effectiveness, since it states the area (in m^2) that can be covered with 1 L of paint applied in a hiding coat.

If the vanishing of the contrast substrate were actually determined with the colorist's eye, hiding power would be a subjective quantity; to derive a functional method of determination, it is therefore necessary to place the procedure on a colorimetric basis.** For achromatic coatings, the "contrast ratio" earlier was used as the hiding criterion; this was the ratio of reflectances on the black and white portions of the substrate for equal film thicknesses. The contrast ratio was specified as 0.98 in order to take account of the 2 % threshold value for brightness perception by the human eye. With colored coatings, brightness contrast is not meaningful, so the hiding criterion was defined some time ago as a certain CIELAB color difference. The criterion $\Delta E^*_{ab} = 1$ is commonly employed. The experimental procedure is as follows: Coatings of various thicknesses are applied to the agreed-on contrast substrate. The color difference ΔE^*_{ab} between the two contrasting fields is then determined by colorimetry for each of the coatings. If the results are plotted versus the reciprocal of film thickness, the hiding power value in m^2/L can be read directly from the diagram at the point corresponding to the criterion $\Delta E^*_{ab} = 1$.

An example is presented in the experimental section (Chapter 7). This determination procedure has, however, two serious deficiencies that can be seen at first glance:
- Several drawdowns with different film thicknesses must be prepared and measured
- The graphical evaluation is imprecise.

---

* For standards see "Coloring Materials, Terms" in Appendix 1.
** Substrates must, of course, be standardized for an exact method of determination. Reflectance values $\varrho_{0b} = 0$ and $\varrho_{0w} = 0.8$ are employed.

102   3 How Spectra Depend on the Scattering and Absorption of Light

It follows that this method must be rationalized, for it would be desirable to use the minimum number of paint drawdowns and also to employ a computer for the evaluation. The following problem now arises: The Kubelka-Munk theory, next to colorimetry the most important principle for rationalizing an optical pigment testing method, applies only to spectral reflectances. On what wavelength should the measurement and calculation be based? It is for this purpose that the "principle of spectral evaluation" was developed as a link from colorimetry back to the Kubelka-Munk theory (Section 3.6).

First, however, achromatic coatings will be analyzed. These lend themselves to a direct application of Kubelka-Munk theory provided certain constraints are observed. This section therefore lays the groundwork for the principle of spectral evaluation.

3.4.2 Achromatic Coatings

In the ideal case (i.e., when the coating contains no colored substances), the small color difference ΔE_{ab}^* remaining between the coating over white and over black can be set equal to the lightness difference ΔL^*. The same holds, however, when there is a slight undertone; for a nearly white or nearly gray specimen, the chroma difference and hue difference between white and black must be of a higher order than the lightness difference. Then, by (2.5) and subsequent equations,

$$\Delta E_{ab}^* = \Delta L^* = 116\, Y_w^* - 16 - (116\, Y_b^* - 16) = 116 \left(\sqrt[3]{\frac{Y_w}{100}} - \sqrt[3]{\frac{Y_b}{100}} \right) \qquad (3.128)$$

where b stands for black and w for white. This equation should be valid for white and light gray coatings, so that the value function for $Y > 0.8856$ has been used here. Further, because of the achromatic character, we can set $Y/100 = \varrho$ (i.e., the reflectance independent of λ).

What is more, we must take into account that the colorist, in the (computationally simulated) application of the hiding power criterion, will select a color matching angle such that the gloss of the specimen surface will not interfere. Thus, for glossy specimens, the colorist "reflects out" the glossy effect by taking a suitable observation angle; this corresponds to a reflectance of $\varrho - gr_o$, where g is the degree of gloss. Now the hiding power criterion can be formulated as

$$\Delta E_{ab}^* = \Delta L^* = 116\, (\sqrt[3]{\varrho_w - gr_o} - \sqrt[3]{\varrho_b - gr_o}) = 1 \qquad (3.129)$$

This condition with equations (3.111), (3.10), and (3.120) forms a closed system that makes it possible to calculate the hiding power value. Specifically, for given auxiliary variables a and b, the product $Sh_D = 1/a$ can be determined. Because S and h in the Kubelka-Munk formulas appear only as their product,

3.4 Hiding Power

if the scattering coefficient S is known along with the value of a, then the hiding power value D can be calculated immediately, for $D = 1/h_D = aS$.

Accordingly, the hiding power value is calculated in a three-step procedure:

1. Determination of auxiliary variables a and b

   This can be done in either of two ways (see Section 3.3.3):

   – ϱ_∞ is available as a measured value; then by equation (3.120)

   $$a = \frac{1}{2}\left(\frac{1}{\varrho_\infty^*} + \varrho_\infty^*\right); \qquad b = a - \varrho_\infty^*$$

   – ϱ_w (along with ϱ_{ow}) and ϱ_b are available as measured values; then by equations (3.123) and (3.27)

   $$a = \frac{1}{2}\left(\varrho_w^* + \frac{\varrho_b^* + \varrho_{ow}^* - \varrho_w^*}{\varrho_b^* \varrho_{ow}^*}\right)$$

   $$b = \sqrt{a^2 - 1}; \qquad \varrho_\infty^* = a - b$$

2. Calculation of scattering coefficient S

   Equation (3.124) can be used for the calculation of S:

   $$S = \frac{1}{bh}\,\text{Arcoth}\,\frac{1 - a\varrho_b^*}{b\varrho_b^*}$$

3. Determination of the factor α

   The system of equations (3.111), (3.114), (3.10), and (3.129) in complete form, with $\alpha = 1/Sh_D$, is

   $$\varrho_b^* = \frac{1}{a + b\coth\dfrac{b}{\alpha}}$$

   $$\varrho_w^* = \frac{1 - \varrho_{ow}^*\left(a - b\coth\dfrac{b}{\alpha}\right)}{a - \varrho_{ow}^* + b\coth\dfrac{b}{\alpha}}$$

   $$\varrho_b = r_o + \frac{(1 - r_o)(1 - r_2)}{1 - r_2\varrho_b^*}\varrho_b^* \qquad \varrho_b^* = \frac{0.36\,\varrho_b^* + 0.04}{1 - 0.6\,\varrho_b^*}$$

   $$\varrho_w = r_o + \frac{(1 - r_o)(1 - r_2)}{1 - r_2\varrho_w^*}\varrho_w^* \qquad \varrho_w^* = \frac{0.36\,\varrho_w^* + 0.04}{1 - 0.6\,\varrho_w^*}$$

   $$\sqrt[3]{\varrho_w - gr_o} - \sqrt[3]{\varrho_b - gr_o} - \frac{1}{116} = 0 \qquad (3.130)$$

3 How Spectra Depend on the Scattering and Absorption of Light

Solution of this system leads to a final equation of at least sixth degree in coth (b/a). Such an equation cannot, however, be solved in elementary functions, and so the function a, which depends on a (or ϱ_∞^*) and g, cannot be written out explicitly.* An iterative solution procedure must therefore be used, for example Newton's method. For this purpose, the derivatives of the individual equations in the system are needed; with $\xi = 1/a$, we have

$$\acute{\varrho}_b^* = \frac{d\varrho_b^*}{d\xi} = \left(\frac{b}{a \sinh b\xi + b \cosh b\xi}\right)^2$$

$$\acute{\varrho}_w^* = \frac{d\varrho_w^*}{d\xi} = b^2 \frac{1 - 2a\varrho_{ow}^* + \varrho_{ow}^{*2}}{[(a - \varrho_{ow}^*) \sinh b\xi - b \cosh b\xi]^2}$$

$$\acute{\varrho}_b = \frac{d\varrho_b}{d\xi} = \frac{(1 - r_o)(1 - r_2)}{(1 - r_2 \varrho_b^*)^2} \acute{\varrho}_b^* = \frac{0.384 \, \acute{\varrho}_b^*}{(1 - 0.6 \, \varrho_b^*)^2}$$

$$\acute{\varrho}_w = \frac{d\varrho_w}{d\xi} = \frac{(1 - r_o)(1 - r_2)}{(1 - r_2 \varrho_w^*)^2} \acute{\varrho}_w^* = \frac{0.384 \, \acute{\varrho}_w^*}{(1 - 0.6 \, \varrho_w^*)^2} \qquad (3.131)$$

Setting the left-hand side of the last equation in (3.130) equal to η (which must be made zero) gives

$$\acute{\eta} = \frac{d\eta}{\delta\xi} = \frac{1}{3}\left[\frac{\acute{\varrho}_w}{(\varrho_w - gr_o)^{\frac{2}{3}}} - \frac{\acute{\varrho}_b}{(\varrho_b - gr_o)^{\frac{2}{3}}}\right] \qquad (3.132)$$

These equations can be used with a desk calculator to prepare a table of a as a function of ϱ_∞^* and g. Such a table, for use as a computational aid, is appended to the first publication (see [24] in Section 3.9). Moreover, the tabulated value can be approximated very closely with a nonlinear regression calculation, so that values of a can be obtained from simple quadratic expressions with two coefficients. This computational aid is included in Chapter 7. There are accordingly two ways of obtaining individual a values without using the somewhat unwieldy formulas from above.

The hiding power value is then

$$\boxed{D = \frac{1}{h_D} = aS} \qquad (3.133)$$

---

* Once again (see also the first footnote in Section 2.3.3), it would have been more advantageous to give the value function a logarithmic form. Then the system of equations could have been solved in elementary functions, so that a could have been written explicitly as a function of ϱ_∞^*.

The following measurements are therefore needed in order to determine the hiding power value:
- The reflectance of the infinitely thick layer (ϱ_∞) or, in its place, the reflectance over white (ϱ_w) for a film thickness the same as over black, together with the reflectance of the white substrate (ϱ_{ow})
- The reflectance over black (ϱ_b)
- The degree of gloss of the infinitely thick layer or the coating over white (g)
- The film thickness over black (h)

In this way, the goal of rationalization stated in Section 3.4.1 has been achieved.

The determination method discussed here is essentially the method of DIN 55 984,* except that the standard is based on high-gloss specimens and does not allow for the actual degree of gloss g. The method just described can therefore be regarded as a refinement of the DIN method. It becomes identical with the DIN method if g is set equal to 1 in all the equations.

3.4.3 Scattering and Absorption Components

The hiding power is entirely due to the cooperation of scattering and absorption, as has already been shown. The "scattering component" and the "absorption component" are those contributions to the hiding power value, expressed as fractions of the total hiding power, that arise from "hiding power due to scattering alone" and "hiding power due to absorption alone." To obtain the equations for these components, we take the Kubelka-Munk equations to the limits $K \to 0$ and $S \to 0$, respectively.

In the assessment of the hiding quality of a coating or pigment, a knowledge of these contributions is important for, say, a white pigment, where only the hiding power contribution due to light scattering is valuable – in this case, the contribution due to absorption is not useful because absorption components in the case of a white pigment are equivalent to loss of quality. As far as possible, therefore, white pigments should cover solely by virtue of their scattering power, not because they also absorb. But this classification can also be significant for colored pigments; in particular, inorganic pigments are often rated better, the more their hiding is due to scattering.

The formulas will now be restated, first for the "hiding power due to scattering only" (equations 3.115 and 3.113):

$$\varrho_b^* = \frac{Sh}{1 + Sh} \quad ; \quad \varrho_w^* = \frac{(1 - \varrho_{ow}^*)\,Sh + \varrho_{ow}^*}{(1 - \varrho_{ow}^*)\,Sh + 1}$$

---

* For further standards see "Hiding Power" in Appendix 1.

106   3 How Spectra Depend on the Scattering and Absorption of Light

and then for the "hiding power due to absorption only" (equation 3.112):

$$\varrho_b^* = 0 \quad \text{(since } \varrho_{ob}^* = 0\text{)}; \qquad \varrho_w^* = \varrho_{ow}^* \, e^{-2Kh}$$

These equations will be used later when the determination of the general hiding power value is discussed.

Again, however, the solution needed for rationalization can be stated immediately for *achromatic* coatings. The "hiding power due to scattering only" D_{sca} is obtained from equation (3.113) with $a_b = 1/Sh_D$:

$$\varrho_b^* = \frac{1}{1 + a_b}; \qquad \varrho_w^* = \frac{1 - \varrho_{ow}^* \,(1 - a_b)}{1 - \varrho_{ow}^* + a_b} \qquad (3.134)$$

These two equations are simply inserted in system (3.130) in place of the first two equations; the iteration then yields D_{sca} directly.

The "hiding power due to absorption only" D_{abs} can be calculated in a similar way. With equation (3.112) and $\beta = 1/Kh_D$,

$$\varrho_b^* = 0; \qquad \varrho_w^* = \varrho_{ow}^* \, e^{-\frac{2}{\beta}} \qquad (3.135)$$

In the case of the absorption component, it is actually possible to solve the system of equations explicitly. We do not need to do this, since it is sufficient to calculate β using equation (3.130) after inserting equations (3.135) in place of the first two equations in the system. Both tables and simple linear formulas are available for the scattering and absorption cases, so that β can be determined; we will become acquainted with these aids in Chapter 7. The "hiding power due to absorption only" D_{abs} is obtained from

$$D_{abs} = \frac{1}{h_D} = \beta K = \beta \, (a - 1) \, S \qquad (3.136)$$

From D_{sca} and D_{abs}, the scattering and absorption components δ_{sca} and δ_{abs} of the hiding power D are calculated as follows:

$$\boxed{\delta_{sca} = \frac{D_{sca}}{D}; \qquad \delta_{abs} = \frac{D_{abs}}{D}} \qquad (3.137)$$

In many practical examples, it has turned out that a residue is still present after these two components are subtracted.

How can this be accounted for? Obviously, that contribution to the hiding power arises from the interaction of scattering and absorption. For if the scattering coefficient increases, the mean distance traveled by light in the layer also grows longer because the light is scattered and rescattered more times. A longer path length, however, necessarily means that more light is absorbed, so there is an additional "scattering/absorption contribution," which would not be present if the layer scattered but did not absorb or if it absorbed but did not scatter. This contribution will be called the "interaction component" δ_{int} and is given by

$$\delta_{int} = 1 - \delta_{sca} - \delta_{abs} \qquad (3.138)$$

The three components given by equations (3.137) and (3.138) can provide much additional information about the hiding power value.

3.5 Transparency

3.5.1 Description and Definition

If several layers are printed on a black substrate, one over another, it is generally seen that the layer increasingly displays a more-or-less weak, bright shimmer in comparison to the unprinted black substrate. This shimmer grows stronger as the process is repeated. The same phenomenon is known with lacquers and paints, provided they are sufficiently transparent. It obviously comes about because scattering particles, commonly pigment particles, are present in the medium (ink or paint). If, in an unprejudiced spirit, one first considers "transparency" as a purely optical phenomenon, it is clear that this scattering effect is opposed to it.

This argument is the basis for the definition of transparency as applied to the field of pigments. The standards* define transparency as the property of a system (not necessarily a pigmented system) to scatter light as little as possible. Accordingly, the color change on application to a black substrate should be as slight as possible. The transparency of the system is greater, the less color change takes place. Given identical specimen preparation, the transparency of the system can be employed for a comparative assessment of transparent pigments.

An expedient approach to the exact definition of a "transparency number," according to the discussion just presented, is to use the CIELAB color difference ΔE^*_{ab} as a function of the film thickness h. Figure 3-10 gives such a plot, showing that the reciprocal slope of the tangent at the origin provides a suitable measure for the transparency number T. We then have

$$T = \tan(90° - \varphi) = \cot \varphi = \frac{1}{\tan \varphi} = \frac{1}{\left.\dfrac{d\Delta E^*_{ab}}{dh}\right|^b_{h=0}} \qquad (3.139)$$

---
* For standards see "Transparency" in Appendix 1.

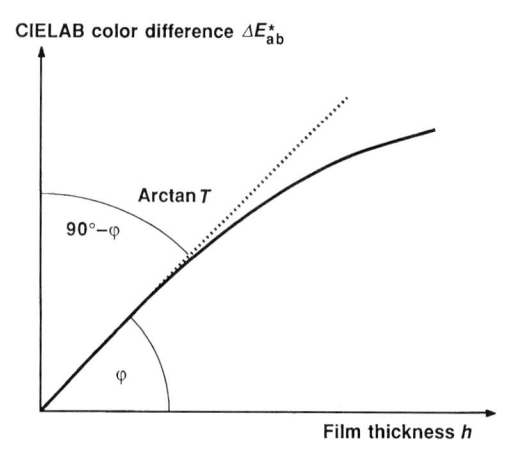

Fig. 3-10. Definition of the transparency number T.

Because the color difference has the dimension 1, expression (3.139) has the units of a length. If the film thickness is measured in mm, the transparency number takes on units of 1 mm = 1 L/m². But what does it mean to state the transparency number T in L/m²? According to the definition in equation (3.139), the number T obviously states how many liters of an ink or paint can be printed or spread on one square meter of a black substrate before ΔE_{ab}^* increases by 1. This interpretation suggests an effectiveness property similar but not identical to the hiding power value. Of two paints, then, the one having the better transparency is the one that can be applied in a thicker coating (i. e., one giving more color) to achieve a given transparency.

Like the definition of the hiding power value, this definition immediately implies a method for determining the transparency number. The curve of ΔE_{ab}^* versus h is plotted through a few experimental points, and then the tangent at the origin is found graphically or by calculation. If one then works close enough to the origin, a single measured point (aside from the origin = black substrate) is generally sufficient.

As in the case of the hiding power, such a procedure lends itself to rationalization. As in the earlier case, one or at most two measurements will be adequate; here, however, the Kubelka-Munk theory, because of its monochromatic character, is not directly applicable. This point will also be taken up in Section 3.6.

3.5.2 Coloring Power

By analogy with the transparency number T, a "coloring power" Φ can be defined. This step is favored by the following argument: of two coatings equal in transparency, the one giving more color is certainly the better. This ability to give color, called coloring power, was not considered in the determination of the transparency number. The determination of color strength must make use of a white substrate;* again a curve of ΔE_{ab}^* versus h is employed. In accordance with the above procedure, the angle of the tangent at the origin is defined as the coloring power; because of the direct proportionality, this is larger the steeper the tangent line (Fig. 3-11).

The coloring power can be read off the plot:

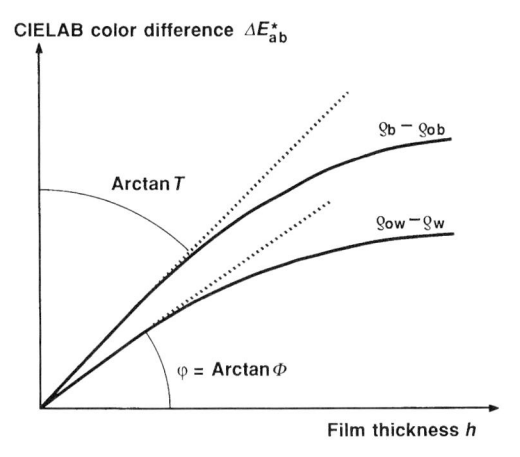

$$\Phi = \tan \varphi = \left. \frac{d\Delta E_{ab}^*}{dh} \right|_{h=0}^{W} \quad (3.140)$$

Fig. 3-11. Definition of coloring power Φ in comparison with transparency number T.

Like the hiding power value, Φ has the dimensions of reciprocal length. Thus it is, once more, an effectiveness property, for $\Phi = 1$ mm^{-1} = 1 m^2/L means that 1 m^2 of the white substrate must be coated with one liter of the paint or ink in order to make the color difference ΔE_{ab}^* relative to the white substrate equal to 1. The dimensions of the transparency number and coloring power are thus reciprocals; accordingly, the product of Φ and T is a quantity having the dimension 1, which we will call the "transparency-coloring power" L. The term "transparency-coloring power" also includes an evaluation of the color-giving property in the sense of the statement, made above, that between two color-giving films of equal transparency the one giving more color is the better. Thus we have

$$L = \Phi T \quad \text{"transparency-coloring power"} \quad (3.141)$$

---

* Standardized as an ideally white substrate, $\varrho_{ow} = 1$.

We must now look ahead to the next section, where it will be shown that the transparency number T as we have defined it is the reciprocal of the colorimetrically evaluated scattering coefficient S, while the transparency-coloring power Φ is directly proportional to the colorimetrically evaluated absorption coefficient K. Accordingly, the product in equation (3.141) contains both these coefficients, even though in colorimetrically evaluated form, precisely as in the numerator and denominator of the Kubelka-Munk function K/S. Conversely, to any K/S value a reflectance can be assigned (see, e.g., eq. 3.116), and so from L we can finally calculate a "transparency-coloring degree" $G = 1 - \varrho^*$, which like the reflectance must lie between 0 and 1:

$$\boxed{G = \sqrt{L(L+2)} - L} \quad \text{"transparency-coloring degree"} \quad (3.142)$$

The quantities L and G are highly suitable for characterizing transparency-coloring action (i.e., along with color-giving action).

3.5.3 Achromatic Coatings

In this section we will show that a model analysis based on achromatic systems can also be meaningful in the understanding of transparency. In practice, such systems play virtually no part; the value of such an analysis lies rather in that it will aid us in acquiring important knowledge.

We proceed as in the definition of T, equation (3.139), following the established pattern (see Section 3.4.2) of replacing ΔE^*_{ab} by $\Delta L^* = L^*_b - L^*_{ob} = L^*_b$; we then have

$$\frac{1}{T} = \acute{L}^*_b = 116\, \acute{Y}^*_b$$

(A prime denotes the result of applying the operator $(d/dh)_{h=0}$ to the unprimed quantity.) For Y^*_b we must use the linear form of equation (2.6) here because of the nearness to the origin. If $Y_b/100 = \varrho_b$,

$$\frac{1}{T} = 116 \cdot 7.787 \cdot \acute{\varrho}_b$$

By the third equation of (3.131), this gives

$$\frac{1}{T} = 116 \cdot 7.787 \cdot 0.384 \cdot \acute{\varrho}^*_b$$

3.5 Transparency 111

Provided the layer does not absorb but only scatters, equation (3.115) applies. The above operator then yields $\acute{o}_b^* = S$, so that

$$\frac{1}{T} = 116 \cdot 7.787 \cdot 0.384 \cdot S = 346.9\ S$$

in other words, the transparency number is inversely proportional to the scattering coefficient of the pigmented layer:

$$T = 0.002883\ \frac{1}{S} \tag{3.143}$$

A similar analysis can be carried out for the coloring power Φ. In definition (3.140), ΔE_{ab}^* is replaced by $\Delta L^* = L_{ow}^* - L_w^* = 1 - L_w^*$, so that

$$\Phi = -\acute{L}_w^* = -116\ \acute{Y}_w^*$$

The cube-root form of equation (2.5) must be used for the value function Y_w^*. Then, with $Y_w/100 = \varrho_w$ and $\varrho_{ow} = 1$,

$$\Phi = -\frac{116}{3\cdot 1}\acute{\varrho}_w = -\frac{116}{3}\cdot \frac{0.384}{(1 - 0.6\ \varrho_{ow}^*)^2}\ \acute{\varrho}_w^*$$

This holds provided the layer absorbs only and does not scatter. Then we have the limit equation (3.112) for ϱ_w^*; the operator gives $\acute{o}_w^* = -2\varrho_{ow}^*\ K$. With $\varrho_{ow}^* = 1$,

$$\Phi = \frac{2\cdot 116\cdot 0.384\cdot 1}{3\ (1 - 0.6\cdot 1)^2}\ K = 185.6\ K \tag{3.144}$$

We thus have the important result that, with the transparency number T and the coloring power Φ as defined in Sections 3.5.1 and 3.5.2,
- T is inversely proportional to the scattering coefficient S in achromatic, nonabsorbing layers
- Φ is directly proportional to the absorption coefficient K in achromatic, nonscattering layers

This close relationship between T and S, on the one hand, and on the other between Φ and K is qualitatively the same for chromatic layers. The example also shows clearly that forming the product $L = \Phi T$ gives an expression proportional to K/S.

Section 3.6 will now show that S and K in these formulas are represented by the colorimetric integrals of $S(\lambda)$ and $K(\lambda)$ in the case of colored coatings.

3.6 Principle of Spectral Evaluation

3.6.1 Description and Significance

In order to work out a rational procedure similar to that for the "achromatic" hiding power (Section 3.4.2), we must face the unanswered question of how to use the Kubelka-Munk theory, which is intrinsically "spectral", in solving the problem. The restriction to a single wavelength is typical for achromatic layers, as has been shown. With colored coatings, however, it is not immediately clear whether the measurement and calculation should be done for a preferred wavelength, although everything speaks for the necessity that the entire spectrum should be included in the calculation. But what approach to evaluation then remains open? These are the questions that will be answered once the principle of spectral evaluation has been discovered.

The procedure can be described as a feed-back between the Kubelka-Munk theory and colorimetry. As has already been illustrated more than once (e.g., in Sections 3.3.3 and 3.4.2), one determines K and S on an infinitely thick layer and on a finite layer of known thickness over black. The infinitely thick layer can be replaced by a layer over white having the same thickness as the layer over black. This determination of K and S is done not once, but several times for distinct wavelengths spread throughout the visible spectrum. As we have seen, $K(\lambda)$ and $S(\lambda)$ as wavelength-specific constants of a coating govern its optical behavior, so that once they are known it is possible to calculate in advance what reflectances over white and black are to be anticipated at any wavelength in the spectrum and for arbitrary film thicknesses. Now the CIE system is used in determining the color stimulus associated with the spectrum just calculated. Two color stimuli are obtained, one over black and one over white. These imply the corresponding CIELAB coordinates, which yield the color difference ΔE_{ab}^*. This color difference between white and black must now be subjected to a suitable iterative procedure to make it equal to 1. Now there is a tool for determining by iterative methods the film thickness such that the color difference between white and black will be exactly 1. By going through these purely computational steps, we have simulated the older sampling method for determining the hiding film thickness; the simulation complies fully with the DIN method discussed in Chapter 7.

But the determination of hiding power of lacquers, paints, and plastics, just briefly sketched, is merely *one* example of the usefulness of the principle of spectral evaluation. Later, we will become acquainted with other applications, such as the determination of transparency and the determination of the relative tinting strength of pigments. The universal applicability of the principle to other problems is implied by the fact that it is the key to rationalizing all test methods based on the visual assessment of a match (or fixed difference) between two color drawdowns.

With this generalization, it is possible to state the principle in pure form, independent of any special application. Figure 3-12 is an explanatory flowchart.

3.6 Principle of Spectral Evaluation

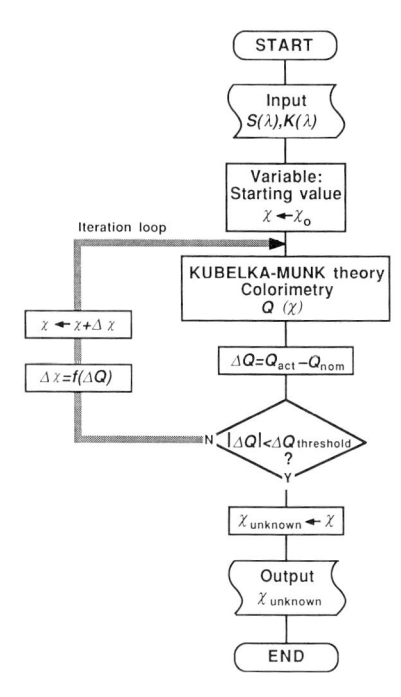

Fig. 3-12. Flowchart explaining the principle of spectral evaluation.

At the start, the scattering and absorption spectra $S(\lambda)$ and $K(\lambda)$ must be known; that is, these quantities have already been determined on suitable coatings or dyed materials. The variable in the problem, here denoted by χ (e.g., film thickness h, scattering coefficient S, absorption coefficient K, pigment volume concentration σ,* etc.), is set to an initial value. From it, the reflection spectrum can be calculated with the Kubelka-Munk formulas, and this in turn, with the aid of colorimetry (CIE and CIELAB systems), yields the value of some colorimetric quantity Q for the two color drawdowns being compared. If there is still a difference between the Q found and the nominal Q, this difference (here denoted by ΔQ) is utilized in calculating a correction to the variable χ. The iteration continues until the nominal and actual Q values are sufficiently close together. The χ value at this time is the desired value of the variable. In order for the correction $\Delta \chi$ to be calculated from ΔQ, additional information (i.e., formulas not contained in the Kubelka-Munk theory) may be needed.

3.6.2 Application to Hiding Power

By the discussion of the achromatic hiding power in Section 3.4.2, it can be stated that the general determination of hiding power with the principle of spectral

---

* If σ is selected as variable, a theoretical notion about the concentration dependence of S or K is generally required; see Chapter 4.

114   *3 How Spectra Depend on the Scattering and Absorption of Light*

evaluation also involves three steps, roughly analogous to those in Section 3.4.2. Specifically, these steps are

- Determination of auxiliary functions $a(\lambda)$ and $b(\lambda)$
- Determination of scattering coefficient $S(\lambda)$ and absorption coefficient $K(\lambda)$
- Determination of ΔE^*_{ab} and iterative calculation to make $\Delta E^*_{ab} = 1$

Now, however, there is the difference that nearly all the quantities to be determined are functions of λ. This takes in $a(\lambda)$, $b(\lambda)$, $S(\lambda)$, $K(\lambda)$, and all the $\varrho(\lambda)$ and $\varrho^*(\lambda)$. In what follows, the λ dependences will not be explicitly shown; the notation "(λ)" will appear only as required for clarity.

1. Determination of auxiliary variables $a(\lambda)$ and $b(\lambda)$

The same formulas as in Section 3.3.3 are used here; thus the two auxiliary variables are determined either from ϱ_∞ or from ϱ_b, ϱ_w, ϱ_{ob}, ϱ_{ow}, all of which must first be transformed into starred quantities by application of the Saunderson correction. In place of equation (3.123), the complete form equation (3.122) is employed, since real black substrates are not always ideally black:

$$a = \frac{(1 + \varrho^*_w \varrho^*_{ow})(\varrho^*_b - \varrho^*_{ob}) + (1 - \varrho^*_b \varrho^*_{ob})(\varrho^*_{ow} - \varrho^*_w)}{2(\varrho^*_b \varrho^*_{ow} - \varrho^*_w \varrho^*_{ob})}$$

As a rule, the λ-dependent quantities are now calculated at intervals of 20 nm, that is, with a total of 16 known points, since this is regarded as sufficient for most applications.

2. Determination of the scattering coefficient $S(\lambda)$ and the absorption coefficient $K(\lambda)$

In this step, again the complete computational formula for S must be used, since the white substrate can be used too for the determination of S:

$$S = \frac{1}{bh} \text{Arcoth} \frac{1 - a(\varrho^* + \varrho^*_o) + \varrho^* \varrho^*_o}{b(\varrho^* - \varrho^*_o)}$$

Depending on whether the measured values are ϱ_b, ϱ_{ob} or ϱ_w, ϱ_{ow}, these are inserted for ϱ^* and ϱ^*_o (after application of the Saunderson correction).

If, however, the values over both substrates are known, it is desirable to use the ϱ^*, ϱ^*_o pair that gives the smaller error of determination for S. To this end, the two errors

- dS_b from ϱ^*_b, ϱ^*_{ob}
- dS_w from ϱ^*_w, ϱ^*_{ow}

must be compared. The formulas needed are derived from the error propagation law (with $d\varrho$ as error of measurement in the reflectance):

3.6 Principle of Spectral Evaluation

$$dS^2 = \left[\left(\frac{\partial S}{\partial \varrho}\right)^2 + \left(\frac{\partial S}{\partial \varrho_o}\right)^2\right] d\varrho^2 \qquad (3.145)$$

The partial derivatives are calculated by the chain rule:

$$\frac{\partial S}{\partial \varrho} = \frac{(1 - r_2 \varrho^*)^2}{(1 - r_o)(1 - r_2)} \frac{\partial S}{\partial \varrho^*} = 2.6 (1 - 0.6 \varrho^*)^2 \frac{\partial S}{\partial \varrho^*} \qquad (3.146)$$

and similarly for $\partial S/\partial \varrho_o$ with ϱ_o^* replacing ϱ^*. Furthermore,

$$\frac{\partial S}{\partial \varrho^*} = -\frac{1}{bh} \frac{1}{y^2 - 1} \frac{\partial y}{\partial \varrho^*} \qquad (3.147)$$

where $y = \dfrac{1 - a(\varrho^* - \varrho_o^*) + \varrho^* \varrho_o^*}{b(\varrho^* - \varrho_o^*)}$

and similarly for $\partial S/\partial \varrho_o^*$ with $\partial y/\partial \varrho_o^*$ replacing $\partial y/\partial \varrho^*$. Finally,

$$\frac{\partial y}{\partial \varrho^*} = -\frac{1 - 2a\varrho_o^* + \varrho_o^{*2}}{b(\varrho^* - \varrho_o^*)^2} \quad \text{and} \qquad (3.148)$$

$$\frac{\partial y}{\partial \varrho_o^*} = \frac{1 - 2a\varrho^* + \varrho^{*2}}{b(\varrho^* - \varrho_o^*)^2} \qquad (3.149)$$

Now both errors of determination dS_b and dS_w can be calculated with equation (3.145); $d\varrho$ can be set equal to 1 because the two errors are only to be compared.

3. Determination of ΔE_{ab}^* and iterative calculation to make $\Delta E_{ab}^* = 1$

The first step is to specify an initial value $h \Leftarrow h_o$ for the film thickness and to calculate ϱ_w^* and ϱ_b^* with the Kubelka-Munk equation (3.111). Both quantities are transformed to uncorrected values with the aid of the inverse Saunderson correction, equation (3.10); at the same time, it is taken into account that the gloss has been reflected away from the observer (see Section 3.4.2), so that

$$\varrho = \frac{(1 - r_o - r_2) \varrho^* + r_o}{1 - r_2 \varrho^*} - gr_o \qquad (3.150)$$

Next, from ϱ_b and ϱ_w, equation (2.3) is used to calculate X, Y, Z by means of the CIE system, for both layers (over white and over black):

$$X = \int \bar{x}(\lambda) \bar{S}(\lambda) \varrho(\lambda) d\lambda \quad \text{etc.} \qquad (3.151)$$

Here $\bar{x}(\lambda) \bar{S}(\lambda)$ stands for the CIE color matching function $\bar{x}$ given the selected

CIE standard illuminant $\bar{S}(\lambda)$.* In this way, X_b, Y_b, Z_b and X_w, Y_w, Z_w are obtained; these are the CIE tristimulus values of the drawdowns over black and over white.

A problem still awaiting solution by the CIE must be mentioned here: how to obtain firm CIE color matching functions $\bar{x}(\lambda)$, $\bar{y}(\lambda)$, $\bar{z}(\lambda)$ for 20 nm intervals. At present, the color matching functions exist only in the form of numerical tables for 5 nm intervals (ISO 7724–1). Without an agreed-on table for 20 nm intervals, it is not certain that two distinct laboratories will obtain the same results. Until 20 nm tables are embodied in standards, it is recommended that the Stearns tables (see Section 3.7; for examples, see Chapter 7) be employed; these may also be suggested to the CIE for adoption, because the interpolation method used is scientifically well justified.

The six CIE tristimulus values are now converted to values X_b^*, Y_b^*, Z_b^* and X_w^*, Y_w^*, Z_w^* by the familiar method (see Section 2.3.3); the cube-root form must be used above the threshold $X/X_n = 0.008856$ etc., the linear form in other cases. Hence, the differences Δa^*, Δb^*, ΔL^* between the layers over black and over white are obtained as differences of the CIELAB color coordinates a_b^*, b_b^*, L_b^* and a_w^*, b_w^*, L_w^*. The value of ΔE_{ab}^* then follows. If this value is not yet close enough to unity, h should be changed in the proper sense and the calculation repeated (iteration loop in Fig. 3-12 with h as variable).

For the iteration proper, any zero-finding technique known from numerical mathematics can be used (to set $\Delta E_{ab}^* - 1$ equal to zero). Because of its many advantages, Newton's method is recommended here again. The main virtue of this technique is fast convergence; considerable amounts of machine time may be required for other procedures. Newton's method does have the drawback that all the derivatives of the variables must be formed, but this laborious manual computation has to be done only once. The equations needed are stated here.

It is useful, instead of taking $\Delta E_{ab}^* - 1$ to zero, to work with the function

$$\eta = \ln \Delta E_{ab}^* \to 0, \qquad (3.152)$$

because the logarithm converges to zero more rapidly. The step-by-step procedure for differentiation with respect to h is as follows:

---

* The notation $\bar{S}(\lambda)$ is used to distinguish this function from the wavelength-dependent scattering coefficient $S(\lambda)$.

3.6 Principle of Spectral Evaluation

$$\frac{\partial \eta}{\partial h} = \frac{1}{\Delta E_{ab}^{*2}} \cdot \left(\Delta a^* \frac{\partial \Delta a^*}{\partial h} + \Delta b^* \frac{\partial \Delta b^*}{\partial h} + \Delta L^* \frac{\partial \Delta L^*}{\partial h} \right)$$

$$\frac{\partial \Delta a^*}{\partial h} = \frac{\partial a_w^*}{\partial h} - \frac{\partial a_b^*}{\partial h}$$

$$\frac{\partial \Delta b^*}{\partial h} = \frac{\partial b_w^*}{\partial h} - \frac{\partial b_b^*}{\partial h}$$

$$\frac{\partial \Delta L^*}{\partial h} = \frac{\partial L_w^*}{\partial h} - \frac{\partial L_b^*}{\partial h}$$

$$\frac{\partial a^*}{\partial h} = 500 \left(\frac{\partial X^*}{\partial h} - \frac{\partial Y^*}{\partial h} \right)$$

$$\frac{\partial b^*}{\partial h} = 200 \left(\frac{\partial Y^*}{\partial h} - \frac{\partial Z^*}{\partial h} \right)$$

$$\frac{\partial L^*}{\partial h} = 116 \frac{\partial Y^*}{\partial h} \tag{3.153}$$

There are two cases:

1) $X/X_n \leq 0.008856$ etc. | 2) $X/X_n > 0.008856$ etc.

$$\frac{\partial X^*}{\partial h} = \frac{7.787}{X_n} \frac{\partial X}{\partial h} \text{ etc.} \quad \bigg| \quad \frac{\partial X^*}{\partial h} = \frac{X^*}{3X} \frac{\partial X}{\partial h} \text{ etc.}$$

$$\frac{\partial X}{\partial h} = \int \bar{x}(\lambda) \bar{S}(\lambda) \frac{\partial \varrho(\lambda)}{\partial h} d\lambda \text{ etc.}$$

$$\frac{\partial \varrho}{\partial h} = \frac{(1 - r_o)(1 - r_2)}{(1 - r_2 \varrho^*)^2} \cdot \frac{\partial \varrho^*}{\partial h} = \frac{0.384}{(1 - 0.6 \varrho^*)^2} \cdot \frac{\partial \varrho^*}{\partial h}$$

$$\frac{\partial \varrho^*}{\partial h} = bS^2 \frac{1 - 2 a \varrho_o^* + \varrho_o^{*2}}{(a - \varrho_o^*) \sinh bSh + b \cosh bSh}$$

The derivatives for the two extremal cases, "scattering only" (ϱ_{sca}^*) and "absorption only" (ϱ_{abs}^*), should also be written out:

$$\frac{\partial \varrho_{sca}^*}{\partial h} = S \left[\frac{1 - \varrho_o^*}{(1 - \varrho_o^*) Sh + 1} \right]^2$$

$$\frac{\partial \varrho_{abs}^*}{\partial h} = -2 K \varrho_o^* e^{-2Kh} = -2 K \varrho_{abs}^* \tag{3.154}$$

So much for the derivatives needed in Newton's method. It goes without saying that these formulas must be worked out back to front; that is, one must begin with the last of the equations just presented, incorporating these into the

3 How Spectra Depend on the Scattering and Absorption of Light

respective steps of the main program if one wishes to write a computer program. An example of these formulas, which may appear somewhat confusing – though only at first glance – appears in Chapter 7 along with a program listing.

3.6.3 Application to Transparency and Coloring Power

The distinctive point about the application of the principle of spectral evaluation to transparency and coloring power is that no iteration is required; the values sought are obtained by direct calculation. The reason is that what is being sought is not an optimal film thickness or pigment concentration; instead, the initial slope (at a film thickness of $h = 0$) of colorimetric quantities versus h is to be determined.

In order for the principle to apply to transparency and coloring power, defining relations (3.139) and (3.140) and the formulas derived from them must be replaced by their partial derivatives with respect to h, then taken to the limit as $h \to 0$. It then becomes clear that the quantities derived from $\Delta E^*_{ab}(h)$ are, for the transparency, the colorimetrically evaluated scattering coefficient and, for the coloring power, the colorimetrically evaluated absorption coefficient.

In the following equations, a prime denotes the result of applying the operator $(\partial/\partial h)_{h \to 0}$ to the unprimed quantity.

$$\Delta \acute{E}^*_{ab} = \left.\frac{\partial \Delta E^*_{ab}}{\partial h}\right|_{h \to 0} = \sqrt{\acute{a}^{*2} + \acute{b}^{*2} + \acute{L}^{*2}}$$

$$\acute{a}^* = 500\,(\acute{X}^* - \acute{Y}^*)$$

$$\acute{b}^* = 200\,(\acute{Y}^* - \acute{Z}^*)$$

$$\acute{L}^* = 116\,\acute{Y}^* \qquad (3.155)$$

This equation is to be used for calculating $\acute{a}^*_s$, $\acute{b}^*_s$, $\acute{L}^*_s$ with $\acute{X}^*_s$, $\acute{Y}^*_s$, $\acute{Z}^*_s$, and for calculating $\acute{a}^*_k$, $\acute{b}^*_k$, $\acute{L}^*_k$ with $\acute{X}^*_k$, $\acute{Y}^*_k$, $\acute{Z}^*_k$; symbols with subscript s are for transparency, those with k for coloring power. As a result, two values are found for $\Delta \acute{E}^*_{ab}$, namely $\Delta \acute{E}^*_{ab,s}$ and $\Delta \acute{E}^*_{ab,k}$.

$$\acute{X}^*_s = 7.787\,\frac{\acute{X}_s}{X_n}$$

$$\acute{Y}^*_s = 7.787\,\frac{\acute{Y}_s}{100}$$

$$\acute{Z}^*_s = 7.787\,\frac{\acute{Z}_s}{Z_n} \qquad (3.156)$$

$$\acute{X}^*_k = \frac{\acute{X}_k}{3\,X_n}\,; \quad \acute{Y}^*_k = \frac{\acute{Y}_k}{3 \cdot 100}\,; \quad \acute{Z}^*_k = \frac{\acute{Z}_k}{3\,Z_n} \qquad (3.157)$$

3.6 Principle of Spectral Evaluation

$$\acute{X}_s = \int \bar{x}(\lambda)\bar{S}(\lambda)(1-r_o)(1-r_2)S(\lambda)\,d\lambda = 0.384\int \bar{x}(\lambda)\bar{S}(\lambda)S(\lambda)\,d\lambda \quad (3.158)$$

(and similarly for $\acute{Y}_s$, $\acute{Z}_s$ with $\bar{x}(\lambda)$ replaced by $\bar{y}(\lambda)$ and $\bar{z}(\lambda)$ respectively). Again $\bar{S}(\lambda)$ denotes the spectrum of the CIE standard illuminant used, and $S(\lambda)$ is the spectral scattering coefficient.

$$\acute{X}_k = 2\int \bar{x}(\lambda)\bar{S}(\lambda)\frac{1-r_o}{1-r_2}K(\lambda)\,d\lambda = 4.8\int \bar{x}(\lambda)\bar{S}(\lambda)K(\lambda)\,d\lambda \quad (3.159)$$

(and similarly for $\acute{Y}_k$, $\acute{Z}_k$ with $\bar{x}(\lambda)$ replaced by $\bar{y}(\lambda)$ and $\bar{z}(\lambda)$ respectively). The expressions to the right of the first equality signs in the last two equations allow for the possibility that the spectra of the two reflection coefficients r_o and r_2 are known.

The equations just stated must, once again, be worked through backwards in the practical calculation, that is, beginning with the last equations. Finally, the transparency number and coloring power are obtained as

$$T = \frac{1000}{\Delta \acute{E}^*_{ab,s}} \quad (\text{mL}\cdot\text{m}^{-2})$$

$$\Phi = \frac{\Delta \acute{E}^*_{ab,k}}{1000} \quad (\text{m}^2\cdot\text{mL}^{-1}) \quad (3.160)$$

The factor 1000 appears twice so that smaller numbers can be used for T and Φ (in derived units). Now T denotes the number of mL that can spread over 1 m² before ΔE^*_{ab} measured against the black substrate becomes equal to 1. Similarly, Φ denotes the number of m² that can be covered with 1 mL so effectively that a ΔE^*_{ab} of 1 is obtained relative to an ideal white.

All the equations needed for transparency and coloring power have now been written down. Similarly to what was done in Section 3.5.2, the transparency-coloring power L and the transparency-coloring degree G can be calculated from T and Φ. All of these equations will be repeated for direct computer processing in the practical portion of Chapter 7. It is now possible to determine hiding power and transparency after a minimum of three spectral reflectance measurements (ϱ_∞, ϱ_b, ϱ_{ob} or ϱ_∞, ϱ_w, ϱ_{ow}). Modern instruments employing the spectral method offer a suitable tool for color measurement.

In conclusion, a further very important point should be noted, again a consequence of the application of the principle of spectral evaluation. The direct determination of the tangent to the curve of ΔE^*_{ab} versus h (Fig. 3-7) is difficult because highly uncertain reflectances close to zero have to be measured. A result obtained by the theoretical analysis based on the principle of spectral evaluation, equations (3.158) and (3.159), namely that $S(\lambda)$ and $K(\lambda)$ in colorimetrically evaluated form satisfy the definitions of T and Φ, makes it possible to determine S and K at higher reflectances (e.g., over a white substrate) where the uncertainties are much smaller. In this way, the procedure allows the transparency number and color strength to be determined with far smaller errors than when they are obtained from the tangents to the curves at the origin.

3.7 List of Symbols Used in Formulas

a, b	auxiliary variables in Kubelka-Munk theory
a^*, b^*, L^*	CIELAB color coordinates
$A = a \sinh x + b \cosh x$	
	abbreviated notation in four-flux theory
$\hat{A} = a \sinh 2x + b \cosh 2x$	
	same as A for doubled film thickness
$\mathbf{A}_j = \{A_{1j}, A_{2j}, \ldots, A_{nj}\}$	
	eigenvectors of many-flux theory
$\alpha = 1/(Sh_D)$	factor for calculation of hiding power
B_1, B_2	constants of integration in four-flux theory
$\beta = 1/(Kh_D)$	factor for determination of hiding power
C	auxiliary variable for coefficient determination in four-flux theory
C_j	constant of integration in many-flux theory
χ	variable in principle of spectral evaluation
D	hiding power value
	with subscripts:
	abs due to absorption only
	sca due to scattering only
δ	components of hiding power value with same subscripts as D, plus int interaction component
$\Delta a^*, \Delta b^*, \Delta L^*$	
	special CIELAB color differences
ΔE^*_{ab}	CIELAB color difference
ΔQ	difference between actual and nominal Q values
$d\varrho$	error of measurement of ϱ
dS	error of measurement of S
η	null function for iterative procedure
η_i	sign factor in many-flux theory
F_{KM}	Kubelka-Munk function
g	degree of gloss
G	transparency-coloring degree
γ	diffuse radiation component
h	film thickness
	with subscript:
	D hiding layer
k'	absorption coefficient for directional radiation
k_j	absorption coefficient in many-flux theory

3.7 List of Symbols Used in Formulas

K	absorption coefficient for diffuse radiation
l	radiance (directional)
l^+, l^-	radiances (directional) in incidence direction and opposite to incidence direction
	with subscripts:
	0 upper boundary of layer
	h lower boundary of layer
L	transparency-coloring power
L^+, L^{++}	radiances (diffuse) in incidence direction
	with same subscripts as l^+, l^-
$L^-, L^=$	radiances opposite to incidence direction
	with same subscripts as l^+, l^-
$\mathbf{L}$	radiance vector
λ_j	eigenvalues in many-flux theory
μ	attenuation coefficient
n	number of channels in many-flux theory
p, P, q, Q	auxiliary variables in four-flux theory
	with subscript:
	sca scattering only
φ	slope angle of function $\Delta E^*_{ab}(h)$
Φ	coloring power
r_o	reflection coefficient at the surface for directional light incident at right angle from outside
r_1	reflection coefficient for diffuse light incident from outside
r_2	reflection coefficient for diffuse light incident on the surface from the interior of the specimen
$R, R^*, \overline{R}^*$	reflection factors: measured, surface-corrected, and surface-corrected after exclusion of directional radiation
	with same subscripts as ϱ
$\overline{R}, \overline{\overline{R}}, \overline{\varrho}$	abbreviated notation in four-flux theory
$\varrho, \varrho^*, \overline{\varrho}^*$	reflectances: measured, surface-corrected, and surface-corrected after exclusion of directional radiation
	with subscripts:
	0 substrate
	abs extreme case "absorption only"
	b black substrate
	dif diffuse irradiation
	ex *ex*clusive of surface reflection
	g directional irradiation
	in *in*clusive of surface reflection
	r perpendicular irradiation and measurement
	w white substrate
	∞ infinitely thick layer
$\overline{\varrho}, \varrho_1, \varrho_2, \varrho_3$	abbreviated notation in four-flux and Kubelka-Munk theories
$\hat{\varrho}$	ϱ for doubled film thickness

s^+, s^-	scattering coefficient for directional light in incidence direction and opposite to incidence direction
S	scattering coefficient for diffuse light
$\overline{S}(\lambda)$	lamp spectrum
s_{ij}, S_{ij}	matrix of coefficients in many-flux theory
$\mathbb{S}$	matrix of scattering coefficients in many-flux theory
t	abbreviated notation in determination of Kubelka-Munk coefficients
T	transparency number
$T, T^*, \overline{T}^*$	transmission factors: measured, surface-corrected, and surface-corrected after exclusion of directional radiation with same subscripts as ϱ
$\tau, \tau^*, \overline{\tau}^*$	transmittances: measured, surface-corrected, and surface-corrected after exclusion of directional radiation with same subscripts as ϱ
$\hat{\tau}$	transmittance for doubled film thickness
u, w	auxiliary variables in four-flux theory
$x = bSh$	abbreviated notation in four-flux and Kubelka-Munk theories
$\overline{x}(\lambda), \overline{y}(\lambda), \overline{z}(\lambda)$	CIE color matching functions

X, Y, Z	CIE tristimulus values with subscripts:
	  b black substrate
	  n CIE standard illuminant
	  w white substrate

X^*, Y^*, Z^*	CIELAB value functions with same subscripts as X, Y, Z
$\xi = 1/\alpha$	variable for calculation of hiding power value
y	abbreviated notation in calculation of hiding power value
z	position coordinate

3.8 Summary

The phenomenological theory of absorption and scattering in turbid media is of fundamental significance for the mathematical treatment of the coloration of lacquers, paints, clouded glasses, etc., and thus for all problems associated with the use and economics of pigments and dyes. The most general form of the phenomenological theory seems to be the many-flux theory, whose mathematical formalism is described in full. Typical for this model is an absorbing and scattering medium bounded by two parallel planes. The boundary conditions can allow for many irradiation geometries; the reflection coefficients at the interfaces can be calculated exactly. A calculation for real layers, however, involves heavy use of the computer and is thus not well-suited to daily practice.

The correction formulas for surface reflection can be written down immediately and are thus an exception to the above statement.

The four-flux theory is derived as an important special case, and its results are written as a collection of explicit formulas. The links between the four-flux theory on the one hand and the theories of Kubelka-Munk (two-flux) and Ryde/Duntley (special case of four-flux) on the other are discussed. The four-flux theory in this generalized form yields equations for the case of irradiation from both sides, for the transmission, and for the particularly important cases of reflection with substrate and reflection with infinitely thick layer, both with mixed (diffuse and directional) irradiation. What is more, the case of a specularly reflecting substrate is treated, as are the extreme cases where absorption is strongly dominant and where scattering is strongly dominant. It is further shown how the five coefficients can be determined experimentally from reflection or transmission measurements, that is, in particular how the poorly accessible coefficients for directional irradiation can be determined exactly. Finally, the equations of the Kubelka-Munk theory and the formulas for obtaining their coefficients are summarized.

The hiding power of scattering and absorbing layers is the most important technical quantity that can be rationalized with the aid of the phenomenological theory. On the basis of the definition in terms of the hiding criterion $\Delta E^*_{ab} = 1$ between the layers over black and white substrates, the simplified case of achromatic layers is considered; equations for the hiding power applicable to this case can be derived in a direct fashion from the Kubelka-Munk formalism. The hiding power so obtained can be split into a scattering component, an absorption component, and an interaction component, so that additional information can be extracted about the quality of the layer.

The transparency number for paints and inks is described and defined. This quantity also makes it possible to characterize the optical performance of transparent pigments. In connection with transparent coatings, a coloring power can also be defined and determined exactly; other quantities describing the transparency-coloring properties can be derived from the transparency number and the coloring power.

Finally, the principle of spectral evaluation is thoroughly discussed as a feedback between colorimetry and the Kubelka-Munk theory. Only in this way can the hiding power of chromatic layers be approached in an exact manner. All the needed formulas, including those for a Newton iteration method and for an improved accuracy in the determination of S, are derived. This principle can also be used to advantage for the transparency number.

3.9 Historical and Bibliographical Notes

The *absorption*-based approach arose out of what was once the most important specialty in physics, an area that received a high degree of early development: astronomy. It was astronomical problems that led to the formulation of what is now known as Lambert's law. LAMBERT, who lived from 1728 to 1777, was also concerned with astronomical topics and was seeking a law to describe the decrease in intensity of starlight as it passed through the atmosphere. The beginnings of this theory may be said to go back to BOUGUER in 1729.

The first worker to take up the phenomenological theory of *scattering* was – we suppose – STOKES [1], but once again, usable formulations came from astronomy. Because both the Earth's atmosphere and interstellar space contain dust particles, a formula based on absorption alone is not always adequate to predict intensities in astronomy. The first closed phenomenological theory of scattering dates from 1905 and was devised by the astronomer SCHUSTER [2]. Here the model of two ray paths appears for the first time; SCHUSTER, who of course was interested only in transmission behavior, obtained a transmittance formula similar to the later ones of RYDE and KUBELKA for diffuse light.

More than a decade later, in 1918, RENWICK [3], [4] confirmed SCHUSTER's results, but went a step further and treated reflectance as well. The formulas were clumsy, however, and not well-suited to calculating the reflectances of scattering layers. In 1927, SILBERSTEIN [5] essentially confirmed SCHUSTER's transmission equation, but this publication contains the first discussion of a model for *directional* irradiation. GUREVIC [6] in turn gave a more thorough treatment of the reflectance.

But all the work cited must be regarded as preliminary to the classic publications of KUBELKA and MUNK [7] and RYDE [8], which came out in 1931. KUBELKA and MUNK, with no knowledge of prior work, analyzed the reflectance for the *diffuse* irradiation case with a substrate, deriving the still-used formulations in terms of hyperbolic functions (these first by Kubelka in 1947 [9]). Their main achievement, however, is the application of the equations to hiding power, on the one hand, and on the other to the reflectance of very ("infinitely") thick layers. The latter aspect is embodied in the Kubelka-Munk function, which found immediate acceptance and is applied to all cases involving scattering layers of high hiding power. The Kubelka-Munk theory gives the first clear expression of the fact that two coefficients are needed in the case of diffuse irradiation ("two-constant theory") and that these coefficients can in fact be determined. The Kubelka-Munk theory fails in the case of a highly dilute pigment or a very thin layer (see Section 3.3.1), because the radiation cannot be ideally diffuse and not every ray collides with a pigment particle ("sieve effect"). With specific application to biological measurements, FUKSHANSKI [10] has devised a theory, but its concepts have not yet been made to work for lacquer and paint coatings.

RYDE [8] analyzed transmittance and reflectance for the case of *directional* irradiation. The two "diffuse" coefficients are complemented by two more scattering coefficients, making a total of four. Later, DUNTLEY [11] extended the theory by introducing a second absorption coefficient for directional irradiation; as a result, five coefficients (including the Kubelka-Munk coefficients) are needed for the phenomenological description of scattering and absorbing layers with directional irradiation. In comparison with the Kubelka-Munk theory, the Ryde theory has not gained such wide adoption.

3.9 Historical and Bibliographical Notes

DUNTLEY [11] was also concerned with surface phenomena. The correction formula proposed by Saunderson [12] dates from the year 1942. A good survey of these formulas can be found in BROCKES [13].

The complete version of the explicit four-flux theory was published in 1962 [14]. Like the Ryde/Duntley theory, it employs five coefficients; it goes further, however, including equations for two-sided irradiation with diffuse and directional light and for layers with substrates. A second part was added in 1964 [15], now describing the determination of the three "directional" coefficients. The equations for specularly reflecting substrates, on the other hand, were not included in the publications cited; they are presented for the first time in this book. The first part of the four-flux theory [14] was confirmed a few years later by BILLMEYER and co-workers [16], who also introduced another (fourth) scattering coefficient. The comprehensive many-flux theory is due to MUDGETT and RICHARDS (1971) [17], whose work marked the completion of the phenomenological theory of scattering and absorption for the time being. The standard work on radiative transfer processes, by CHANDRASEKHAR [18], should also be mentioned here. Recommended monographs, particularly on the use of the Kubelka-Munk theory, are those of KORTÜM [19] and JUDD [20] (in English). Experimental applications of the many-flux theories have been described by only a few authors, for example PAULI and EITLE [21], [22], who applied a three-flux theory to color formulation problems. Chapter 6 will give an account of the results obtained by MUDGETT and RICHARDS [17] and by ATKINS and BILLMEYER [23].

The first equations for the *hiding power* based on scattering and absorption coefficients for diffuse irradiation were included by KUBELKA and MUNK in their renowned work [7]. In present-day terms, they used the lightness contrast ratio as hiding criterion; as it then turned out, however, this criterion is applicable only to achromatic layers. Standardized methods for the achromatic case were accordingly worked out in the 1950s and 60s and have been used since that time. Not until the advent of the CIELAB system could the desire for more suitable hiding criteria be fulfilled. First, in 1981, the introduction of the criterion $\Delta E_{ab}^* = 1$ made possible a computational procedure for achromatic layers based on the Kubelka-Munk theory [24]; this duly led to a standard. At the same time, the first description was given of how gloss could be handled in color matching and how the scattering and absorption components of the hiding power could be calculated. It had already been shown that the determination of hiding power from ϱ_∞^*, ϱ_b^* is more accurate than the determination from ϱ_b^*, ϱ_{ob}^*, ϱ_w^*, ϱ_{ow}^* [25]. A pneumatic apparatus for measuring film thickness was developed by KÄMPF [26] for use in the determination of scattering coefficients, absorption coefficients, hiding power, and transparency. Economic aspects of the hiding power were analyzed by KEIFER [27]; these will be taken up in Section 7.4.4.

Not until 1964, however, was the *principle of spectral evaluation* [28], [29] applied in order to make the Kubelka-Munk theory applicable to colored layers. The first procedure used was a graphical one, with the lightness as hiding criterion. GALL [30] then showed how the principle could be rationally applied (by computer programming) when other criteria (color difference) were used in determining the hiding power of inks and paints. STROCKA [31] demonstrated that 16 chosen wavelengths (20 nm intervals) would be sufficient for most applications. The tables of STEARNS [32] can be expediently employed for the CIE color matching functions (see Chapter 7 for examples). Further improvements followed [33]: iteration with Newton's method, calculation of S for smaller error of determination. On the last point, it should be remarked that the calculation of the error dS in the scattering coefficient as done in Section 3.6.1 is preferable to the computational scheme in [33] because it is simpler.

With regard to *transparency*, the definition generally accepted today was not written until 1976 [34]. This definition provided the basis for a standardized test procedure. The same publication [34] suggested the use of the principle of spectral evaluation in order to create a rational and sufficiently accurate determination method using the Kubelka-Munk coefficients, especially when the calculation can be done in a single pass after the hiding power has been determined [33]. Reference should also be made to a report on experience with diverse methods of hiding power determination [35] and to a summary of the Kubelka-Munk theory in pigment testing [36].

References

[1] G. G. Stokes, *Proc. Roy. Soc.* **11** (1860) 545.
[2] A. Schuster, *Astrophys. J.* **21** (1905) 1.
[3] F. F. Renwick, *Phot. J.* **58** (1918) 140.
[4] F. F. Renwick, H. J. Channon, B. V. Storr, *Proc. Roy. Soc., Ser. A*, **94** (1918) 222.
[5] L. Silberstein, *Phil. Mag.* **4** (1927) 7.
[6] M. Gurevic, *Phys. Z.* **31** (1930) 753.
[7] P. Kubelka, F. Munk, *Z. techn. Phys.* **12** (1931) 593.
[8] J. W. Ryde, *Proc. Roy. Soc., Ser. A*, **131** (1931) 451.
[9] P. Kubelka, *J. opt. Soc. Am.* **38** (1948) 448.
[10] L. Fukshanski, *J. Math. Biol.* **6** (1978) 177.
[11] S. Q. Duntley, *J. opt. Soc. Am.* **32** (1942) 61.
[12] J. L. Saunderson, *J. opt. Soc. Am.* **32** (1942) 727.
[13] A. Brockes, *Die Farbe* **9** (1960) 53.
[14] H. G. Völz, VI. *Fatipec*-Kongress Wiesbaden 1962, Kongr.-B. (Verlag Chemie, Weinheim) 98.
[15] H. G. Völz, VII. *Fatipec*-Kongress Vichy 1964, Kongr.-B. (Verlag Chemie, Weinheim) 194.
[16] J. K. Beasley, J. T. Atkins, F. W. Billmeyer, Jr., *Proc. II. Interdisc. Conf. Electromagnet. Scatt.* 1967, Congress B. (Gordon & Breach, New York) 765.
[17] P. S. Mudgett, L. W. Richards, *Appl. Opt.* **10** (1971) 1485.
[18] S. Chandrasekhar, *Radiative Transfer,* Clarendon Press, Oxford, 1950.
[19] G. Kortüm, *Reflexionsspektroskopie,* Springer, New York, 1969.
[20] D. B. Judd, *Color in Business, Science and Industry,* 2. Edit., J. Wiley & Sons, New York 1969.
[21] H. Pauli, D. Eitle, 2. *AIC*-Congress York 1973, Congress B. (A. Hilger, London) 423.
[22] D. Eitle, H. Pauli, XIV. *Fatipec*-Congress, Budapest 1978, Congress B. 209.
[23] J. T. Atkins, F. W. Billmeyer, *Color. Eng.* **6** Nr. 3 (1968) 40.
[24] H. G. Völz, *Die Farbe* **29** (1981) 297.

[25] H. G. Völz, *Die Farbe* **19** (1970) 231.
[26] G. Kämpf, H. G. Völz, *farbe + lack* **77** (1971) 629.
[27] S. Keifer, „Deckvermögen" in BAYER-Ringbuch „Anorganische Pigmente für Anstrich- und Beschichtungsstoffe", Leverkusen 1973.
[28] H. G. Völz, VII. *Fatipec*-Congress Vichy 1964, Congress B. (Verlag Chemie, Weinheim) 366.
[29] H. G. Völz, *farbe + lack* **71** (1965) 725.
[30] L. Gali, *farbe + lack* **72** (1966) 955 and 1973.
[31] D. Strocka, 2. *AIC*-Congress, York 1973, Congress B. (A. Hilger, London) 453.
[32] E. I. Stearns, *Rev. Ind. Management Text. Sci.* **14** (1975) 1.
[33] H. G. Völz, *Progr. Org. Coat.* **15** (1987) 99.
[34] H. G. Völz, *Dtsch. Farb. Z.* **30** (1976) 392 and 454.
[35] M. Cremer, *farbe + lack* **90** (1984) 548.
[36] H. G. Völz, *Progr. Org. Coat.* **13** (1985) 153.

4 How Light Scattering and Absorption Depend on the Content of Coloring Material (Beer's Law, Scattering Interaction)

4.1 Introduction

4.1.1 Description and Significance of the Concentration Dependence

The complex of ideas to be discussed next occupies a position that is not just central in the interlinked system of fundamental theories (Chapter 1) but is also very significant in practice. Now that we have become acquainted with the absorption and scattering of light as the root cause of the spectral distribution, we must ask how the concentration of dyes or pigments affects scattering and absorption and thus, ultimately, spectra. The link with practical concerns can be understood immediately because this topic relates to all aspects of producing color recipes: not just formulation, but also correction or rematching. The manipulation of pigment concentrations is the formulating colorist's most vital tool.

In addition to the concentration dependences of absorption and scattering, however, this chapter also deals with that of the Kubelka-Munk function, especially with regard to special pigment mixtures such as those with white and black pastes. These are particularly relevant to colorimetric pigment testing, because they form the basis for determining the tinting strength and lightening power.

The last two concepts will be taken up in the pertinent sections; here it should be noted that the tests are often used for incoming inspection of coloring materials, since they can be performed in a relatively short time.

Colorimetric tests of coloring material generally presuppose that dyes are dissolved and pigments are dispersed in a medium. The resulting pigment dispersion should contain the pigments in homogeneously distributed form (optimally dispersed); "dispersion" as the term is used in colloid chemistry must therefore be dealt with first, and this means providing some essential information about particle size. A section on measures of pigment content then follows.

4.1.2 Pigment Particle Size

Along with the optical constants, the key physical features of pigments include geometrical information such as the particle size and shape. The effect of particle size on color properties was mentioned briefly in Section 1.2, which made the point that both the absorption and the scattering properties are strongly influenced by the size and distribution of pigment particles. The terminology of pigment particles will be outlined briefly here as set forth in standards.*

The following terms are defined (see also Fig. 4-1):

- *Particle:* A delimitable unit of a pigment, which can have any form and any structure.
- *Primary particle, single particle:* A particle that can be identified as an entity with the help of suitable physical methods (optical or electron microscopy). In many cases, it may be a single crystal, but it may also consist of a number of crystallites that coherently scatter x rays.
- *Aggregate:* An intergrown combination of primary particles in surface contact with one another such that the surface area is smaller than the sum of the surface areas of the primary particles.
- *Agglomerate:* A combination of primary particles or aggregates in contact with one another such that the components of the agglomerate are not inter-

Fig. 4–1. Primary particles, aggregates, and agglomerates, after HONIGMANN and STABENOW.

---

* See "Particle-Size Analysis" in Appendix 1.

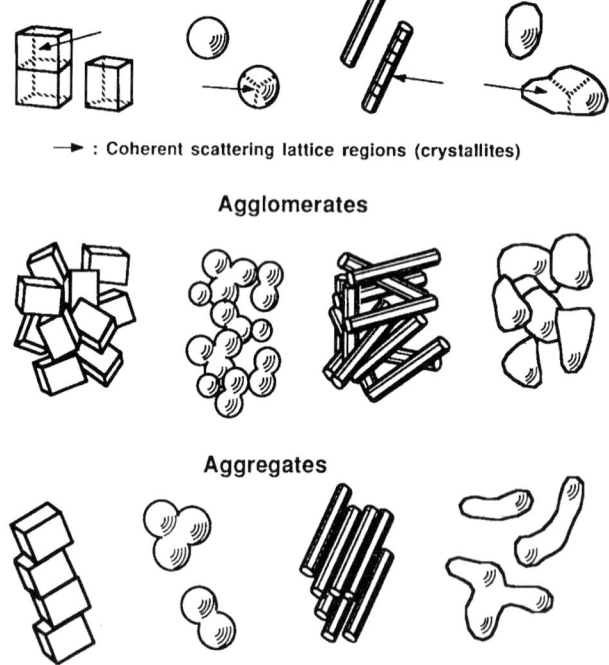

grown with one another and the total surface area is not substantially different from the sum of the individual surface areas.
- *Flocculate:* An agglomerate that is formed in pigment/binder systems and can be broken down again by low shear forces.

The *particle size* denotes the geometric dimension used to characterize the spatial extent of a particle. Such measures can be

- The particle diameter, that is, the diameter in the case of spherical particles or those bounded by regular figures
- The equivalent diameter, that is, the diameter of a sphere having the same volume as the particle
- The particle surface area, that is, the external surface area (without voids) or internal surface area (including voids) of a particle
- The particle volume, that is, the effective volume (without voids) or apparent volume (including voids)
- The particle mass, that is, the mass of the particle determined by weighing

Tables 4-1 and 4-2 present some information on the particle shape and most frequent diameter of pigment particles.

The granulometric composition of a pigment is described by its particle-size distribution. Topics relating to this aspect will be discussed in Chapter 5.

4.1.3 Dispersing of Pigments

In general, the testing of pigments requires their incorporation into a binder or a plastic material; this holds, in particular, for the color properties. The look of the dry pigment powder, as a rule, affords only a very incomplete picture of the actual color qualities. In order to gain a complete knowledge of the colorimetric properties of a pigment, one would have to work it into a multitude of coating materials or plastics. The impulse toward rationalization means that such tests must be limited in number. However, it is often sufficient to do tests in typical systems, those that promote pigment dispersion and allow the color quality of interest to manifest itself fully.

A pigment/binder dispersion is a suspension before drying, a solid sol after drying. Accordingly, the concepts and relations known from colloid chemistry apply to the description of pigment/binder systems. The dispersing of pigments and fillers in binders is a highly complex process involving subprocesses, some of which have to precede or follow others, some of which can overlap in time. These subprocesses can be systematized and classified under the following three heads:

- *Wetting:* the displacement of air from the particle surface and the formation of a layer of solvent molecules.
- *Disintegration:* the disruption and dissipation of agglomerates, involving energy supplied externally (i. e., by dispersing equipment).

Table 4–1. Particle-size properties of some inorganic pigments.

	Particle shape	Modal diameter $\overline{D}_{hv}$, μm
Titanium dioxide, rutile (aftertreated)	almost spherical	0.18–0.35
Titanium dioxide, anatase (untreated)	almost spherical	0.10–0.20
Zinc white	almost cubic to acicular	0.7 –1.0
Red iron oxide	spherical	0.1 –0.7
Yellow iron oxide	acicular	0.1 × 0.3 to 0.2 × 0.8
Cadmium sulfoselenides	cubic to spherical	0.2 –0.4

Table 4–2. Particle sizes of some organic pigments.

	Modal diameter $\overline{D}_{hv}$, μm
Azo (monoazo) pigments	0.50–0.55
Azo (disazo) pigments	0.05–0.21
Phthalocyanines, blue	0.06–0.10
Phthalocyanines, green	0.04–0.08
Polycyclic pigments	0.05–0.30

– *Stabilization:* the establishment of repulsive forces between particles (e.g., through solvate shells) in order to maintain the dispersed state already achieved. These repulsive forces must be stronger than the attractive van der Waals forces, which would otherwise cause the pigment to flocculate.

As a rule, the quality of dispersing has a crucial effect on the properties of the system. The color properties, in particular the full shade, tinting strength, gloss, and gloss haze, show an especially strong dependence on the state of dispersing. In systems made up of various pigments, demixing can occur, and a change in the optical aspect (floating) can result. Floating can be controlled by preliminary coflocculation of the pigments with suitable wetting agents, or by the addition of thickeners.

A frequent departure from ideal dispersion takes the form of flocculation. As defined by KEIFER, this means any departure from statistical uniformity of the pigment distribution in the medium. Deviations from uniformity commonly result in a loss of optical and coloristic performance. Flocculation also takes in the cases where pigment particles are drawn close together by attractive forces (but do not agglomerate), and where the pigment distribution is distorted by foreign particles. In particular, this always occurs when such particles are larger than the pigment particles, and so any addition of substances having a larger particle size (e. g., fillers) is always problematic.

At higher pigment contents, there is no longer any certainty that the voids between pigment particles will be completely filled by the binder. Above a given concentration (the "critical pigment volume concentration"), characteristic anomalies are seen, leading to disrupted and sometimes singular behavior and thus to deterioration of the pigment properties. The color properties are also affected.

4.1.4 Measures of Pigment Content

From the standpoint of colloid chemistry, a pigmented medium (lacquer, paint, plastic) is a dispersion, that is, a multiphase system. The terminology and notation originally developed for mixed-phase systems* can be applied to the quantitative description of such systems. Thus the portions of the two components in such a dispersion are described in terms of:

– the *mass m* (unit: g or kg)
– the *volume V* (unit: cm³ or L)

In this book, composition-related symbols will always appear without subscripts if they refer to pigment, and with the subscript M if they refer to (unpigmented) media. The same will apply to all material constants of the pigment and of the medium, respectively. Accordingly, we define:

– the *pigment mass portion (PMP) w* as

$$w = \frac{m}{m + m_M} \quad \text{(unit: 1)} \quad (4.1)$$

– the *pigment mass concentration (PMC)* β as

$$\beta = \frac{m}{V + V_M} \quad \text{(unit: g cm}^{-3}\text{)} \quad (4.2)$$

– the *pigment volume concentration (PVC)* σ as

$$\sigma = \frac{V}{V + V_M} \quad \text{(unit: 1)} \quad (4.3)$$

---

* See "Mixtures, Composition" in Appendix 1.

4 How Light Scattering and Absorption Depend on the Content

The PMP and PVC are expressed as fractions, often in % (example for PVC: $\sigma = 0.0295 = 2.95\,\%$). If the volumes in this formula are replaced by the masses m, m_M and the respective densities* ϱ, ϱ_M (unit: g cm^{-3}), the PVC becomes:

$$\sigma = \frac{\dfrac{m}{\varrho}}{\dfrac{m}{\varrho} + \dfrac{m_M}{\varrho_M}} = \frac{m}{m + \mu} = \frac{w}{w + \acute{\mu}} \qquad \text{(unit: 1)} \tag{4.3}$$

where

$$\mu = \varrho\, V_M = m_M \frac{\varrho}{\varrho_M} \qquad \text{(unit: g)} \tag{4.4}$$

is *the pigment mass equivalent to the medium volume (PVM)* and

$$\acute{\mu} = \frac{\varrho}{\varrho_M}(1-w) \tag{4.5}$$

is *the pigment mass equivalent to the medium volume of the mass portion (PMM)*. For constant portions by weight of the medium, μ is a constant (provided the same pigment is used); thus *the pigment/medium volume ratio (PMV)* defined as

$$\psi = \frac{V}{V_M} = \frac{m}{\mu} = \frac{w}{\acute{\mu}} = \frac{\sigma}{1-\sigma} \qquad \text{(unit: 1)} \tag{4.6}$$

is an especially convenient measure, since it can be calculated easily from m and μ or from σ alone. Table 4-3 lists the equations for the quantities just defined. Conversion relations for these measures are stated in matrix form in Table 4-4. If one wishes, for example, to do the conversion $w \rightarrow \sigma$, one begins in the second row ($w \rightarrow$) and follows across to the fourth column ($\rightarrow \sigma$) to find equation (4.3).

---

* The use of the same symbol in the standard notation for the density, ϱ, and the spectral reflectance, $\varrho(\lambda)$, is all right because there is no danger of confusion. The same holds for the symbol σ, which represents both standard deviation and PVC.

4.1 Introduction

Table 4–3. Measures of pigment content in media, with their relationships to masses and densities of pigment and medium.

(4.4)	Pigment mass equivalent to the medium volume (PVM)	$\mu = m_M \dfrac{\varrho}{\varrho_M} = \varrho V_M$
(4.1)	Pigment mass portion (PMP)	$w = \dfrac{m}{m + m_M}$
(4.2)	Pigment mass concentration (PMC)	$\beta = \varrho \dfrac{m}{m + \mu}$
(4.3)	Pigment volume concentration (PVC)	$\sigma = \dfrac{m}{m + \mu}$
(4.6)	Pigment/medium volume ratio (PMV)	$\psi = \dfrac{m}{\mu}$

Table 4–4. Conversion relations between measures of pigment content in media, equations (4.7) to (4.15).

	$\to w$	$\to \beta$	$\to \sigma$	$\to \psi$
Pigment mass portion (PMP) $w \to$	w	(4.7) $\beta = \varrho \dfrac{w}{w + \acute{\mu}}$	(4.3) $\sigma = \dfrac{w}{w + \acute{\mu}}$	(4.6) $\psi = \dfrac{w}{\acute{\mu}}$
Pigment mass concentration (PMC) $\beta \to$	(4.8) $w = \acute{\mu} \dfrac{\beta}{\varrho - \beta}$	β	(4.9) $\sigma = \dfrac{\beta}{\varrho}$	(4.10) $\psi = \dfrac{\beta}{\varrho - \beta}$
Pigment volume concentration (PVC) $\sigma \to$	(4.11) $w = \acute{\mu} \dfrac{\sigma}{1 - \sigma}$	(4.12) $\beta = \varrho \sigma$	σ	(4.6) $\psi = \dfrac{\sigma}{1 - \sigma}$
Pigment/medium volume ratio (PMV) $\psi \to$	(4.13) $w = \acute{\mu} \psi$	(4.14) $\beta = \varrho \dfrac{\psi}{1 + \psi}$	(4.15) $\sigma = \dfrac{\psi}{1 + \psi}$	ψ
Pigment mass equivalent to the medium volume of the mass portion (PMM) $\acute{\mu} =$	(4.5) $\acute{\mu} = \dfrac{\varrho}{\varrho_M}(1 - w) = \dfrac{\mu}{m + m_M}$			

This book will make exclusive use of the measures described here in quantifying the pigment portions in pigmented media. Among them, the pigment volume concentration (PVC) σ is the most important, because the optical pigment properties (chiefly the absorption and scattering properties) are referred to volume (not weight, cross section, or surface area). As a consequence, physical quantities such as the scattering and absorption coefficients often display a linear or quasi-linear dependence on the volume fractions. Related to the PVC is the *pigment volume portion (PVP,* unit : 1) φ; it is defined in terms of the partial volumes *before* mixing, in contrast to σ, which is defined in terms of the volumes V and V_M *after* mixing. This measure is less interesting as far as colorimetric test methods are concerned; in any case, $\varphi = \sigma$ if there is no volume reduction in the mixing process, as in nondrying pastes.

Other measures are seen occasionally, such as the *pigment surface concentration* (in m²/L), which states the amount of pigment particle/medium interfacial area per unit volume in the dispersion. This quantity is sometimes used with latex paints, because it provides a measure of the free surface area exposed not to the medium but to the air, which leads to increased scattering power by virtue of the higher refractive index (e.g., TiO_2 in rutile form has a refractive index of 2.75 relative to air but 1.9 relative to binder).

The problem arises now and again of converting volumes in finished pigment/medium mixtures to masses, or masses to volumes. The *density of the dispersion* ϱ_D is needed for this purpose. If there is no volume reduction in the mixing process, the mixture density can be calculated from the components:

$$\frac{1}{\varrho_D} = \frac{V + V_M}{m + m_M} = \frac{\frac{m}{\varrho} + \frac{m_M}{\varrho_M}}{m + m_M} = \frac{w}{\varrho} + \frac{1-w}{\varrho_M} \quad \text{(unit of } \varrho \text{: g cm}^{-3}) \quad (4.16)$$

A comparison with equations (4.5) and (4.7) shows that the mixture density is very easy to obtain if w and σ or w and β are known:

$$\varrho_D = \frac{\varrho}{w + \mu} = \frac{\beta}{w} = \frac{\varrho\sigma}{w} \quad \text{(unit: g cm}^{-3}) \quad (4.17)$$

In general, it should be kept in mind that quantities expressed in other units (e.g., β in g/L) must be multiplied by the appropriate powers of ten to get them in the stated units before equations (4.1) to (4.17) are applied.

4.2 Absorption and Content of Coloring Material

4.2.1 Dyes

The absorption coefficients defined in Chapter 3, $k'(\lambda)$ for directional irradiation and $K(\lambda)$ for diffuse irradiation, are not actually constants, for they are also affected by the concentration of the coloring matter used. We now inquire into the dependence on concentration.

Absorbing substances present as molecules or ions in the solution show a strictly linear dependence of absorption on solute concentration. This statement applies to dissolved *dyes* as well, so that

$$k' = k''_\beta \beta \quad \text{and} \quad K = k_\beta \beta \tag{4.18}$$

Here β is the concentration by mass in g/L, and the two new quantities k''_β and k_β are constants for a given dye, called "absorption coefficients per unit mass concentration". If k' and K are measured in m²/L, they have the unit m²/g. Equations (4.18) constitute Beer's law. Deviations from the behavior described are seen only when the solution departs from ideality, that is, when reactions occur between the individual solute molecules or between dye and solvent molecules.

Equation (4.18) is sometimes directly combined with (3.48) to yield

$$k''_\beta \beta = \frac{1}{h} \ln \frac{1}{T^*} \tag{4.19}$$

(where h is the layer thickness and T^* the transmission factor), the whole expression being referred to as the Lambert-Beer law. This is somewhat confusing, because the equation does not describe a single effect discussed by two authors, but two fundamentally separate insights into absorption phenomena in transmission measurements. Equation (4.19), however, states that for a given cell thickness h if the concentration by mass β is known then the absorption coefficient per unit mass concentration k''_β can be determined by transmission measurements, and conversely if k''_β is known then β can be determined. The latter technique in particular has found wide adoption, e.g., spectroscopic analytical methods.

4.2.2 Pigments

If a pigmented medium is now taken in place of the dissolved dye, it seems at first glance that the ideal conditions existing in the solution are now violated.

138   4 *How Light Scattering and Absorption Depend on the Content*

However, the total observed absorption is precisely the sum of all individual absorption contributions of the pigment particles, just as the absorption of the dye solution was the sum of all absorption contributions of the molecules or ions. For this reason Beer's law also holds rigorously for *pigments* up to high concentrations. Figure 4-2 clearly shows the linear dependence on PVC. The plotted measured values were obtained from the Kubelka-Munk function $F = K/S$; care was taken to hold the scattering coefficient S constant, since the red iron oxide pigments display scattering of the same order as the TiO_2 pigments used in the white paste. As the concentration of the colored pigment was increased, it was therefore necessary to lower the concentration of the white pigment.

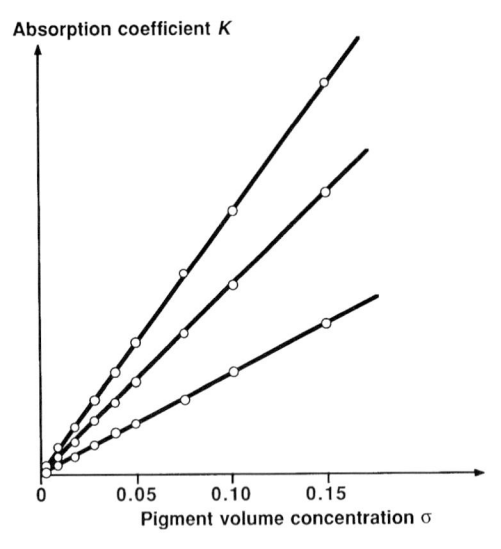

Fig. 4–2. Absorption coefficient K for diffuse irradiation (arbitrary units) versus PVC for three red iron oxide pigments (measured at $\lambda = 550$ nm).

Thus we can write

$$k' = k''\sigma; \qquad K = k\sigma \qquad (4.20)$$

for the absorption of pigments; this is again Beer's law, but with the PVC σ as the pigment portion. The reason for making this change will become quite evident in the next chapter (see p. 210). The new absorption coefficients referred to the PVC thus have the unit m^2/L, since PVC itself has the unit 1.

An important remark should be made on the newly-defined "referred" coefficients. For both dissolved dyes and dispersed pigments, these coefficients can be formed out of contributions by several species. If there is no interaction between the components, multicomponent systems can be described by simply adding the individual referred coefficients, so that for n distinct pigments we have:

$$k' = \sum_{i=1}^{n} k_i'' \sigma_i; \qquad K = \sum_{i=1}^{n} k_i \sigma_i \qquad (4.21)$$

Thus in, say, a three-component mixture, one can measure $k'(\lambda)$ at three different wavelengths λ. If k_i'' of the individual components are known, three equations can be written; therefrom the concentrations of the three components in the mixture can be calculated. It is expedient to perform these measurements at wavelengths corresponding to the absorption maxima of the components or, when the absorption curves overlap greatly, in ranges that permit each component to be distinguished as clearly as possible from the others.

What was already said about dyes also holds for the validity of Beer's law with regard to the absorption coefficients of pigmented systems. Deviations from ideal dispersion of the pigment/binder system can again result in departures from the linear concentration dependence. The initiating factors are of the same type as in the other case but differ in significance and nomenclature. Specifically, phenomena such as flocculation and agglomeration are implicated, especially at higher pigment concentrations; anomalous effects are also seen near the critical pigment volume concentration.

4.3 Scattering and Pigment Content

4.3.1 Scattering Interaction

Because scattering can occur only in the presence of pigments, dyes can now be left out of the analysis. The situation with scattering is somewhat more involved than with absorption. Beer's law, that is, the proportionality between the scattering coefficient and the pigment volume concentration, holds only for low pigment contents.* At higher PVC values, because of the smaller and smaller distances between pigment particles, a *scattering interaction* occurs, which can be interpreted as particles hindering each other in light scattering. In the literature, "multiple scattering" and "dependent scattering" are distinguished by whether the hindrance is due to the same pigment particles or to others. As a result, there are departures from linearity as the measured scattering coefficient falls below the straight line. The curve passes through a maximum and then falls off more-or-less steeply (Fig. 4-3). In other words, simple summing over

---

* Up to ca. $\sigma \leq 10\%$.

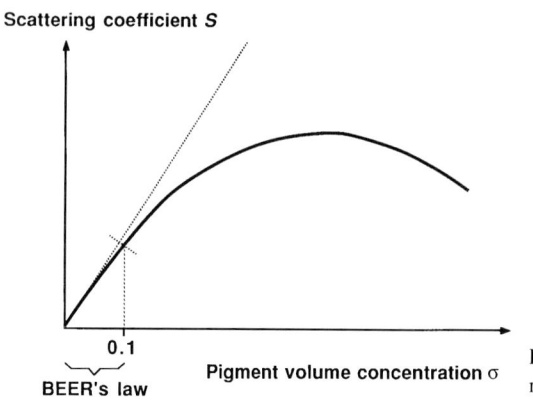

140   4 *How Light Scattering and Absorption Depend on the Content*

Fig. 4-3. Scattering of light versus pigment volume concentration.

single-particle scattering properties, as specified by Beer's law, no longer works as the particles come to be nearer together. Physically, this problem can be interpreted as the scattering of light on multiple particles, and the parallelism with other multiparticle problems in physics makes it understandable that the problem of scattering interaction cannot be solved generally.

The only help that can be expected here must come from theoretical descriptions based on simplified geometrical models. We begin by considering an isodisperse collection of particles regularly distributed in a medium. "Regularly distributed" means that all particles must have the same distance to all their nearest neighbors. This assumption leads to the arrangement taken up by the sphere centers in a closest packing of spheres. In Figure 4-4, the three-dimen-

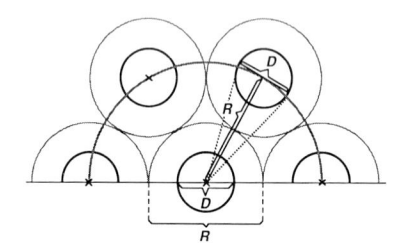

Fig. 4-4. Geometry of the scattering interaction.

sional relationships are shown in a cross section. In deriving an equation to describe the scattering interaction, the quantity of interest is the ratio of the sphere diameter D to the distance R between spheres. As shown in the drawing, the ratio of the volume of the spheres (diameter D) to the volume of the closest-packed spheres (diameter R) is exactly equal to the pigment volume concentration σ divided by $\pi\sqrt{2}/6$, the fraction of space filled by closest-packed spheres, so that

4.3 Scattering and Pigment Content

$$\frac{\frac{1}{6}D^3\pi}{\frac{1}{6}R^3\pi} = \frac{\sigma}{\frac{\pi}{6}\sqrt{2}}$$

from which the ratio D/R is

$$\frac{D}{R} = \sqrt[3]{\frac{3\sqrt{2}}{\pi}\sigma} = 1.105\sqrt[3]{\sigma}$$

If one examines the central particle in Figure 4-4, one sees immediately that the scattering interaction occurs because the light backscattered into the upper half-space can no longer travel unhindered past the nearest neighbors. Instead, the scattered light is reduced by some fraction, here – as a first approximation – taken proportional to the area fraction covered by the particle.

This area fraction p of the individual particle can be read directly from Figure 4-4; it is given approximately by the ratio of the cross-sectional area of the hindering spheres (diameter D) to the area of the hemisphere described by R:

$$p \approx \frac{\frac{D^2}{4}\pi}{2R^2\pi} = \frac{1}{8}\left(\frac{D}{R}\right)^2$$

For closest-packed spheres, there are twelve nearest neighbors. Because only the back half-space is of interest here, six spheres must be considered, so the fraction p must be multiplied by six (giving $6p$) in order to obtain a measure of the hindrance.

What is more, it must be taken into account that the extent of scattering hindrance must be proportional to the area fraction but not necessarily equal to it. For example, it would certainly be better to replace the geometric cross section of the hindering spheres by the scattering contribution of the individual particle, which must be proportional to the PVC-referred scattering coefficient s (mathematically, this corresponds to multiplying by s). Further, an attenuation factor η characteristic of the pigment can be inserted in the interaction term; the total amount by which the scattering interaction reduces total scattering is then represented by $6p\eta s$. By the last two equations for p and D/R, this expression can be rewritten as

$$6\,p\eta s = \gamma s\sigma^{\frac{2}{3}} \tag{4.22}$$

where the abbreviated notation

$$\gamma = \frac{3}{4}\left(\frac{3\sqrt{2}}{\pi}\right)^{\frac{2}{3}} \cdot \eta$$

has been used. The scattering contribution that remains after the hindrance, according to equation (4.22), is thus

$$1 - \gamma s \sigma^{\frac{2}{3}}$$

Because Beer's law is taken to hold at small σ, the scattering coefficient is given by:

$$S = s\sigma (1 - \gamma s \sigma^{\frac{2}{3}}) \qquad (4.23)$$

This relation is not affected by the fact that the next nearest (and so on) spheres contribute to the hindrance. Their effect can be described by some change in the value of the free parameter γ (analogous to the Madelung factors used in crystallographic lattice-energy calculations).

The quantity s appearing for the first time in equation (4.22) is the scattering coefficient per unit PVC, which is defined (more rigorously than in the absorption case) as:

$$s = \left.\frac{dS}{d\sigma}\right|_{\sigma \to 0}$$

The coordinates of the maximum on the S versus σ curve are obtained as follows from the condition $dS/d\sigma = 0$:

$$\sigma_{max} = \left(\frac{3}{5\,\gamma s}\right)^{\frac{3}{2}}$$

$$S_{max} = \frac{2}{5} \sqrt{\frac{1}{s}\left(\frac{3}{5\gamma}\right)^3} = \frac{2}{5} s\sigma_{max} \qquad (4.25)$$

The possibility of the dependent scattering when several pigments are present means that S cannot generally be expressed as a simple sum of contributions due to the various species as was done for K (eq. 4.21). A sum formulation can be used only when one of the summands dominates and the other contributions are small in comparison. In all other cases, the dependent scattering must be taken care of by adding another interaction term to the sum.

4.3.2 Experimental Test of an Empirical Formula

Manipulation of equation (4.23) yields:

$$\frac{S}{\sigma} = s\,(1 - \gamma s \sigma^{\frac{2}{3}}) \tag{4.26}$$

If one makes the change of variable $S/\sigma = y$ and $\sigma^{2/3} = x$, one obtains the equation of a straight line:

$$y = s - \gamma s^2 x \tag{4.27}$$

In this way, the S versus σ curve of Figure 4-3 is linearized; equation (4.27) is therefore especially well-suited to an experimental test.

In order to verify the applicability of this formula, a series of lacquers with increasing pigment content was prepared. The pigment volume concentrations σ lay between 5% and 20%. Several titanium dioxide pigments differing in scattering properties were used, so that the test would cover the widest possible range of properties. From each pigment, a master batch with a σ of roughly 0.45 was prepared by grinding for several hours in a vibratory ball mill. Vehicle was added to this stock to bring the PVC to the desired final value; then drawdowns on black/white paper were made with an automatic Bird applicator. The procedure described in Section 7.2.1 was next followed in order to determine S values for each PVC. When the values obtained were plotted versus the PVC, the typical S versus σ curves shown in Figure 4-5 resulted.

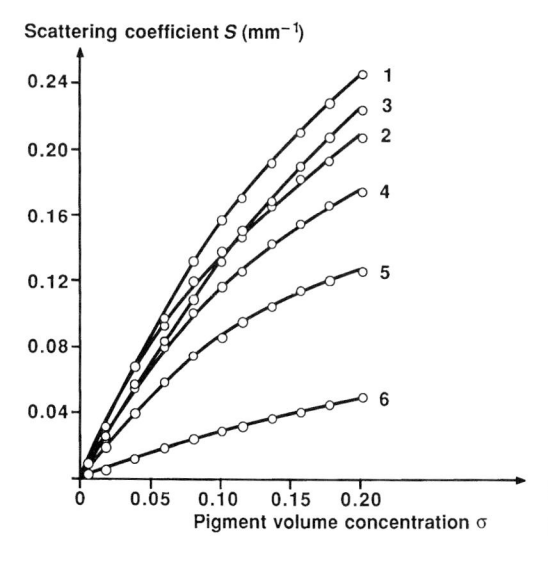

Fig. 4–5. S versus σ plots for six white pigments.

4 How Light Scattering and Absorption Depend on the Content

$\frac{S}{\sigma}$ (mm^{-1})

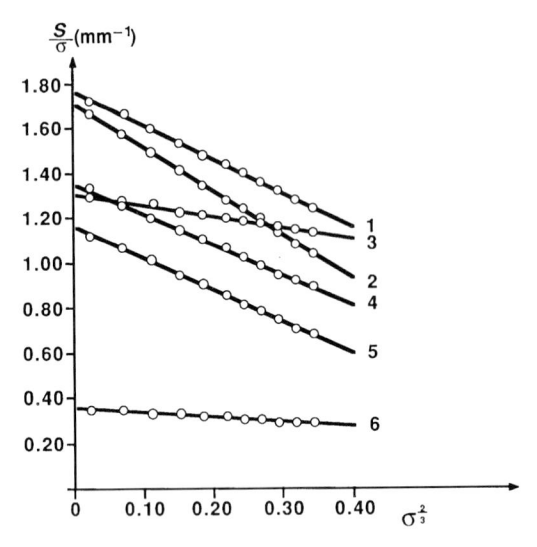

Fig. 4–6. Linearization of the S versus σ curves of Figure 4–5.

An x,y plot as described by equation (4.27) was also prepared, and the optimal lines were calculated by a least-squares method. Figure 4-6 presents the results. The measured points fall quite well on straight lines, providing satisfactory confirmation of the geometrical-empirical formula to the anticipated accuracy.

From the intercepts $y(0) = y_o$ and $x|_{y \to 0} = x_o$, equation (4.27) gives the parameters s and γ:

$$s = y_o; \qquad \gamma = \frac{1}{x_o y_o} \qquad (4.28)$$

A knowledge of x_o and y_o makes it quite simple to find the coordinates of the maximum.*

Equation (4.23) is of fifth degree in $\sigma^{1/3}$ and, for this reason, cannot be solved explicitly for σ. We do not have to propose a graphical solution procedure here, for any desktop computer can be programmed to solve for σ by an iterative method given S, γ, and s.

It must be stated that the simple theoretical model performs well and has been confirmed again and again to within the claimed accuracy. As required by the Kubelka-Munk theory, the relation derived here for the example of white pigments essentially describes pure scattering behavior – regardless (in the terms of our discussion up to now) of whether absorption takes place along with the scattering. It can therefore be expected that the derived equation will also be valid for layers with colored pigments.

---

* Equations (4.25), (4.27), and (4.28) immediately yield $x_{max} = 0.6\, x_o$, $y_{max} = 0.4\, y_o$. In the linearized plot, in other words, the maximum of the S versus σ curve lies at 60 % of the x axis intercept and at 40 % of the y axis intercept.

4.4 Systematic Treatment of Pigment/Achromatic Paste Mixing

4.4.1 Standard Methods of Pigment/Paste Mixing

The mixing of pigments with a black or white paste and the colorimetric assessment of the mixture are governed by certain laws, which can be formulated quite generally, regardless of whether the case involves black and colored pigments in a white paste or white pigments in a black paste. This general relationship has not been discussed a great deal; it is based on physical and colorimetric principles that are common to all pigment/paste mixtures. This section will propose a systematic description of mixing methods and state a conservation law for each method. Systematization is needed all the more as a large number of determination methods are employed in practice (see Chapter 8) and no overarching relationship has been seen between these. But such a generalization would seem necessary to the further rationalization of pigment testing methods.

The development begins with a system ("paste") that already contains a black or white pigment ("achromatic pigment"). The problem is to mix or grind this paste with another pigment ("test pigment") in such a way that a series of increasing or decreasing pigment volume concentrations of the test pigment are obtained. Several approaches can be taken to this problem; these can be unambiguously delimited and thus distinguished. These distinctions are not always manifest in the literature, and both vagueness and wrong conclusions have sometimes resulted.

Just three methods of pigment/paste mixing are of practical interest. These are, in a sense, three prototypes to which the few remaining methods can also be reduced in principle. At present, however, the three standard methods have no names. So that they can be told apart in the further discussion, the following names are therefore introduced:

– Constant volume method
– Adding pigment method
– Dilution method

Every mixture of test pigment and paste is a three-component system comprising the test pigment, the achromatic pigment, and the medium. The new concepts will be explained in what follows, and characteristic features and typical conservation laws for these methods will be stated. The following terminology will be employed:

σ = PVC of test pigment
σ_{UP} = volume concentration of achromatic paste $\Big\}$ in the *mixture*
σ_{U} = PVC of achromatic pigment ($\sigma_{U} = \sigma_{UP}\sigma_{Uo}$)
σ_{Uo} = PVC of achromatic pigment in the *achromatic paste*

Constant volume method. In this method, care is taken to make the PVC of the achromatic pigment equal in all mixtures. On moving to a lower PVC of the test pigment in the mixture, the reduction in pigment volume is always made up by the same volume of medium. In other words, the total volume of the mixture is held constant by adding a calculated quantity of binder (Fig. 4-7a). The conservation law is thus

$$\boxed{\sigma_U = \text{const}} \qquad (4.29)$$

Adding pigment method. An equal quantity of the achromatic paste is measured out every time in this method, and increasing masses of the test pigment are added to such equal-sized portions. The three-component system is reduced to a binary system in this way, and the conservation law can be drawn directly from Figure 4-7b:

$$\boxed{\sigma + \sigma_{UP} = 1} \qquad (4.30)$$

Dilution method. The third way of preparing mixtures with decreasing PVC of the test pigment is to take an already ground mixture of pigment with achromatic paste and dilute it down by adding vehicle. Neither the PVC of the achro-

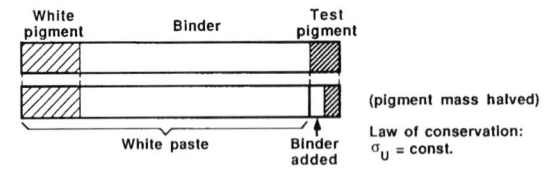

a) Constant volume method
(Example: Colored pigment mixed with white paste)
White pigment | Binder | Test pigment
(pigment mass halved)
White paste | Binder added
Law of conservation: σ_U = const.

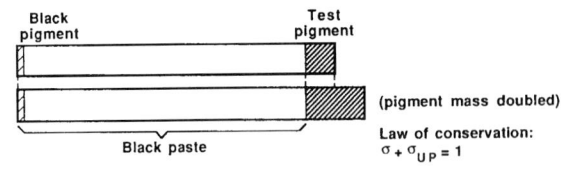

b) Adding pigment method
(Example: White pigment mixed with black paste)
Black pigment | Test pigment
(pigment mass doubled)
Black paste
Law of conservation: $\sigma + \sigma_{UP} = 1$

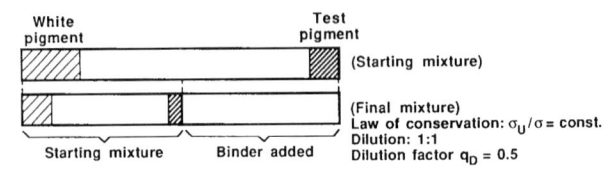

c) Dilution method
(Example: Colored pigment mixed with paste)
White pigment | Test pigment (Starting mixture)
Starting mixture | Binder added
(Final mixture)
Law of conservation: σ_U/σ = const.
Dilution: 1:1
Dilution factor $q_D = 0.5$

Fig. 4–7. The three standard methods of pigment/paste mixing.

4.4 Systematic Treatment of Pigment/Achromatic Paste Mixing

matic pigment nor the PVC of the test pigment is held constant. What does remain constant is the ratio between the PVC values of the achromatic pigment and the test pigment (Fig. 4-7c). Accordingly, the conservation law for the dilution method is:

$$\frac{\sigma_U}{\sigma} = \text{const} = \frac{\sigma_{UA}}{\sigma_A} \tag{4.31}$$

where subscript A identifies initial values (before dilution).

4.4.2 Importance of the Methods

The *adding pigment method* plays a prominent role because it forms the basis for defining the tinting strength and lightening power (Sections 4.6 and 4.7). Weighed quantities of the test pigment are added to equal-sized portions of an achromatic substrate until the mixture visually matches a reference mixture. When this is not done, for example when pigment/achromatic paste mixtures for tinting strength or lightening power determination are prepared by the constant volume method, the values obtained can be expected to differ from the classical values according to the adding pigment method. International standards do not consistently embody a rigorous enough treatment of this point. In view of its importance, the adding pigment method is also discussed very thoroughly in the subsequent analyses of tinting strength and lightening power.

The other two methods have gained some importance in testing practice; this is particularly true of the *constant volume method*, because it alone permits the PVC of the achromatic pigment to be held constant (conservation law, equation 4.29). This point is important for the relation, derived from the Kubelka-Munk function, between $F = K/S$ and the PVC σ of the test pigment. When testing is done with white paste, the constant PVC of the white pigment means that the scattering coefficient S remains virtually unchanged in the mixtures, because the white pigment is dominant. As a result, a plot of F versus σ gives a straight line, because the absorption coefficient K is linear in σ up to high PVC, as was shown earlier (see Fig. 4-2). Similarly, when white pigments are tested in black paste, the constant PVC of the black pigment in the mixture means that the absorption coefficient K remains unchanged; if $1/F = S/K$ is plotted versus σ, the curve has practically the same shape as the S versus σ curve, but this is linear only for low PVC ("Beer's law range"; see Fig. 4-3).

The *dilution method* is also liked by practitioners because it involves only one pigment/paste grinding operation. Here again, however, it must be kept in mind that matching of two such paste mixtures for tinting strength or lightening power may give a somewhat different result from the match obtained with the adding pigment method.

Not much remains to be added about other methods for pigment/paste mixing. By analogy with the constant volume method, in which the PVC of the *achromatic pigment* is held constant, one might seek a method in which the PVC of the *test pigment* remains constant. This is possible if one adds decreasing portions of paste to equal-weight portions of the pigment, adding vehicle to make up for each decrement in paste volume. The method just described is rather uninteresting, because with σ = const the PVC of the pigment under study can no longer be varied. The quantity varied is the PVC of the achromatic pigment, so that the technique is a reciprocal variant of the constant volume method with the achromatic and test pigments interchanged. The discussion implies a further possibility, if fixed paste portions and variable pigment portions are replaced by fixed pigment portions and variable paste portions added to them. A closer examination, however, reveals that this method is not a novel one but is identical with the adding pigment method, since once again a two-component system is involved while the mass balance (conservation law eq. 4.30) is unchanged.

Other methods, for instance one using no achromatic stock paste – weighed portions of achromatic pigment, test pigment, and binder being brought together – are likewise of little interest and so will not be discussed separately. Practical experience shows that colorimetric data for the tinting strength or lightening power can be determined with certainty only if at least the achromatic pigment is incorporated into the medium in a separate operation. What is more, such methods also lead away from the classical tinting strength match and may give rise to dubious results, e.g., the Reynolds method for lightening power, formerly much used, in which $\sigma + \sigma_U$ = const. Individually, such methods can be related back to one of the three prototypes; the Reynolds method is then seen to be a derivative of the adding pigment method, though not identical with it.

4.5 Kubelka-Munk Functions of Pigment/Paste Mixture

4.5.1 General Formula

The discussion in Sections 4.2 and 4.3 leads directly to a formula for the Kubelka-Munk function of a pigment/paste mixture if the components of K and S are known:

$$F(\lambda) = \frac{K(\lambda)}{S(\lambda)} = \frac{K_F(\lambda) + K_W}{S_F(\lambda) + S_W} \tag{4.32}$$

in which subscript F refers to the colored or black pigment and W to the white pigment. At this point, it is unnecessary to include a further interaction term in

4.5 Kubelka-Munk Functions of Pigment/Paste Mixture

the denominator to take account of dependent scattering (see Section 4.3), provided one of the two scattering contributions is small in comparison with the other. Table 4-5 compares the magnitudes of the several terms for typical pigment/paste mixtures. K_F in the numerator and S_W in the denominator are seen to be the dominant terms. This fact suggests that the Kubelka-Munk function might be calculated with the dominant terms (K_F/S_W) and the deviations taken care of by perturbation terms depending on the weaker terms K_W and S_F. The λ dependences shown in equation (4.32) hold only for colored pigments; they can be ignored for black pigments, so that F itself becomes independent of λ.

From equation (4.32) it is now possible to derive the Kubelka-Munk functions for pigment/paste mixtures obtained by the three standard mixing methods, both for white pigments in black paste and for black or colored pigments in white paste. The equations of Sections 4.2 and 4.3 for the concentration dependences of K and S are used for the purpose. Aside from the scattering and absorption coefficient per unit concentration, the equations to be derived contain only the concentration of the test pigment; first and second approximations for the three standard mixing methods are obtained from them.

Table 4–5. Typical values of the quantities in equation (4.32).

	Carbon black pigments, $\sigma_b = 0.01$		Colored pigments, $\sigma = 0.05$	TiO$_2$ pigments, $\sigma_w = 0.2$	
	In black paste	Tested in white paste, $\sigma_w = 0.2$, $s_w = 2000$ mm^{-1}	Tested in white paste, $\sigma_w = 0.2$, $s_w = 2000$ mm^{-1}	In white paste	Tested in black paste, $\sigma_b = 0.01$, $k_b = 20,000$ mm^{-1}
Absorption coefficient referred to PVC, k, mm^{-1}	~20 000	~20 000	200…20 000	5	~20 000
Scattering coefficient referred to PVC, s, mm^{-1}	10	~2000	~2000	~2000	~2000
Absorption coefficient K, mm^{-1}	200	200	10…1000	1	200
Scattering coefficient S, mm^{-1}	0.1	400	400	400	400
Kubelka-Munk function F	2000	0.5	0.02…2	0.002	0.5
Reflectance (corrected) $\varrho^*(\lambda)$	0	0.4	0.2…0.8	0.9	0.4

150   4 How Light Scattering and Absorption Depend on the Content

The approximations are arrived at by expanding the Kubelka-Munk function as a power series in the test-pigment concentration and then truncating after the linear and quadratic terms, respectively. Because the later terms are much smaller than the linear term of the first approximation, they take on the character of perturbations and can be handled as corrections to the linear leading term. The result is an advantage for the experimental determination of the perturbation coefficients: the correction terms in perturbation calculations do not have to be very accurate, i.e., they can exhibit a relative error greater than that of the leading term without impairing the value of the correction.

4.5.2 Black Pigments Mixed with White Paste

For black pigments,* by equations (4.20) and (4.23),

$$K_F(\lambda) = k\sigma; \qquad S_F(\lambda) = s\sigma\,(1 - g\sigma^{\frac{2}{3}}) \tag{4.33}$$

where $g = \gamma s$. The Kubelka-Munk functions for the standard mixing methods are obtained by applying the conservation law for each method.

Constant volume method. The conservation law in equation (4.29) is σ_w = const, so that by the equations (4.20) and (4.23) K_w = const and S_w = const. Substitution in equation (4.32) yields the following expression for the Kubelka-Munk function $F_{b,c}$ (the first subscript b denotes black pigment, the c stands for the constant volume method):

$$F_{b,c} = \frac{k\sigma + K_w}{s\sigma\,(1 - g\sigma^{\frac{2}{3}}) + S_w} = \frac{K_w}{S_w} \cdot \frac{\dfrac{k}{K_w}\sigma + 1}{1 + \dfrac{s}{S_w}\sigma\,(1 - g\sigma^{\frac{2}{3}})}$$

$K_w/S_w = F_w$ is the Kubelka-Munk function of the white paste after the addition of medium to restore constant volume (i.e., the system without black pigment). The development in the preceding section implies that the second summand in the denominator is very small compared to 1, so that the quotient can be truncated after the second-order term to obtain:

$$F_{b,c} = F_w\left(\dfrac{k}{K_w}\sigma + 1\right) \cdot \left[1 - \dfrac{s}{S_w}\sigma + g\,\dfrac{s}{S_w}\sigma^{\frac{5}{3}} + \left(\dfrac{s}{S_w}\right)^2\sigma^2\ldots\right]$$

---

* The following convention should always be observed: symbols with no subscript refer to the pigment being tested.

4.5 Kubelka-Munk Functions of Pigment/Paste Mixture

Multiplying out the brackets and neglecting the terms of order higher than 2 yields:

$$F_{b,c} = F_W \left[\left(\frac{k}{K_w} - \frac{s}{S_w} \right) \sigma + g \frac{s}{S_w} \sigma^{\frac{5}{3}} - \frac{s}{S_w} \left(\frac{k}{K_w} - \frac{s}{S_w} \right) \sigma^2 + 1 \right]$$

and, finally, the *PVC-dependent Kubelka-Munk function for black pigment/white paste mixtures prepared by the constant volume method*:

$$F_{b,c}^{(2)} - F_W = c_{b_1} \sigma \left(1 + c_{b\frac{5}{3}} \sigma^{\frac{2}{3}} - c_{b_2} \sigma \right) \quad (4.34)$$

(second approximation)

The coefficients in this equation have the following meanings:

$$c_{b_1} = \frac{1}{S_w}(k - F_W s) = \frac{k}{S_w}\left(1 - \frac{F_W}{F_{b,PU}}\right) \approx \frac{k}{S_w}$$

$$c_{b\frac{5}{3}} = gF_W \frac{s}{k - F_W s} = g \frac{F_W}{F_{b,PU} - F_W} \approx g \frac{F_W}{F_{b,PU}}$$

$$c_{b_2} = \frac{s}{S_w} \quad (4.35)$$

where $F_{b,PU} = k/s = (K/S)|_{\sigma \to 0}$ is the Kubelka-Munk function of the black pigment full shade for PVC in the Beer's law range.

Similarly, the first approximation is obtained from equation (4.34) by neglecting terms in higher powers than 1 (which are always very small in comparison with 1) as well as the additive term F_W (which is always very small in comparison with $F_{b,c}$):

$$F_{b,c}^{(1)} = c_{b_1} \sigma \approx \frac{k}{S_w} \sigma \quad \text{(first approximation)} \quad (4.36)$$

Thus, from well-defined premises, we have rigorously derived the well-known linear dependence of the Kubelka-Munk function (first approximation) on PVC for black pigment/white paste mixtures. A far simpler way to reach the same result, of course, is to ascribe all scattering by the paste mixture to the white paste and all absorption by the mixture to the black pigment, as indicated by the right-hand expression in equation (4.36). This procedure, however, leaves it unclear what simplifications have been made and whether these are reasonable.

4 How Light Scattering and Absorption Depend on the Content

Adding pigment method. The conservation law for this method, from equation (4.30), is $\sigma + \sigma_{WP} = 1$ (where σ_{WP} is the volume concentration of the white paste after addition of the black pigment). The PVC of the white pigment σ_w in the mixture must be calculated from σ_{WP}; the formula is

$$\sigma_w = \sigma_{ow}\, \sigma_{WP} \tag{4.37}$$

where σ_{ow} is the PVC of the white pigment in the white paste. By the conservation law, it follows that:

$$K_W = k_w\, \sigma_w = k_w\, \sigma_{ow}\, \sigma_{WP} = k_w\, \sigma_{ow}\, (1-\sigma)$$

$$S_W = s_w\, \sigma_w\, (1 - g_w\, \sigma_w^{\frac{2}{3}}) = s_w\, \sigma_{ow}\, (1-\sigma)\, [1 - g_w\, \sigma_{ow}^{\frac{2}{3}} (1-\sigma)^{\frac{2}{3}}]$$

$$= s_w\, \sigma_{ow}\, (1-\sigma) \cdot \left[1 - g_w\, \sigma_{ow}^{\frac{2}{3}}\left(1 - \frac{2}{3}\sigma - \frac{1}{9}\sigma^2 \ldots\right)\right]$$

After substitution in equation (4.32), the numerator and denominator are divided by $1 - \sigma$; this causes the pigment/medium volume ratio $\psi = \sigma/(1-\sigma)$ (PMV, eq. 4.6) to appear in the numerator and denominator. The PMV thus becomes the appropriate measure of the pigment portion for the adding pigment method, and so the remaining PVC expressions are also replaced by $\sigma = \psi/(1+\psi) \approx \psi\,(1-\psi)$ (eq. 4.15). Now $F_{b,a}$ (subscript a for adding pigment method), once again with terms of higher order neglected, becomes

$$F_{b,a} = \frac{k\psi + k_w\, \sigma_{ow}}{s\psi\,(1 - g\psi^{\frac{2}{3}}) + s_w\, \sigma_{ow}\left[1 - g_w\, \sigma_{ow}^{\frac{2}{3}}\left[1 - \frac{2}{3}\psi\,(1-\psi) - \frac{1}{9}\psi^2\right]\right]}$$

Here $k_w \sigma_{ow} = K_{WP}$ in the numerator is the absorption coefficient of the white paste, while the isolable expression $s_w \sigma_{ow}\,(1 - g_w \sigma_{ow}^{2/3}) = S_{WP}$ is the scattering coefficient of the white paste. The quotient $K_{WP}/S_{WP} = F_{WP}$ denotes the Kubelka-Munk function of the white paste, so that

$$F_{b,a} = F_{WP} \cdot \frac{\dfrac{k}{K_{WP}}\psi + 1}{1 + \dfrac{1}{S_{WP}}\left(s + \dfrac{2}{3} g_w s_w \sigma_{ow}^{\frac{5}{3}}\right)\psi - g\,\dfrac{s}{S_{WP}}\psi^{\frac{5}{3}} - \dfrac{5}{9} g_w\, \dfrac{s_w}{S_{WP}}\, \sigma_{ow}^{\frac{5}{3}}\, \psi^2}$$

Once again, the quotient expression can be truncated after the quadratic term, because all the summands after the 1 in the denominator are much smaller than 1. Thus the *Kubelka-Munk function for black pigment/white paste mixtures prepared by the adding pigment method* is

4.5 Kubelka-Munk Functions of Pigment/Paste Mixture

$$F_{b,a}^{(2)} - F_{WP} = z_{b_1}\, \psi\, (1 + z_{b5/3}\, \psi^{\frac{2}{3}} - z_{b_2}\, \psi) \quad (4.38)$$

(second approximation)

Rather than write exact expressions for the coefficients, we write them after dropping terms that contribute practically nothing:

$$z_{b_1} \approx \frac{k}{S_{WP}}$$

$$z_{b5/3} \approx g\, \frac{S}{k}\, F_{WP} \approx g\, \frac{F_{WP}}{F_{b,PU}}$$

$$z_{b_2} \approx \frac{S}{S_{WP}}\left(1 + \frac{2}{3} g_w\, \frac{S_w}{S}\, \sigma_{ow}^{5/3}\right) \approx \frac{S}{S_{WP}} \quad (4.39)$$

(approximation for black pigments with high light-scattering power, with $F_{b,PU} = k/s = (K/S)|_{\sigma \to 0}$, where $F_{b,PU}$ is the Kubelka-Munk function of the black pigment full shade in the Beer's law range). About the first approximation, the same statements can be made as in the case of the constant volume method, and the same conclusions drawn, but referred to the PMV:

$$F_{b,a}^{(1)} = z_{b_1}\, \psi = \frac{k}{S_{WP}}\, \psi \quad \text{(first approximation)} \quad (4.40)$$

A comparison of these equations with equations (4.34) and (4.36) for the constant volume method shows that the expressions are completely identical (aside from the reference to PMV), even up to equality of the coefficients (though after simplification).*

Dilution method. The conservation law here is $\sigma_w/\sigma = \text{const} = \sigma_{wA}/\sigma_A$ (where the subscript A marks initial values before dilution). Accordingly, equation (4.32) can be written again with σ canceled out of the numerator and denominator:

---

* Here it must be noted that $F_{b,c}$ and $F_{b,a}$ are nonetheless two distinct functions; $F_{b,c}$ is still the Kubelka-Munk function of the constant volume method, even when σ is replaced by $\sigma = \psi/(1 + \psi)$, and similarly for $F_{b,a}$. Likewise, F_W and F_{WP} are not the same quantities.

$$\frac{1}{F_{b,d}} = \frac{s\sigma(1-g\sigma^{\frac{2}{3}}) + s_w \frac{\sigma_{wA}}{\sigma_A} \sigma \left[1 - g_w \left(\frac{\sigma_{wA}}{\sigma_A}\right)^{\frac{2}{3}} \sigma^{\frac{2}{3}}\right]}{k\sigma + k_w \frac{\sigma_{wA}}{\sigma_A} \sigma}$$

$$= \frac{s\sigma_A \left[1 - g\sigma_A^{\frac{2}{3}} \left(\frac{\sigma}{\sigma_A}\right)^{\frac{2}{3}}\right] + s_w \sigma_{wA} \left[1 - g_w \sigma_{wA}^{\frac{2}{3}} \left(\frac{\sigma}{\sigma_A}\right)^{\frac{2}{3}}\right]}{k\sigma_A + k_w \sigma_{wA}}$$

To find the initial value of the Kubelka-Munk function, one sets $\sigma \to \sigma_A$:

$$\frac{1}{F_A} = \frac{s\sigma_A (1 - g\sigma_A^{\frac{2}{3}}) + s_w \sigma_{wA} (1 - g_w \sigma_{wA}^{\frac{2}{3}})}{k\sigma_A + k_w \sigma_{wA}}$$

With $q_D = \sigma/\sigma_A$, dividing by the first equation and neglecting higher-order terms gives the *Kubelka-Munk function for black pigment/white paste mixtures prepared by the dilution method*:

$$\frac{F_{b,d}}{F_A} = \frac{1 - v_b}{1 - v_b q_D^{\frac{2}{3}}} \approx 1 - v_b (1 - q_D^{\frac{2}{3}}) \qquad \text{(first approximation)} \qquad (4.41)$$

where $q_D = \sigma/\sigma_A$ is the dilution ratio by volume and F_A is the initial value before dilution. This formula has just one coefficient:

$$v_b = \frac{g s \sigma_A^{\frac{5}{3}} + g_w s_w \sigma_{wA}^{\frac{5}{3}}}{s \sigma_A + s_w \sigma_{wA}} \approx g_w \sigma_{wA}^{\frac{2}{3}} \qquad (4.42)$$

The simplification at right is obtained because we always have $s_w \sigma_{wA} \gg s\sigma_A$ (see also Table 4-5).

Equation (4.41) is a "2/3" approximation implied by the second approximation. The zero-th approximation is obtained for $\sigma \to 0$, that is, infinite dilution:

$$\frac{F_{b,d\infty}}{F_A} = 1 - v_b$$

and is of little interest.

4.5.3 White Pigments Mixed with Black Paste

By equations (4.20) and (4.23), the following hold for white pigments:*

$$K_W = k\sigma; \qquad S_W = s\sigma(1 - g\sigma^{\frac{2}{3}}) \qquad (4.43)$$

where $g = \gamma\sigma$. The Kubelka-Munk functions for the standard mixing methods are once more found by imposing the conservation laws.

Constant volume method. The conservation law $\sigma_b = $ const now implies $K_b = $ const and $S_b = $ const. For the Kubelka-Munk function $F_{w,c}$, equation (4.32) then yields

$$\frac{1}{F_{w,c}} = \frac{s\sigma(1 - g\sigma^{\frac{2}{3}}) + S_b}{k\sigma + K_b}$$

(where it is advantageous to use the reciprocal). The procedure is now similar to that in Section 4.5.2: Extraction of the Kubelka-Munk function of the black paste (after addition of medium to restore constant volume), $K_b/S_b = F_B$; division by the denominator; truncation after the second-order term. The result is:

$$\frac{1}{F^{(2)}_{w,c}} - \frac{1}{F_B} = c_{w_1}\sigma\left(1 - c_{w_5}\frac{\sigma^{\frac{2}{3}}}{3} - c_{w_2}\sigma\right) \qquad \text{(second approximation)} \qquad (4.44)$$

the *PVC-dependent Kubelka-Munk function for white pigment/black paste mixtures prepared by the constant volume method.* As in the preceding section, the first approximation is obtained by dropping the higher powers and the $1/F_B$ term:

$$\frac{1}{F^{(1)}_{w,c}} = c_{w_1}\sigma = \frac{s}{K_b}\sigma \qquad \text{(first approximation)} \qquad (4.45)$$

The coefficients now have the following meanings:

---

* Quantities relating to the white pigment have no subscripts here.

4 How Light Scattering and Absorption Depend on the Content

$$c_{w_1} = \frac{1}{F_B}\left(\frac{s}{S_b} - \frac{k}{K_b}\right) \approx \frac{s}{K_b}$$

$$c_{w\frac{5}{3}} = g\,\frac{F_B}{F_B - F_{w,PU}} \approx g$$

$$c_{w_2} = \frac{k}{K_b} \tag{4.46}$$

where $F_B = K_b/S_b$ and $F_{w,PU} = k/s = (K/S)|_{\sigma \to 0}$ (the Kubelka-Munk functions of, respectively, the black paste and the white pigment full shade in the Beer's law range). Equation (4.45), the first approximation of the Kubelka-Munk function for white pigment/black paste mixtures, expressed as a function of concentration, is – like equation (4.36) – one of the most frequently used formulas.

Adding pigment method. With the conservation law $\sigma_{BP} + \sigma = 1$ and the PVC equation for the black pigment,

$$\sigma_b = \sigma_{BP}\sigma_{ob} \tag{4.47}$$

where σ_{ob} is the PVC of the black pigment in the black paste, the derivation of the approximations is similar to the development in Section 4.5.2. The *Kubelka-Munk function for white pigment/black paste mixtures prepared by the adding pigment method* is

$$\boxed{\frac{1}{F_{w,a}^{(2)}} - \frac{1}{F_{BP}} = z_{w_1}\psi\,(1 - z_{w\frac{5}{3}}\psi^{\frac{2}{3}} - z_{w_2}\psi)} \quad \text{(second approximation)} \tag{4.48}$$

or, after dropping higher powers and the $1/F_{BP}$ term,

$$\boxed{\frac{1}{F_{w,a}^{(1)}} = z_{w_1}\psi = \frac{s}{K_{BP}}\psi} \quad \text{(first approximation)} \tag{4.49}$$

where the coefficients (approximate values) are

$$z_{w_1} \approx \frac{s}{K_{BP}}\left(1 + \frac{2}{3} g_b \frac{s_b}{s} \sigma_{ob}^{\frac{5}{3}}\right) \approx \frac{s}{K_{BP}}$$

$$z_w \frac{5}{3} \approx g$$

$$z_{w_2} \approx \frac{k}{K_{BP}} + \frac{5}{9} g_b \frac{s_b}{s} \sigma_{ob}^{\frac{5}{3}} \approx \frac{k}{K_{BP}} \quad (4.50)$$

(approximations for black paste pigments with low light-scattering power, such as carbon black; K_{BP} is the absorption coefficient of the black paste).

Dilution method. The conservation law for this method is $\sigma_b/\sigma = \text{const} = \sigma_{bA}/\sigma_A$ (subscript A referring to initial mixture before dilution), so a derivation similar to that in Section 4.5.2 yields the *Kubelka-Munk function for white pigments in black paste*:

$$\boxed{\frac{F_{w,d}}{F_A} \approx 1 - v_w (1 - q_D^{\frac{2}{3}})} \quad (4.51)$$

where F_A is the initial value and $q_D = \sigma/\sigma_A$ is the dilution ratio by volume; the coefficient is

$$v_w = \frac{g s \sigma_A^{\frac{5}{3}} + g_b s_b \sigma_{bA}^{\frac{5}{3}}}{s \sigma_A + s_b \sigma_{bA}} \approx g \sigma_A^{\frac{2}{3}} \quad (4.52)$$

The simplified form at the extreme right holds because we always have $s\sigma_A \gg s_b \sigma_{bA}$.

4.5.4 Colored Pigments Mixed with White Paste

The λ dependence of absorption and scattering has to be considered for colored pigments. The formulas can be found in Section 4.5.2, where the λ-dependent quantities are identified as such.

Constant volume method. From equations (4.32) to (4.36), the *spectral Kubelka-Munk function for colored pigments* is

$$\boxed{F_c^{(2)}(\lambda) - F_W = c_1(\lambda) \sigma \left(1 + c_{\frac{5}{3}}(\lambda) \sigma^{\frac{2}{3}} - c_2(\lambda) \sigma\right)} \quad (4.53)$$

(second approximation)

4 How Light Scattering and Absorption Depend on the Content

$$\boxed{F_c^{(1)}(\lambda) = c_1(\lambda)\,\sigma \qquad \text{(first approximation)}} \tag{4.54}$$

where the coefficients (higher-order terms being dropped) are

$$c_1(\lambda) \approx \frac{k(\lambda)}{S_W}$$

$$c_{\frac{5}{3}}(\lambda) \approx g(\lambda)\,\frac{F_W}{F_{PU}(\lambda)}$$

$$c_2(\lambda) = \frac{s(\lambda)}{S_W} \tag{4.55}$$

With regard to the Kubelka-Munk function of colored pigments, however, there is a second aspect of great interest: the question whether Kubelka-Munk functions can be written for the CIE tristimulus values X, Y, Z. If so, there will be an appreciable benefit, for these "broad-band" quantities* could be directly represented as functions of the pigment concentration. This would be wholly similar to what was done for the spectral Kubelka-Munk functions. That this can indeed be done will now be demonstrated for the CIE tristimulus value Y.

We start with the λ-dependent quantities of equations (4.53) to (4.55):

$$F_c(\lambda) - F_W = F_c(\lambda)\left(1 - \frac{F_W}{F_c(\lambda)}\right) = c_1(\lambda)\,\sigma\,(1 + c_{\frac{5}{3}}(\lambda)\,\sigma^{\frac{2}{3}} - c_2(\lambda)\,\sigma \ldots) \tag{4.56}$$

For an experimental determination of the Kubelka-Munk function to the required accuracy, the spectral reflectance $\varrho(\lambda)$ of the colored pigment/white paste mixture (see Section 3.3.2 and Fig. 3-9) must lie in the range of $0.20 \le \varrho(\lambda) \le 0.75$. With this restriction and except for the upper part of the range ($0.55 \le \varrho_{in} \le 0.75$), equation (3.118) implies that $\ln F = A - B\varrho^*$. Further, the additive term $F_W/F_c(\lambda)$ is small enough compared to 1 that $F_c(\lambda)$ in the formula can be replaced by the first approximation $F_c^{(1)}(\lambda) = c_1(\lambda)\sigma$. Taking logarithms on both sides of the equation leads to

$$A - B\varrho^*(\lambda) - \frac{F_W}{c_1(\lambda)\,\sigma} = \ln c_1(\lambda) + \ln \sigma + c_{\frac{5}{3}}(\lambda)\,\sigma^{\frac{2}{3}} - c_2(\lambda)\,\sigma \ldots \tag{4.57}$$

---

* The CIE tristimulus values X, Y, Z (or the reflectometer values R'_x, R'_y, R'_z) are not spectral values now, as required by the Kubelka-Munk theory; instead, each corresponds to a broad band of wavelengths from which it was derived.

4.5 Kubelka-Munk Functions of Pigment/Paste Mixture

On both the left-hand and right-hand sides, the approximate rule $\ln(1 + \varepsilon) \approx \varepsilon$ (for small quantities) has been used. In order to get to the CIE tristimulus value Y, the colorimetric integral $\int \bar{y}_N(\lambda)\ldots d\lambda$ is applied on both sides of the equation (where $\bar{y}_N(\lambda)$ denotes the CIE color matching function $\bar{y}(\lambda)$ normalized to the CIE standard illuminant N, that is, already divided by 100). Upon integration, quantities independent of λ remain constant, since $\int \bar{y}_N(\lambda)\, d\lambda = 1$.

$$A - B\frac{Y^{**}}{100} - \frac{F_W}{\sigma} \int \frac{\bar{y}_N(\lambda)}{c_1(\lambda)}\, d\lambda$$

$$= \int \bar{y}_N(\lambda) \cdot \ln c_1(\lambda)\, d\lambda + \ln \sigma + \sigma^{\frac{2}{3}} \int \bar{y}_N(\lambda)\, c_{\frac{5}{3}}(\lambda)\, d\lambda - \sigma \int \bar{y}_N(\lambda)\, c_2(\lambda)\, d\lambda \ldots$$

Here Y^{**} is the CIE tristimulus value in the interior of the paste (actually Y^*, but this notation has already been claimed by the value function in the CIE-LAB system). The three integrals on the right-hand side are now written by turns as $\ln \bar{c}_1$, $\bar{c}_{5/3}$, $\bar{c}_2$. Now $1/\bar{c}_1 = \exp(-\int y_N(\lambda) \cdot \ln c_1(\lambda)\, d\lambda)$; because $\exp(-\ln x) = 1/x$, an obvious substitution for the remaining integral on the left is

$$\int \frac{\bar{y}_N(\lambda)}{c_1(\lambda)}\, d\lambda = \frac{1}{\bar{c}_1} + \varepsilon$$

where ε is a perturbation term. If this expression is substituted in the equation, the resulting term $\varepsilon F_W/\sigma$ will certainly be small of higher order than the already very small term $F_W/(\bar{c}_1 \sigma)$ and thus can be neglected. Then

$$A - B\frac{Y^{**}}{100} - \frac{F_W}{\bar{c}_1 \sigma} = \ln \bar{c}_1 + \ln \sigma + \bar{c}_{\frac{5}{3}}\, \sigma^{\frac{2}{3}} - \bar{c}_2\, \sigma$$

A comparison with equation (4.57) reveals full agreement if the following substitutions are made: $\varrho^*(\lambda)$ by $Y^{**}/100$, $c_1(\lambda)$ by $\bar{c}_1$, $c_{5/3}(\lambda)$ by $\bar{c}_{5/3}$, and $c_2(\lambda)$ by $\bar{c}_2$. It is thus possible to go back through the steps from equation (4.57) to (4.56) in reverse order. An equation like (4.56) with the above substitutions is obtained.

Now the same arguments can be repeated for the other two CIE tristimulus values X and Z in order to obtain similar formulas. Listing all three equations together, we then have the *Kubelka-Munk functions of the CIE tristimulus values for colored pigment/white paste mixtures prepared by the constant volume method*:

160   4 How Light Scattering and Absorption Depend on the Content

$$\bar{F}_{Xc}^{(2)} - F_W = \bar{c}_{X_1}\sigma \left(1 + \bar{c}_{X\frac{5}{3}}\sigma^{\frac{2}{3}} - \bar{c}_{X_2}\sigma\right)$$

$$\bar{F}_{Yc}^{(2)} - F_W = \bar{c}_{Y_1}\sigma \left(1 + \bar{c}_{Y\frac{5}{3}}\sigma^{\frac{2}{3}} - \bar{c}_{Y_2}\sigma\right)$$

$$\bar{F}_{Zc}^{(2)} - F_W = \bar{c}_{Z_1}\sigma \left(1 + \bar{c}_{Z\frac{5}{3}}\sigma^{\frac{2}{3}} - \bar{c}_{Z_2}\sigma\right) \tag{4.58}$$

(second approximation)

and

$$\bar{F}_{Xc}^{(1)} = \bar{c}_{X_1}\sigma; \quad \bar{F}_{Yc}^{(1)} = \bar{c}_{Y_1}\sigma; \quad \bar{F}_{Zc}^{(1)} = \bar{c}_{Z_1}\sigma \tag{4.59}$$

(first approximation)

where the coefficients are

$$\bar{c}_{X_1} = e^{\int \bar{x}_N(\lambda)\,\cdot\,\ln c_1(\lambda)\,d\lambda}$$

$$\bar{c}_{Y_1} = e^{\int \bar{y}_N(\lambda)\,\cdot\,\ln c_1(\lambda)\,d\lambda}$$

$$\bar{c}_{Z_1} = e^{\int \bar{z}_N(\lambda)\,\cdot\,\ln c_1(\lambda)\,d\lambda}$$

$$\bar{c}_{X\frac{5}{3}} = \int \bar{x}_N(\lambda)\, c_{\frac{5}{3}}(\lambda)\, d\lambda$$

$$\bar{c}_{Y\frac{5}{3}} = \int \bar{y}_N(\lambda)\, c_{\frac{5}{3}}(\lambda)\, d\lambda$$

$$\bar{c}_{Z\frac{5}{3}} = \int \bar{z}_N(\lambda)\, c_{\frac{5}{3}}(\lambda)\, d\lambda$$

$$\bar{c}_{X_2} = \int \bar{x}_N(\lambda)\, c_2(\lambda)\, d\lambda$$

$$\bar{c}_{Y_2} = \int \bar{y}_N(\lambda)\, c_2(\lambda)\, d\lambda$$

$$\bar{c}_{Z_2} = \int \bar{z}_N(\lambda)\, c_2(\lambda)\, d\lambda \tag{4.60}$$

$\bar{x}_N(\lambda)$, $\bar{y}_N(\lambda)$, and $\bar{z}_N(\lambda)$ being the CIE color matching functions normalized to CIE standard illuminant N.

The Kubelka-Munk functions described by equations (4.58) and (4.59) are not identical to the colorimetric integrals over $F(\lambda)$; in accordance with equations (4.57) ff. they are defined as $\bar{F}_x = \exp(\int \bar{x}_N(\lambda) \cdot \ln F(\lambda) d\lambda)$, etc. These are related in the following way to the CIE tristimulus values from the corrected reflectance:

4.5 Kubelka-Munk Functions of Pigment/Paste Mixture

$$\bar{F}_{Xc} = \frac{\left(1 - \frac{X^{**}}{X_n}\right)^2}{2 \cdot \frac{X^{**}}{X_n}}$$

$$\bar{F}_{Yc} = \frac{\left(1 - \frac{Y^{**}}{100}\right)^2}{2 \cdot \frac{Y^{**}}{100}}$$

$$\bar{F}_{Zc} = \frac{\left(1 - \frac{Z^{**}}{Z_n}\right)^2}{2 \cdot \frac{Z^{**}}{Z_n}} \qquad (4.61)$$

where X_n and Z_n are the CIE tristimulus values for the standard illuminant and

$$\frac{X^{**}}{X_n} = \int \bar{x}_N(\lambda)\, \varrho^*(\lambda)\, d\lambda$$

$$\frac{Y^{**}}{100} = \int \bar{y}_N(\lambda)\, \varrho^*(\lambda)\, d\lambda$$

$$\frac{Z^{**}}{Z_n} = \int \bar{z}_N(\lambda)\, \varrho^*(\lambda)\, d\lambda \qquad (4.62)$$

These results complete the proof that the concentration-dependent Kubelka-Munk functions of the CIE tristimulus values can be described by the same formulas commonly used for the spectral values or black pigments (see the above argument and Section 4.5.2), provided the restriction $0.1 < \varrho_{in}(\lambda) < 0.55$ is imposed. But this is not an onerous restriction, since in striving for accuracy one will always make certain to take quantities of paste and pigment such that $\varrho_{in}(\lambda)$ will lie in this range.

Adding pigment method. First, *the spectral Kubelka-Munk function for colored pigments* is

4 How Light Scattering and Absorption Depend on the Content

$$F_a^{(2)}(\lambda) - F_{WP} = z_1(\lambda)\,\psi\,(1 + z_{\frac{5}{3}}(\lambda)\,\psi^{\frac{2}{3}} - z_2(\lambda)\,\psi) \qquad (4.63)$$

(second approximation)

$$F_a^{(1)}(\lambda) = z_1(\lambda)\,\psi \qquad \text{(first approximation)} \qquad (4.64)$$

where the coefficients (without higher-order terms) are

$$z_1(\lambda) \approx \frac{k(\lambda)}{S_{WP}}$$

$$z_{\frac{5}{3}}(\lambda) \approx g(\lambda)\,\frac{F_{WP}}{F_{PU}(\lambda)}$$

$$z_2(\lambda) \approx \frac{s(\lambda)}{S_{WP}}\left(1 + \frac{2}{3}\,g_w\,\frac{S_w}{s(\lambda)}\,\sigma_{ow}^{\frac{5}{3}}\right) \approx \frac{s(\lambda)}{S_{WP}} \qquad (4.65)$$

(approximation for colored pigments with high light-scattering power).

With regard to the Kubelka-Munk functions of the CIE tristimulus values for the *adding pigment method*, the same arguments as cited above for the constant volume method lead to similar results, the *Kubelka-Munk functions of the CIE tristimulus values of colored pigment/white paste mixtures prepared by the adding pigment method*:

$$\bar{F}_{Xa}^{(2)} - F_{WP} = \bar{z}_{X_1}\psi\,(1 + \bar{z}_{X\frac{5}{3}}\psi^{\frac{2}{3}} - \bar{z}_{X_2}\psi)$$

$$\bar{F}_{Ya}^{(2)} - F_{WP} = \bar{z}_{Y_1}\psi\,(1 + \bar{z}_{Y\frac{5}{3}}\psi^{\frac{2}{3}} - \bar{z}_{Y_2}\psi)$$

$$\bar{F}_{Za}^{(2)} - F_{WP} = \bar{z}_{Z_1}\psi\,(1 + \bar{z}_{Z\frac{5}{3}}\psi^{\frac{2}{3}} - \bar{z}_{Z_2}\psi) \qquad (4.66)$$

(second approximation)

and

$$\bar{F}_{Xa}^{(1)} = \bar{z}_{X_1}\psi; \quad \bar{F}_{Ya}^{(1)} = \bar{z}_{Y_1}\psi; \quad \bar{F}_{Za}^{(1)} = \bar{z}_{Z_1}\psi \qquad (4.67)$$

(first approximation)

where the coefficients are

4.5 Kubelka-Munk Functions of Pigment/Paste Mixture

$$\bar{z}_{X_1} = e^{\int \bar{x}_N(\lambda) \cdot \ln z_1(\lambda) \, d\lambda}$$

$$\bar{z}_{Y_1} = e^{\int \bar{y}_N(\lambda) \cdot \ln z_1(\lambda) \, d\lambda}$$

$$\bar{z}_{Z_1} = e^{\int \bar{z}_N(\lambda) \cdot \ln z_1(\lambda) \, d\lambda}$$

$$\bar{z}_{X\frac{5}{3}} = \int \bar{x}_N(\lambda) \, z_{\frac{5}{3}}(\lambda) \, d\lambda$$

$$\bar{z}_{Y\frac{5}{3}} = \int \bar{y}_N(\lambda) \, z_{\frac{5}{3}}(\lambda) \, d\lambda$$

$$\bar{z}_{Z\frac{5}{3}} = \int \bar{z}_N(\lambda) \, z_{\frac{5}{3}}(\lambda) \, d\lambda$$

$$\bar{z}_{X_2} = \int \bar{x}_N(\lambda) \, z_2(\lambda) \, d\lambda$$

$$\bar{z}_{Y_2} = \int \bar{y}_N(\lambda) \, z_2(\lambda) \, d\lambda$$

$$\bar{z}_{Z_2} = \int \bar{z}_N(\lambda) \, z_2(\lambda) \, d\lambda \tag{4.68}$$

Equations (4.61) and (4.62) with appropriate substitutions hold for $\overline{F}_{Xa}$, $\overline{F}_{Ya}$, and $\overline{F}_{Za}$.

Dilution method. The *spectral Kubelka-Munk function of colored pigment/white paste mixtures*, similarly to equation (4.41), are

$$\frac{F_d(\lambda)}{F_A(\lambda)} \approx 1 - v(\lambda)(1 - q_D^{\frac{2}{3}}) \tag{4.69}$$

where $q_D = \sigma/\sigma_A$, $F_A(\lambda)$ being the initial value before dilution, and the coefficient is

$$v(\lambda) = \frac{g(\lambda) \, s(\lambda) \, \sigma_A^{\frac{5}{3}} + g_w \, s_w \, \sigma_{wA}^{\frac{5}{3}}}{s(\lambda) \, \sigma_A + s_w \, \sigma_{wA}} \approx g_w \, \sigma_{wA}^{\frac{2}{3}} \tag{4.70}$$

Now the Kubelka-Munk functions of the CIE tristimulus values will again be derived. Starting with equation (4.69), we proceed as we did before, taking logarithms to get to the proven equality $\ln F = A - B\varrho^*$, we once more apply the approximations for small quantities on the right-hand side:

$$B(\varrho^*(\lambda) - \varrho_A^*(\lambda)) = (1 - q_D^{\frac{2}{3}}) \, v(\lambda) \tag{4.71}$$

Application of the colorimetric integral $\int \bar{y}_N(\lambda) \ldots d\lambda$ leads immediately to

$$\frac{B}{100}(Y^{**} - Y_A^{**}) = (1 - q_D^{\frac{2}{3}}) \int \bar{y}_N(\lambda) \, v(\lambda) \, d\lambda$$

and formal identity with equation (4.71).

From here, the steps are repeated in reverse order, yielding the *Kubelka-Munk functions of the CIE tristimulus values for colored pigment/white paste mixtures prepared by the dilution method*:

$$\frac{\bar{F}_{Xd}}{\bar{F}_{A,Xd}} \approx 1 - \bar{v}_X \left(1 - q_D^{\frac{2}{3}}\right)$$

$$\frac{\bar{F}_{Yd}}{\bar{F}_{A,Yd}} \approx 1 - \bar{v}_Y \left(1 - q_D^{\frac{2}{3}}\right)$$

$$\frac{\bar{F}_{Zd}}{\bar{F}_{A,Zd}} \approx 1 - \bar{v}_Z \left(1 - q_D^{\frac{2}{3}}\right) \tag{4.72}$$

$$(q_D = \sigma/\sigma_A)$$

where $\bar{F}_{A,Xd}$, $\bar{F}_{A,Yd}$, and $\bar{F}_{A,Zd}$ are the Kubelka-Munk functions of the initial CIE tristimulus values and the coefficients are given by

$$\bar{v}_X = \int \bar{x}_N(\lambda)\, v(\lambda)\, d\lambda \approx g_w\, \sigma_{wA}^{\frac{2}{3}}$$

$$\bar{v}_Y = \int \bar{y}_N(\lambda)\, v(\lambda)\, d\lambda \approx g_w\, \sigma_{wA}^{\frac{2}{3}}$$

$$\bar{v}_Z = \int \bar{z}_N(\lambda)\, v(\lambda)\, d\lambda \approx g_w\, \sigma_{wA}^{\frac{2}{3}} \tag{4.73}$$

Equations (4.61) and (4.62), suitably modified, hold for $\bar{F}_{Xd}$, $\bar{F}_{Yd}$, and $\bar{F}_{Zd}$.

Thus the concentration dependence of the Kubelka-Munk functions of CIE tristimulus values for colored pigments by the dilution method is described by the same law as for black pigments.

4.6 Tinting Strength

4.6.1 Significance and Definition

Tinting strength is a fundamental characteristic of coloring materials. It describes one of the most important optical qualities of dyes and pigments in the form of an "effectiveness" criterion and is continually used for optical performance comparisons of pigments and dyes. The assessment of the optical performance of pigments is commonly based on the tinting strength and hiding power in a stated medium. When the price is then taken into consideration along with the optical performance, conclusions can be drawn about the economics of a pig-

4.6 Tinting Strength

ment ("money value"; see also Section 7.4.4). As a rule, however, the determination of hiding power involves appreciable experimental effort. It appears much easier to determine the tinting strength, and a first examination reveals no problems in doing so; this approach is normally preferred, even if tinting strength and hiding power do not necessarily lead to the same economic results.

Tinting strength is defined as "a measure of the ability of a coloring matter to impart color, through its absorbing power, to other substances."* If one understands a definition as simultaneously being an instruction, telling how to carry out a determination implied by the definition, one comes to a surprising conclusion here: current testing specifications, without exception, define (i.e., tell how to determine) only a "relative tinting strength" relative to a reference coloring matter. The justified question of absolute tinting strength remains largely unanswered, and existing standards say little about it. We can only return to this interesting question further on and seek an answer to it then.

The concept and definition of relative tinting strength are worded differently for dyes and pigments and at first glance they do not seem to be identical. For dyes, the relative tinting strength is the ratio of the extinction coefficients for the test and reference dyes measured at the absorption maxima of the solutions:

$$P_\text{d} = \frac{k''_\beta}{k''_{\beta,\text{R}}} \qquad (4.74)$$

(Because there is no scattering, the extinction coefficients are equal to the absorption coefficients per unit concentration by mass k''_β.) The absorption curves for both dyes should be quite similar.

In the pigment field, tinting strength applies only to black and colored pigments, and accordingly there has to exist an analog for white pigments: the "lightening power".

The procedural instruction (the "standard experiment") based on the definition of tinting strength or lightening power runs as follows: each of the pigments to be compared is mixed into a white paste (into a black or blue paste for lightening power), and the portion of one of the two pigments is varied until the two drawdowns match, i.e., look the same to an observer. The tinting strength or lightening power is the weight ratio of reference pigment (mass m_R) and test pigment (mass m) needed to achieve the same color quality in a white system or the same lightening in a colored system. Thus the defining equation for tinting strength and lightening power is

---

* For standards see "Tinting Strength, Relative" in Appendix 1.

$$P = \left.\frac{m_R}{m}\right|_{\substack{\Delta Q = 0 \\ m_{FP} = \text{const}}} \tag{4.75}$$

where $\Delta Q = 0$ denotes a match and m_{FP} is the constant portion of white or black paste. Tinting strength and lightening power describe true effectiveness properties, since a tinting strength a factor of two greater signifies that only half as much of this pigment is needed to achieve the same optical effect as with the reference pigment.

It is common to supplement the notion of tinting strength with that of "tinting equivalent." This is the reciprocal of the tinting strength and thus tells how many parts of a test pigment will have a color-imparting effect equivalent to that of 1 part by weight of the reference pigment. Another way to explain this property is to say that the tinting equivalent is lower for more strongly coloring pigments and higher for less strongly coloring ones, corresponding to the smaller and larger portions of these pigments required.

In a formulation calculation context, the tinting strength of a pigment is sometimes linked with its coloristic properties in an arbitrary pigment mixture. Thus one attempts to interpret the tinting strength as the fundamental pigment property for formulation, namely as the substitution ratio when a formula is to be converted. A definition of tinting strength in this sense would not, however, be possible unless the system itself or some general principle of formulation systems were included in the definition. Even if it were, results under this definition would necessarily differ from those under the classical definition. In practice, this means that substitution ratios in general formulas need not be identical with the relative tinting strength (substitution ratio in white). It is essential to keep this point in mind when tinting strength measurements are used to answer questions in formulation calculations.

Equation (4.75) provides a basis for a test specification. The standards here have retained the decades-old interpretation of the tinting strength concept as well as its definition, both of which are far older than the standards themselves. As a consequence of the definition, the test is performed in a light-scattering substrate, but for each substrate the vehicle as well as the white pigment must be specified. Today, nondrying vehicles containing a titanium dioxide pigment are common use. A method of tinting strength determination usable in practice must, of course, cover some further points, for example the test pigment concentration, because the relative tinting strength depends not only on the white pigment used and its concentration, but also on the concentration of the colored pigment in the test. The procedure ultimately used may differ in a variety of ways from the standard experiment described above. This fact may well conceal the latent presence of defining equation (4.75) and the system of the standard experiment.

This holds, in particular, for photometric techniques derived from the standard experiment. In these methods, a rationalization profit is achieved because only one mixture of white paste with the test pigment need be prepared. As in the case of the hiding power, this economy is the great motivation for using colorimetric methods to determine tinting strength; but the problem of a colorimetric matching criterion arises – the problem to be discussed now in Section 4.6.2.

4.6.2 Coloristic Matching Criteria for Pigments

As has been shown, methods of determining the tinting strength for pigments specify that the reference and test mixtures with white paste are to be "matched" through variation of the pigment portions; when the match is completed, they should appear alike to the colorist. Matching does not present any difficulties to the colorist when the pigments involved have very similar spectra, that is, when the differences in hue and saturation are small. When these differences are greater, criteria based on coloristic experience are applied; the uncertainty of the assessment unavoidably grows larger as the color difference increases. On the other hand, as the spectral curves become less and less similar, a point is soon reached where the determination of relative tinting strength ceases to be meaningful, because the test pigment is no longer such that it can replace the reference in a color formulation. It is thus reasonable to restrict tinting strength determinations and studies to small color differences.

The question of a matching criterion to be used becomes acute (if indeed it has not done so already) when visual matching is abandoned in favor of a colorimetric method. The only way for the colorimeter, and the computer fed by it, to draw on experience comparable to that of the colorist is to render objective the criterion used by colorists. Not until the color quality that the colorist utilizes has been identified and quantified can instructions for the computer be embodied in program form.

The problematic aspect of tinting strength determination can be expressed in the question: What colorimetric or physiological criterion governs the optimal replacement of reference by test pigment? There are cases in which the pigment being assessed is fully capable of replacing the reference pigment at a certain concentration; i.e., the color difference between the mixtures of the two pigments with white paste vanishes. With inorganic colored pigments and some organic pigments, however, this is the exception, because in most such cases the differences in tinting strength can mostly be attributed to differences in particle size and particle-size distribution, with the inevitable result of shifts in hue. Because these color differences between test and reference pigments do not result from differences in concentration or unequal tinting strengths, there is no concentration at which the color difference between the two will vanish. The question then arises: What criterion can be applied in order to find the mass ratio at which the reference pigment can be optimally replaced? Is the matching of absorption coefficients (or reflectances) at the absorption maximum (that is the reflection minimum) the proper criterion, as older methods suggest? Or, in the case of pigments differing in color nuance, is it better to seek the concentration such that the color difference between reference and test pigments is a minimum? The applicability test for these criteria is the agreement on the visual match among many colorists.

If a working technique is specified and the portion of the test pigment is progressively increased, the resulting colors describe a three-dimensional curve in color space. This curve is called the "tinting characteristic" of the pigment. If the method is to add a colored pigment to a white paste, say by the constant volume method, then as expected the three-dimensional curve begins near the white locus (color locus of white paste; PVC of colored pigment $\sigma = 0$) and ends close to the colored pigment full shade ($\sigma \gg \sigma_w$). The nature of the curve between these two points may vary greatly from one pigment to another and from one color range to another. For the CIELAB color space, the tinting characteristic for mixtures with a white preparation ("reducing white") can even be formulated analytically (though not explicitly). From the PVC-dependent reflection spectra $\varrho(\lambda, \sigma)$ of reducing whites, the CIE tristimulus values $X(\sigma)$, $Y(\sigma)$, $Z(\sigma)$ can be obtained, and from these the CIELAB color loccus $\mathbf{V}(\sigma) = \{a^*(\sigma), b^*(\sigma), L^*(\sigma)\}$ can be determined. If the components of a vector depend exclusively on a single variable (σ), a three-dimensional curve always exists. Figure 4-8 presents one example of a tinting characteristic for reducing whites.

In equation (4.75), the match is signified by $\Delta Q = 0$. The quantity Q, the color difference between test and reference pigments, which is to be made equal to zero, is called the "tinting strength criterion." The selected tinting strength criterion can usually be represented by equipotential surfaces in color space (see Fig. 4-9 for L^* as tinting strength criterion). One of the pigments to be compared (constant portion by weight) fixes a certain equipotential surface, and the tinting characteristic of the other pigment (variable portion by weight) intersects this surface at a point. The point is the location of the match, and the pigment concentration at this point of the three-dimensional curve is the desired concentration at the match. It should be clear that the value of the tinting strength is more-or-less strongly influenced by what tinting strength criterion is chosen. Accordingly, one can no longer speak of just the "tinting strength"; there are as many values of tinting strength as there are tinting strength criteria in use. What follows is a survey of criteria.

Tinting Strength Criteria

1) Criteria definable in physiological and simultaneously colorimetric terms

– *Lightness L^* (or CIE tristimulus value Y)*. This criterion does not require any further explanation. Both quantities, L^* and Y, lead to the identical match, since $\Delta L^* = 0$ and $\Delta Y = 0$ have the same meaning. Figure 4-9 shows the CIELAB color space with equi-lightness surfaces drawn in (planes in this instance). The figure also indicates how the tinting characteristic of a colored pigment intersects the plane of the reference.

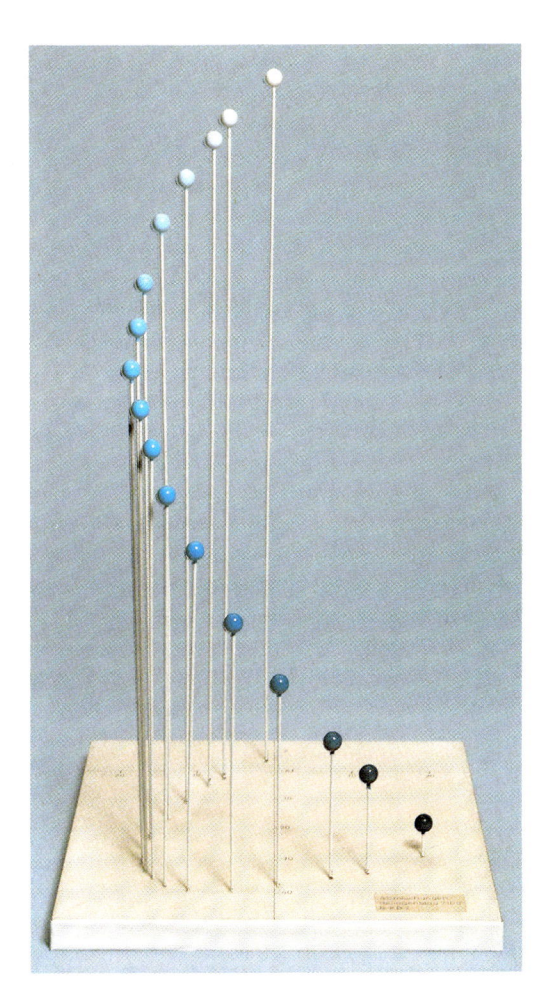

Fig. 4–8. Tinting characteristic of a copper phthalocyanine pigment, after a model due to KEIFER. (The highest color locus corresponds to the white paste; the negative b^* axis points toward the observer.)

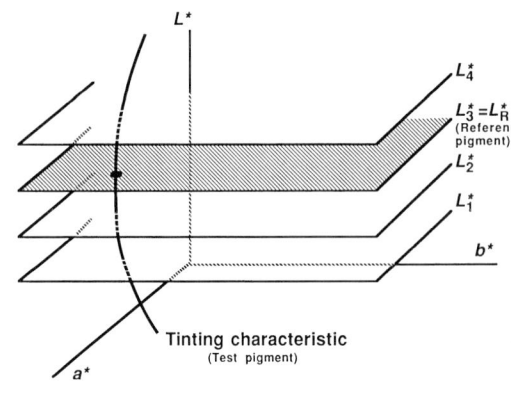

170   4 How Light Scattering and Absorption Depend on the Content

Fig. 4–9. Equipotential surfaces of lightness (tinting strength criterion L^*) with tinting characteristic of a colored pigment in CIE-LAB color space.

- *Color depth T^*.* The color depth is a measure of the "coloredness" (intensity) of a color perception.* It increases with increasing saturation, but decreases with increasing lightness. Accordingly, in the CIELAB color space, it can be approximately represented as the distance from the white locus ($L^*_w = 100$):

$$T^* = \sqrt{a^{*2} + b^{*2} + (100 - L^*)^2} = \sqrt{C^{*2}_{ab} + (100 - L^*)^2} \tag{4.76}$$

Surfaces of constant color depth are thus spherical shells centered on the white locus, as shown in Figure 4-10. The above formula is, however, too rough for practical use.

- *The smallest of the CIE tristimulus values X, Y, Z.* This is the CIE tristimulus value that is most strongly governed by the absorption.

- *The smallest of the reflectometer values R'_x, R'_y, R'_z.* This measure provides a simplification when measurement is done by the tristimulus method.

- *Smallest possible color difference ΔE^*_{ab}.* This tinting strength criterion is already a difference, but it is to be minimized instead of made equal to zero. In the CIELAB color space, the mixture of the reference pigment with white represents a point (color locus). On the tinting characteristic of the test pigment, the point is sought that is nearest to the reference color locus.

- *Chroma difference ΔC^*_{ab}.* This criterion is of little importance but is included here to complete the systematic order.

---

* The term originated in textile dyeing and was coined for use in fastness comparisons, which are to be done not at an equal dye concentration but at an equal "depth" of coloration.

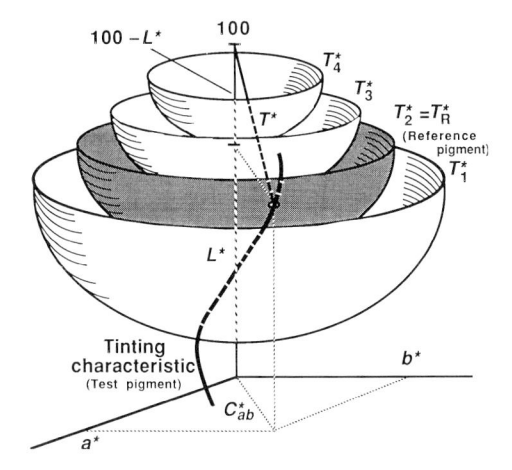

Fig. 4–10. Equipotential surfaces for color depth (tinting strength criterion T^*) with tinting characteristic of a colored pigment, plotted in the CIELAB color space.

2) Criteria derived from the Kubelka-Munk function F

– *Maximum of the spectral Kubelka-Munk function $F_{max}(\lambda)$ or minimum of the reflection spectrum $\varrho_{min}(\lambda)$.*

This is a variant of the "smallest CIE tristimulus value" criterion. $\Delta F_{max}(\lambda) = 0$ and $\Delta \varrho_{min}(\lambda) = 0$ lead to the same match; $\Delta \varrho_{min}(\lambda)$ is also the tinting strength criterion for dyes in solution.*

– *Sum of the colorimetrically evaluated Kubelka-Munk functions $\overline{F}_{x,y,z}$.*

This criterion is formulated as

$$\overline{F}_{x,y,z} = \int (\overline{x}(\lambda) + \overline{y}(\lambda) + \overline{z}(\lambda)) \, F(\lambda) \, d\lambda \tag{4.77}$$

The sum of the CIE color matching functions, which is multiplied by $F(\lambda)$, has the form of a double-peaked curve (see Fig. 2–4) with a deep notch at 500 nm. It is not immediately obvious why the wavelengths around 500 nm are thus discriminated against in the summation. The criterion originated in the textile field and is used there for tests of dye strength (on the fiber).

– *Averaged spectral Kubelka-Munk function $\overline{F}$.*

---

* Thus the definitions of tinting strength for dyes, equation (4.74), and for pigments, equation (4.75), are identical for this tinting strength criterion.

This is given by

$$\overline{F} = \int F(\lambda)\,d\lambda \qquad (4.78)$$

and should take care of the objections raised in connection with the criterion $\overline{F}_{x,y,z}$. The integral must be formed by summation; the magnitude of $d\lambda$ is not crucial provided the mixtures with the test and reference pigments are formulated with the same $d\lambda$.

- Smallest possible variance sum of the spectral Kubelka-Munk functions $\overline{\Delta F^2}$ *(min)*, calculated as

$$\overline{\Delta F^2}\,(min) = \int \Delta F^2(\lambda)\,d\lambda = \int (F(\lambda) - F_R(\lambda))^2\,d\lambda \to Min. \qquad (4.79)$$

4.6.3 Color Matching Studies

One may naturally ask which, among these tinting strength criteria, is the "right" one – or which comes closest to the coloristic match. Two studies have been published that shed some light on this problem; their results will appreciably aid in the selection of a suitable tinting strength criterion.

KEIFER and the author pursued the question of what criteria colorists apply to assess the tinting strength of inorganic pigments. Six pairs of inorganic pigments corresponding to the principal color directions red, yellow, blue, green, and in addition orange and black were chosen. In each pair, mixtures of the two pigments in reduction displayed a hue difference $|\Delta H^*_{ab}| < 3$ after matching. This procedure insured the "replaceability" of one pigment by the other in a coloration. Mixtures of the pigments with a white paint were prepared; the quantity of one pigment ("test pigment") was incremented 10 mg at a time, while the quantity of the other ("reference pigment") was held constant. From each test pigment mixture and the reference pigment mixture (the same every time), side-by-side drawdowns were made with a Bird applicator; after these had dried, matching trials and color measurement were performed. Half of the evaluators were experienced colorists skilled in the determination of tinting strength of inorganic pigments; the other half were persons who did not meet this qualification but who had done visual color matching before. One result can be stated immediately: The difference between the two groups was not significant. The overall result is therefore not influenced just by coloristic learning processes; it holds some universal significance. The color matchers were given the following question to answer: "Which drawdown seems to you to be the best match between reference and test pigments?" For the colorimetric evaluation, the following tinting strength criteria were tested in the study: L^*, ΔE^*_{ab}, and $\varrho_{min}(\lambda)$. Two forms of the color depth were used, T^* (eq. 4.75) and T_S according to Gall's improved color depth formula (Section 4.6.5).

Figures 4-11 to 4-16 present the results in graphical form, with the various tinting strength criteria identified as in Section 4.6.2. From these graphs it is

4.6 Tinting Strength

quite clear that the criterion $\Delta L^* = 0$ gives the best agreement with the visual assessment (heavy arrow) for each of the six hues tested. The plots also show how much the tinting strength value obtained under the other criteria, which is proportional to the quantity of pigment, can diverge from the visual match. The larger deviations include those of the tinting strength criteria $\varrho_{min}(\lambda)$, T^*, and T_S. Even ΔE^*_{ab} comes off poorly with regard to both the positions of the minima and the uncertainty (due to the weakly marked minima).

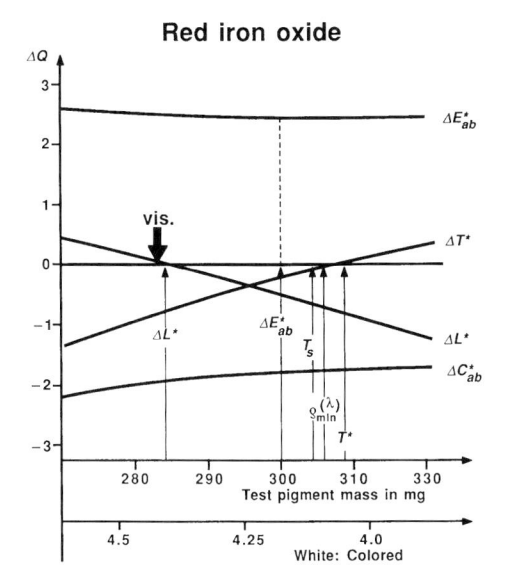

Fig. 4-11. Comparison of tinting strength criteria with visual assessment (heavy arrow) for two red iron oxide pigments.

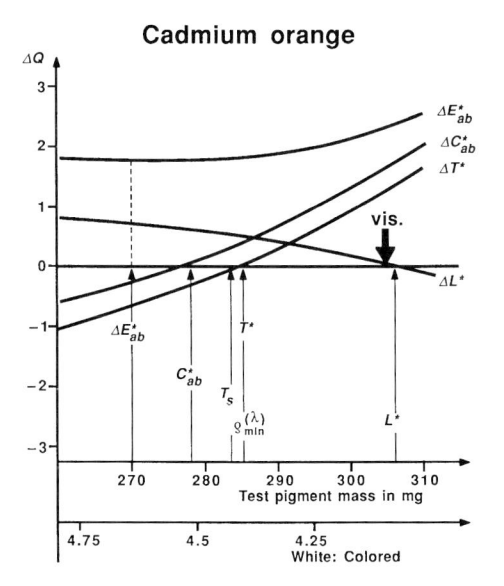

Fig. 4-12. Same as Figure 4-11, for two cadmium orange pigments.

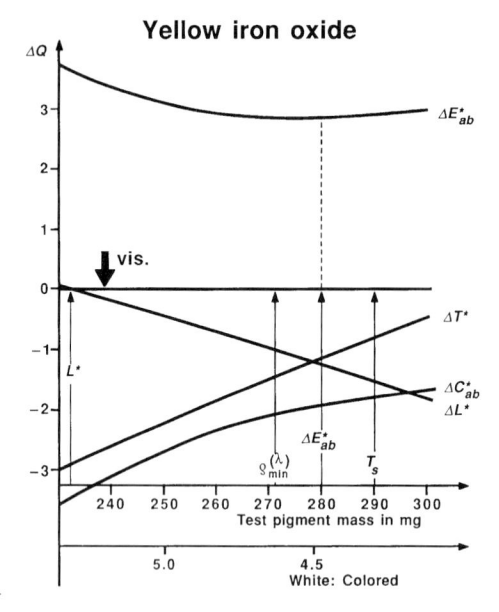

174   4 *How Light Scattering and Absorption Depend on the Content*

Fig. 4–13. Same as Figure 4–11, for two yellow iron oxide pigments.

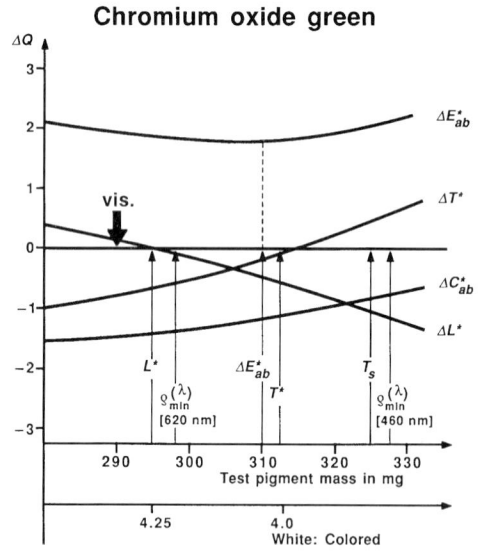

Fig. 4–14. Same as Figure 4–11, for two chromium oxide green pigments [both minima labeled $\varrho_{min}(\lambda)$].

4.6 *Tinting Strength* 175

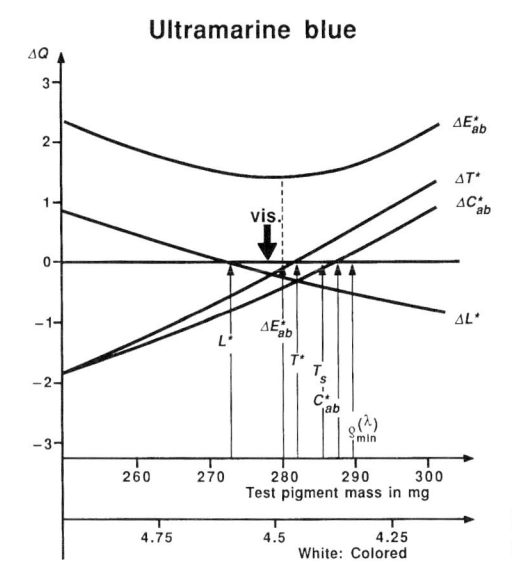

Fig. 4-15. Same as Figure 4–11, for two ultramarine blue pigments.

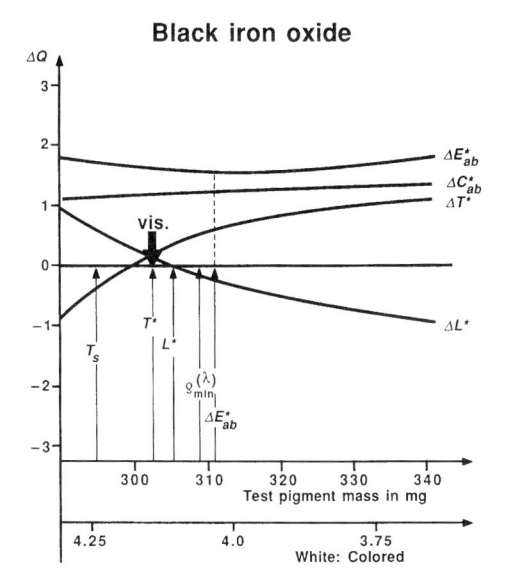

Fig. 4-16. Same as Figure 4–11, for two black iron oxide pigments.

To find out about the reliability of coloristic judgment, the frequency distributions for test pigment selection in color matching trials were calculated and plotted. Two of the resulting curves are presented here. For chromium oxide green (Fig. 4-17), the distribution is relatively narrow (under ±1% of the tinting strength); for the more highly saturated cadmium pigment (Fig. 4-18), it is markedly broader (over ±2%). According to these results, coloristic judgment appears to be more reliable for unsaturated colors in reduction.

The other color matching study that needs to be discussed here is that carried out and reported by GALL, an investigation performed with great care and based on profound knowledge. It differs in several points from the Uerdingen study just described:

– Organic pigments were studied

– Color depth was taken as the typical tinting strength criterion for the pigments in the study.

– In accordance with the objective, reducing whites were ordered by the specified levels of color depth.

– The color matchers were practiced in working with organic pigments (i.e., pigments commonly having higher color saturation).

The samples comprised 18 or 22 pigments covering roughly the entire color circle. For every hue, there was a selection from a series of gradations in the concentration of colored pigment in reduction. A mixture (standard) made with a copper phthalocyanine pigment in reduction, adjusted to the specified textile color depth standard, was to be rematched in color depth with the other pigments. The task was as follows: "From each concentration series, select a reducing white that best agrees in color depth

– with the blue standard and

– with the samples already color-matched to the standard."

The textile standards were identified, in the traditional way, with fractions of the "standard depth of shade" (R.T.). The levels studied were 1/1, 1/3, 1/25, and 1/200 R.T. Mixtures with white at 1/3 R.T. made with a further 18 pigments (five red, four yellow, four green, five blue) and lying on "hue profiles" (i.e., each having the same hue but different saturation) were also prepared and matched. The purpose of doing this was to determine the extent to which the colorist rates a decrease in saturation as an increase in tinting strength.

Table 4-6 presents the matching results. The last column gives the probability of the assessments falling in a concentration range of ±10% about the average value. It can be seen that the reliability of coloristic judgment is appreciably higher for saturated samples (i.e., the ones with higher CIE tristimulus values)

4.6 Tinting Strength 177

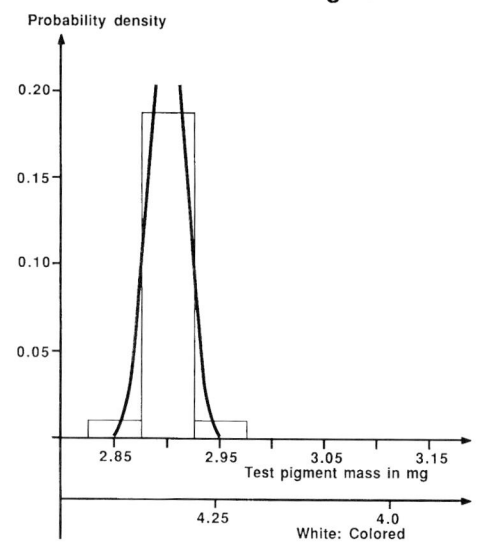

Fig. 4-17. Statistical analysis of the visual matching of chromium oxide pigments.

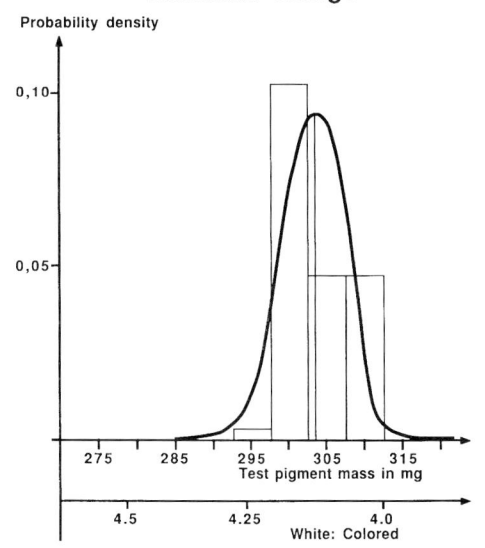

Fig. 4-18. Statistical analysis of the visual matching of cadmium orange pigments.

than for unsaturated samples*. The table also shows that the luminance (tristimulus value Y) differs significantly between these samples, which were selected to have equal color depth; the differences are particularly marked in the case of yellow.

Table 4–6. Color matching results along the 1/3 R. T. hue profile, after GALL.

Color	X	Y	Z	H
Red, standard	48.18	32.38	9.89	
A	41.83	29.38	9.76	80 %
B	36.96	26.28	9.48	67 %
C	32.49	23.36	9.40	69 %
D	28.07	21.00	9.49	58 %
E	25.26	19.12	9.66	59 %
Green, standard	15.81	29.23	24.46	
A	15.89	29.01	24.03	94 %
B	16.04	28.85	24.24	98 %
C	15.84	27.90	23.88	98 %
D	15.69	27.25	23.53	91 %
Blue, standard	17.38	20.70	63.82	
A	16.63	19.86	60.59	98 %
B	16.41	19.78	57.72	96 %
C	15.83	19.04	55.93	92 %
D	15.26	18.88	50.98	86 %
E	13.94	18.72	45.76	78 %
Yellow, standard	62.05	64.76	15.35	
A	58.25	61.39	15.97	88 %
B	56.43	59.74	15.83	85 %
C	54.75	57.92	15.75	78 %
D	50.84	53.97	15.78	64 %

These studies also included matching of the same samples by "uneducated" observers, who arrived at virtually the same results as experienced colorists. The goal of this investigation was the colorimetric establishment of the various color-depth levels. From the color coordinates of the experimentally determined color depths corresponding to the same level, improved formulas for color depth were obtained.

---

* The terms "brilliant," "pure," and "hazy" have purposely not been applied to pigments here because they are ambiguous. Further, they appear to contain hidden value judgments.

The results of these two painstaking color matching studies of tinting strength may at first appear contradictory. How is it possible that one group of colorists uses lightness as the tinting strength criterion over a wide range of color depths, while the other group adjudges samples differing greatly in lightness to have the same color depth in the context of a tinting strength determination?

To clear up the apparent contradiction, it is necessary to return to the assumptions and constraints in the two studies and the differences that result from them. Table 4-7 briefly sums up the key points for both studies. Different pigments were used: in one case, inorganic pigments, relatively unsaturated, mostly dark in full shade; in the other, organic pigments, relatively saturated in reduction. The inorganic pigments had smaller color differences between test and reference than did the organic pigments. Of the latter, Gall's colorists gave more reliable matches of the more saturated reductions than of the less saturated ones; it was just the other way around with the Uerdingen colorists, and in addition their judgment (because of the smaller color differences) was on the whole more reliable than that of the other group.

There would appear to be no arguing with the view that pigments in reduction giving higher saturations are well represented by the color depth as tinting

Table 4–7. Summary of differences between the color matching studies.

Study Aspect	GALL	KEIFER/VÖLZ
Tested pigments	Organic pigments	Inorganic pigments
ΔE_{ab}^* test/reference in reduction after matching	Generally >3	Generally <3
Colorists	Experienced in matching organic pigments	inorganic pigments
Task	"Equal color depth"	"Best match"
Objective	To improve color depth formulas	To identify best tinting strength criterion
Result	Improved formulas	Lightness
Reliability of judgment	Better on saturated samples than unsaturated	Better on unsaturated samples than saturated; generally much higher than in GALL'S study

strength criterion, while pigments in reduction giving lower saturations are better evaluated by the lightness as the tinting strength criterion. Another point in favor of this conclusion is that "uneducated" matchers in both studies came to almost the same results as experienced colorists – which appears to prove that the assessment criteria used by the colorists cannot be purely learned ones. It must be kept in mind that the matching in GALL's study was done by colorists whose special experience, not just with organic pigments but also with the problems of formula conversion and rematching, was greater than that of the Uerdingen colorists. Accordingly, it cannot be ruled out that formulation experience influenced the formation of a "color depth" judgment. This would mean that too much weight was placed on color nuance because it is commonly a serious problem in formula conversions. In turn, this leads too far in the substitution ratio direction in formulas and can lead away from the tinting strength, as was already stated in Section 4.6.1.

One important point must not be neglected here: The more unsaturated the color of a pigment in reduction, in other words the nearer its color locus is to the lightness axis in color space, the closer are the two equipotential surfaces of lightness (horizontal plane) and color depth (spherical shell), which ultimately become identical on the lightness axis.

Thus it has become usual to use lightness as the tinting strength criterion in evaluating inorganic pigments and color depth as the criterion for organic pigments. In the case of highly saturated yellow pigments, which scatter strongly, both criteria appear to fail; the reason may, however, be an expectation of the tester, who ordinarily uses inorganic yellow pigments less for coloring than for lightening, a function that is of course not regarded in the tinting strength evaluation.

Finally, it should be asked why ΔE_{ab}^* showed such poor performance in the Uerdingen study* when, as the smallest color difference, it should in fact be the decisive criterion. It may again be that this expectation failed because of inadequacies in the CIELAB system. It has already been established that CIELAB regularly understates the lightness and overstates the saturation. A glance at any of Figures 4-11 to 4-16 will reveal what changes come about if ΔL^* is multiplied by a quite reasonable factor of 4. The ΔL^* curve becomes far steeper, so that the ΔE_{ab}^* curve is much more strongly dominated by ΔL^*; as a consequence, the minimum of ΔE_{ab}^* moves closer to the zero crossing of ΔL^* and also becomes more strongly marked. Thus tinting strength criterion min(ΔE_{ab}^*) practically coincides with $\Delta L^* = 0$.

---

* According to a communication from GALL, no other known ΔE formula makes a suitable tinting strength criterion either.

4.6.4 Lightness Matching

In a rational method for tinting strength determination, a desirable feature would be that the value can be obtained by a simple calculation based on a single reducing white. Because measurements by the tristimulus method directly yield the measurement $R'_y = Y$, the tinting strength criterion $\Delta L^* = \Delta Y = 0$ indeed satisfies this wish.

This section begins by deriving the equation of the relative tinting strength for the criterion Y as a function of the mass of test pigment. The fundamental equation (4.75) requires a relation between m and m_R after matching to $\Delta Y = 0$. The Kubelka-Munk function expressed in terms of the PMV ψ is a suitable way to get m_R as a function of m. This approach is particularly convenient, since the tinting strength criterion $\Delta Y = 0$ can obviously be replaced by the corrected values $\Delta Y^{**} = 0$ and finally by the Kubelka-Munk functions $\Delta F = 0$, that is, by setting $F = F_R$ (test and reference). Unfortunately, this chain, which looks so plausible, has a blemish in the form of a weak link. The problem lies in the step from $Y = Y_R$ to $Y^{**} = Y^{**}{}_R$, for

$$\Delta Y = Y - Y_R = \int \bar{y}(\lambda)(\varrho(\lambda) - \varrho_R(\lambda)) \, d\lambda = \int \bar{y}(\lambda) \Delta\varrho(\lambda) d\lambda = 0 \quad (4.80)$$

does not necessarily imply that

$$\Delta Y^{**} = Y^{**} - Y^{**}_R = \int \bar{y}(\lambda)(\varrho^*(\lambda) - \varrho^*_R(\lambda)) \, d\lambda$$

$$= \int \bar{y}(\lambda) \Delta\varrho^*(\lambda) \, d\lambda \approx \int \bar{y}(\lambda) \frac{d\varrho^*}{d\varrho} \Delta\varrho(\lambda) \, d\lambda \quad (4.81)$$

vanishes. This becomes clear if substitution (3.11) is made for ϱ^*, ϱ^*_R.

The integral on the right would vanish identically with $\Delta Y = 0$ if the derivative $d\varrho^*/d\varrho$ were constant. This also is nearly given: the factor lies between 0.4 and 2.6, and so there are contributions to the integral, but the crucial point is that these contributions, because of the Y matching performed, must have different signs in different regions of the spectrum, so that they practically cancel out. Figure 8-4 shows the spectra of two red iron oxide pigments in reduction after Y matching. The $\Delta\varrho$ are positive for $\lambda = 420$ to 540 nm but negative for $\lambda = 580$ to 700 nm. A quantitative analysis of this example shows that $\Delta Y^{**} < 0.03\%$, which is sufficiently close to zero.

Though selected on purpose, this example is typical enough for reflection spectra of mixtures with white after Y matching that it can be generalized: ΔY^{**} must always be small of higher order than Y^{**} and Y^{**}_R as ΔY goes to zero. Thus, at a practical level of accuracy, there is justification for setting the two quantities equal, and there is no further obstacle to using the criterion $\Delta F = 0$.

The equations for the Kubelka-Munk functions in Sections 4.5.2 (black pigments) and 4.5.4 (colored pigments) have the same form and can therefore be handled together. The CIE tristimulus value Y for the adding pigment method is described by

$$F - F_{WP} = z_1 \psi (1 + z_{\frac{5}{3}} \psi^{\frac{2}{3}} - z_2 \psi) \tag{4.82}$$

Here $\psi = m/\mu$ is the volume ratio of pigment to white paste, in accordance with equation (4.6), with μ given by equation (4.4):

$$\mu = m_{WP} \frac{\varrho}{\varrho_{WP}} \tag{4.83}$$

where m_{WP} and ϱ_{WP} are the mass and density of the white paste and ϱ is the density of the test pigment. The notations F for the Kubelka-Munk function and z_i for the coefficients appear in the following forms:

$F^{(2)}_{b,a}$, z_{bi} for black pigments

$\overline{F}^{(2)}_{Ya}$, $\overline{z}_{Yi}$ for colored pigments

A similar equation then describes the reduced reference pigment:

$$F_R - F_{WP} = z_{R_1} \psi_R (1 + z_{R\frac{5}{3}} \psi_R^{\frac{2}{3}} - z_{R_2} \psi_R) \tag{4.84}$$

where $\psi_R = m_R/\mu_R$,

$$\mu_R = m_{WP} \frac{\varrho_R}{\varrho_{WP}} \tag{4.85}$$

and ϱ_R is the density of the reference pigment. For easily discerned reasons, the desired function $m_R(m)$, or in its place $\psi_R(\psi)$, must have a structure like that of $F(\psi)$:

$$\psi_R \big|_{\Delta F = 0} = p_o \psi (1 + p_{\frac{2}{3}} \psi^{\frac{2}{3}} - p_1 \psi)$$

After this expression is substituted in equation (4.84), coefficients can be compared with equation (4.82), since F is supposed to be equal to F_R. The following are therefore obtained for the coefficients p_i:

$$p_o = \frac{z_1}{z_{R_1}}$$

$$p_{\frac{2}{3}} = z_{\frac{5}{3}} - p_o^{\frac{2}{3}} z_{R\frac{5}{3}}$$

$$p_1 = z_2 - p_o z_{R_2} \tag{4.86}$$

The fundamental equation for the tinting strength can now be written down. By equations (4.83) and (4.85), the requirement $m_{WP} = $ const implies that $\mu_R/\mu = \varrho_R/\varrho$, so that

4.6 Tinting Strength

$$P^{(1)} = \left.\frac{m_R}{m}\right|_{\substack{\Delta F=0 \\ m_{WP}=\text{const}}} = \left.\frac{\mu_R}{\mu}\right|_{m_{WP}=\text{const}} \cdot \left.\frac{\psi_R}{\psi}\right|_{\Delta F=0}$$

$$= \frac{\varrho_R}{\varrho} p_o \left[1 + p_{\frac{2}{3}}\left(\frac{m}{\mu}\right)^{\frac{2}{3}} - p_1 \frac{m}{\mu}\right] = \frac{\varrho_R}{\varrho} p_o \left[1 + p_{\frac{2}{3}}\left(\frac{\sigma}{1-\sigma}\right)^{\frac{2}{3}} - p_1 \frac{\sigma}{1-\sigma}\right]$$

(first approximation) (4.87)

This is the desired equation for the relative tinting strength as a function of the quantity m or the PVC σ of the test pigment. The zero-th approximation is obtained by setting higher-order terms equal to zero:

$$P^{(0)} = \frac{\varrho_R}{\varrho} p_o \quad \text{(zero-th approximation)} \quad (4.88)$$

The coefficients are described again by equations (4.86). They can be experimentally determined by mixing three suitable quantities of pigment with equal quantities of white paste and matching the reference pastes or, more simply, from the Kubelka-Munk functions (4.82) and (4.84) of the test and reference pigments. In testing practice, they are always determined empirically rather than by the laborious calculation from the $z_i(\lambda)$. Nonetheless, for the sake of completeness and to identify their source, they must be written out exactly. By equations (4.68) in Section 4.5.4 and (4.86),

$$p_o = \frac{z_1}{z_{R_1}} = \frac{\bar{z}_{Y_1}}{\bar{z}_{RY_1}} = e^{\int \bar{y}_N(\lambda) \cdot \ln \frac{z_1(\lambda)}{z_{R_1}(\lambda)} d\lambda} \approx \int \bar{y}_N(\lambda) \frac{k(\lambda)}{k_R(\lambda)} d\lambda$$

$$p_{\frac{2}{3}} = z_{\frac{5}{3}} - p_o^{\frac{2}{3}} z_{R\frac{5}{3}} = F_{WP} \int \bar{y}_N(\lambda) \left(\frac{g(\lambda)}{F_{PU}(\lambda)} - p_o^{\frac{2}{3}} \frac{g_R(\lambda)}{F_{PU,R}(\lambda)}\right) d\lambda$$

$$p_1 = z_2 - p_o z_{R_2} = \frac{1}{S_W} \int \bar{y}_N(\lambda) (s(\lambda) - p_o s_R(\lambda)) d\lambda \quad (4.89)$$

Here $k(\lambda)$, $k_R(\lambda)$ and $s(\lambda)$, $s_R(\lambda)$ are the absorption and scattering coefficients per unit PVC; $g(\lambda)$ and $g_R(\lambda)$ are the attenuation coefficients of the scattering interaction. The first of each pair is for the test pigment, the second for the reference. The formulas for black pigments simplify because the λ dependences and thus the integrals vanish; $p_o \approx k/k_R$ and so forth.

Equation (4.87), however, is not yet well-suited to the rational determination of tinting strength, for it requires a new determination of the p_i for every

184   4 *How Light Scattering and Absorption Depend on the Content*

new test pigment, or at least of p_o if one is working with the zero-th approximation. An approach to further rationalization is offered by the use and direct analysis of the Kubelka-Munk functions, which are quickly derived from Y measurements. For this purpose, the functions $\hat{F}$ (test pigment) and $\hat{F}_R$ (reference) *for equal masses* $m = m_R = \hat{m}$ are found from equations (4.82) and (4.84), and the quotient of the functions is taken. With the approximations for small quantities, neglecting higher-order terms, one then obtains

$$\hat{f} = \frac{\hat{F} - F_{WP}}{\hat{F}_R - F_{WP}} = \frac{z_1}{z_{R_1}} \frac{\varrho_R}{\varrho} \left[1 + \left(\frac{z_{\frac{5}{3}}}{\mu^{\frac{2}{3}}} - \frac{z_{R\frac{5}{3}}}{\mu_R^{\frac{2}{3}}} \right) \hat{m}^{\frac{2}{3}} - \left(\frac{z_2}{\mu} - \frac{z_{R_2}}{\mu_R} \right) \hat{m} \right] \quad (4.90)$$

Now one forms the ratio to the tinting strength in equation (4.87) for the same quantity $\hat{m}$, so that

$$\frac{P^{(1)}}{\hat{f}} = 1 + \left[\left(\frac{\varrho}{\varrho_R} \right)^{\frac{2}{3}} - p_o^{\frac{2}{3}} \right] z_{R\frac{5}{3}} \left(\frac{\hat{m}}{\mu} \right)^{\frac{2}{3}} - \left(\frac{\varrho}{\varrho_R} - p_o \right) z_{R_2} \frac{\hat{m}}{\mu} \quad (4.91)$$

It is useful to consider the effects of the two perturbation terms on the right-hand side. Equation (4.65) shows that the second perturbation term (containing z_{R_2}) is influenced by the scattering power of the colored pigment while the first (containing $z_{R_{5/3}}$) is influenced only by its much slighter scattering interaction. Accordingly, the first perturbation term can be neglected; the second, on the other hand, may be appreciable in the case of strongly scattering colored pigments. What is more, it must be taken into account that the reference pigment (thus z_{R_2}) and the mass of pigment $\hat{m}$ are fixed in the practical determination of relative tinting strength, and further that p_o can be replaced by $\hat{f} \cdot \varrho / \varrho_R$ when higher-order terms are neglected. Finally, then,

$$\boxed{P^{(1)} = \hat{f}\,[1 + a\,(\hat{f} - 1)]} \qquad \text{(first approximation)} \quad (4.92)$$

Here

$$\hat{f} = \frac{\hat{F} - F_{WP}}{\hat{F}_R - F_{WP}} \; ; \quad a = z_{R_2} \frac{\hat{m}}{m_{WP}} \frac{\varrho_{WP}}{\varrho_R} \quad (4.93)$$

and

$$\boxed{P^{(o)} = \frac{\hat{F}}{\hat{F}_R}} \qquad \text{(zero-th approximation)} \quad (4.94)$$

The coefficient a is a constant and needs to be determined just once as long as the masses and the reference pigment are not changed. In addition, $\hat{F}_R$ does not have to be determined anew for every tinting strength test. The effect of the bracketed expression in equation (4.92) is that high tinting strength values are further increased while low ones are reduced.

Another important point should be added. When the Kubelka-Munk functions are calculated in order to determine the relative tinting strength with equations (4.92) to (4.94), the surface-corrected CIE tristimulus values Y^{**} and Y_R^{**} are to be obtained from equation (4.62). In tristimulus measurements, however, the spectral reflectances needed are not available. The Y values with the Saunderson corrections applied directly must therefore be used:

$$\frac{Y^{++}}{100} = \frac{\dfrac{Y}{100} - 0.04}{0.36 + 0.6 \dfrac{Y}{100}} \tag{4.95}$$

(To distinguish these from the exact values, we use the notation Y^{++}.) But this direct correction has no further consequences, for in this case the chain of equalities $\Delta Y = \Delta Y^{++} = \Delta F = 0$ always holds rigorously.

The zero-th approximation in equation (4.94) is often sufficient for determining the relative tinting strength. Even for pigments with strong intrinsic scattering, it can always be employed for relative tinting strengths in the range of $80\% \leq P \leq 120\%$; the difference between $P^{(o)}$ and $\hat{f}$ due to dropping F_{WP} is less than 0.2%. For this reason, the zero-th approximation is suitable especially for production control, where only small departures from 100% relative tinting strength are seen. Because of its continual use all over the world, this is the most commonly employed equation in colorimetric pigment testing.

Finally, the use of lightness matching $\Delta Y = \Delta L^* = 0$ as the tinting strength criterion makes it possible to answer the question of Section 4.6 as to the *absolute* tinting strength: The *coloring power* Φ introduced in Section 3.5.2 can serve this function. The coloring power was defined for the white substrate as the rate of increase in ΔE_{ab}^* with film thickness h, beginning at $h = 0$, that is, at the intrinsic lightness L_{ow}^* of the substrate. Because of the lightness criterion for the tinting strength, only ΔL^* matters, and so ΔE_{ab}^* can be replaced by ΔL^*; now, with equations (3.140), (3.155), and (3.157), the Y analog of equation (3.159), and the PVC-dependent absorption coefficient $K(\lambda) = \sigma\, k(\lambda)$ of the pigment, one gets

$$\Phi = \left.\frac{\partial \Delta L^*}{\partial h}\right|_{h \to 0} = 116\, \acute{Y}_k^* = \frac{116}{3 \cdot 100}\, \acute{Y}_k$$

$$= \frac{116 \cdot 4.8}{3 \cdot 100} \int \bar{y}_N(\lambda)\, K(\lambda)\, d\lambda = 1.856\, \sigma \int \bar{y}_N(\lambda)\, k(\lambda)\, d\lambda$$

186   4 How Light Scattering and Absorption Depend on the Content

A pigmented medium is used for the determination of coloring power, a pigment mixture with white for the determination of relative tinting strength. For the two results to be comparable, the pigment concentration must be the same in both cases. If one takes the same quantity $\hat{m}$ of test pigment as in the tinting strength test, then since $V \ll V_M$

$$\sigma = \frac{V}{V_M + V} \approx \frac{V}{V_M} = \frac{\hat{m}}{\varrho V_M}\bigg|_{V_M = V_{WP}}$$

In other words, the volume V_M of the medium used in determining Φ should be equal to the volume of white paste in the tinting strength test. Then

$$\Phi = 1.856 \, \frac{\hat{m}}{\varrho V_M}\bigg|_{V_M = V_{WP}} \cdot \int \bar{y}_N(\lambda) \, k(\lambda) \, d\lambda \tag{4.96}$$

For Φ to be accepted as the absolute tinting strength, it must be shown that the ratio of two absolute tinting strengths (Φ relative to Φ_R) gives the same relative tinting strength P obtained from equation (4.88). Since $\hat{m}$ and V_M are fixed,

$$\frac{\Phi}{\Phi_R} = \frac{\varrho_R}{\varrho} \, \frac{\int \bar{y}_N(\lambda) \, k(\lambda) \, d\lambda}{\int \bar{y}_N(\lambda) \, k_R(\lambda) \, d\lambda} \tag{4.97}$$

On the other hand, the relative tinting strength is given to a good approximation by equations (4.88) and (4.89):

$$P^{(0)} = \frac{\varrho_R}{\varrho} \, p_o = \frac{\varrho_R}{\varrho} \int \bar{y}_N(\lambda) \, \frac{k(\lambda)}{k_R(\lambda)} \, d\lambda \tag{4.98}$$

In general, the quotient of integrals in equation (4.97) is not exactly equal to the integral of the quotient in equation (4.98). But under the assumptions taken in tinting strength determination (only small hue differences allowed, hence similar reference and sample spectra), the two expressions are always so similar that they are nearly equal up to a numerical factor close to 1. Thus if $k(\lambda) = \text{const} \cdot k_R(\lambda)$, Φ/Φ_R and $P^{(0)}$ come out to the same value; the same is true if $k(\lambda)$ and $k_R(\lambda)$ are independent of λ (black pigments).

Now the coloring power defined by equation (4.96) can indeed be regarded as the absolute tinting strength, provided the quantity m_M is such that $V_M = m_M/\varrho_M = m_{WP}/\varrho_{WP} = V_{WP}$. It should be noted that the absolute tinting strength has as its units not 1 but mm^{-1} = m^2/L (see Section 3.5.2) and, as a consequence, presents itself as yet another effectiveness measure.

4.6.5 Color Depth Matching

On the basis of results obtained in the color depth study that was described in Section 4.6.3, GALL was able to work out new formulas for the color depth. The T^* formula (eq. 4.76) had proved quite inadequate for practical use, but the color depth formulas initially taken over from the textile field also failed to reproduce the results of color matching tests.

Typically, the formulas used today* span three discrete surfaces of constant color depth; if a color locus x, y, Y is specified, they give the deviation ΔT_S from these color depth level surfaces. Once again, the surfaces are labeled with fractions, this time of the standard color depth SDS (= *standard depth of shade*) instead of the older standard depth R.T. Thus the standard gives formulas for $\Delta T_{S,1/3}$, $\Delta T_{S,1/9}$, and $\Delta T_{S,1/25}$ (using the notation $B_{1/3}$, $B_{1/9}$, and $B_{1/25}$). The reference is always adjusted to the selected SDS, and the quantity m_R of the reference is recorded. The test pigment mass m is varied in order to achieve a match (i.e., to make $\Delta T_S = 0$).

The formula

$$\Delta T_{S,i} = B_i = \sqrt{Y} \, (s \cdot a_i(\varphi) - 10) + b_i \qquad (4.99)$$

gives ΔT_S values for the several color depth levels. Here

i   denotes the color depth level (1/3, 1/9, or 1/25 times SDS)

Y is the CIE tristimulus value Y

s is a measure of the saturation, equal to ten times the linear distance between the CIE color locus x, y of the test pigment and the equi-energy stimulus

$$s = 10 \sqrt{(x - x_n)^2 + (y - y_n)^2} \qquad (4.100)$$

$a_i(\varphi)$ is a factor used in evaluating the saturation measure; it depends on the hue angle $\varphi = \text{Arctan}\,[(y - y_0)/(x - x_0)]$ and, to a slight degree, on the SDS

b_i are constants: $b_{1/3} = 29$, $b_{1/9} = 41$, $b_{1/25} = 56$

The $a_i(\varphi)$ are tabulated in DIN 53235 Part 1 (enter the tables with i and φ). The convenient formulas (third-degree polynomials) presented in Chapter 8 are preferred for computer calculations.

---

* For standards see "Color Depth" in Appendix 1.

188   4 How Light Scattering and Absorption Depend on the Content

The levels defined by these formulas are again surfaces in color space resembling spherical shells or goblets (Fig. 4-19). To gain some idea of the situation, one can calculate from the b_i where these surfaces intersect the lightness axis in CIELAB color space. For $\Delta T_S = s = 0$, equation (4.99) gives

$$Y_{0,\frac{1}{3}} = \left(\frac{\frac{b_1}{3}}{10}\right)^2 = 2.9^2 = 8.41$$

$$Y_{0,\frac{1}{9}} = 16.81$$

$$Y_{0,\frac{1}{25}} = 31.36$$

The respective lightness values are $L^*_{0,1/3} = 34.8$, $L^*_{0,1/9} = 48.0$, and $L^*_{0,1/25} = 62.8$; thus the lightness range of interest is well spanned.

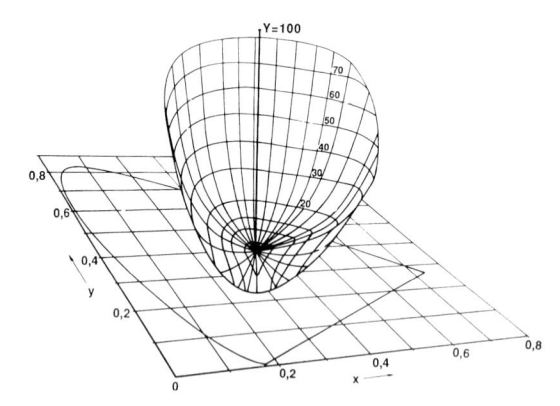

Fig. 4–19. Surface of constant color depth T_S ($T_S = (1/3)SDS$) in the CIE system, after GALL.

For the sake of rationalization, it is desirable if the test can be performed with a single weighed quantity of pigment. Bringing ΔT_S to zero requires an iteration. The new quantity of pigment mass to add is

$$m_{i+1} = m_i \, 0.9^{\Delta T_S} \tag{4.101}$$

while the m dependence of the CIE tristimulus values Y used in equation (4.99) can be obtained from equation (4.66) or (4.67); the coefficients are determined by appropriate pigment mass additions.

It is still better to apply the principle of spectral evaluation immediately. The function $F(\lambda)$ can be obtained from equation (4.63) for individual λ values; the needed coefficients z_{Ri} for the reference can then be found with one to two pigment quantities. On the other hand, the tristimulus method can also be employed. The basis for this approach comprises equations (4.66) and (4.67) for the Kubelka-Munk functions of the CIE tristimulus values X, Y, Z, or the corresponding equations for the reflectometer values R'_x, R'_y, R'_z. The coefficients here are easily determined with two pigment quantities.

4.7 Special Problems

4.7.1 Lightening Power

As defined by standards, lightening power is "the ability of a pigment to increase the lightness of a colored, gray, or black medium." The lightening power is the relative tinting strength of white pigments, and the definition accords with that of the general relative tinting strength of pigments, equation (4.75). The problem, once again, is to derive a simple relation suitable for rationalizing the procedure; the tool used, once again, is the Kubelka-Munk function.

Dispersions of white pigments in black pastes yield reflection spectra virtually independent of λ, so that $\varrho(\lambda)$ can be treated as equal to $Y/100$. Thus $\Delta Y = \Delta Y^{++} = \Delta F = 0$ rigorously.

Similarly to the development in Section 4.6.4, the Kubelka-Munk functions of the test and reference pigments are written out in accordance with equation (4.48). Next, an expression is sought for the function $\psi_R(\psi)$ at matching, which leads in the end to the lightening power equation:

$$P_w^{(1)} = \frac{\varrho_R}{\varrho} p_{wo} \left[1 - p_w \tfrac{2}{3} \left(\frac{m}{\mu} \right)^{\frac{2}{3}} - p_{w1} \frac{m}{\mu} \right]$$

$$= \frac{\varrho_R}{\varrho} p_{wo} \left[1 - p_w \tfrac{2}{3} \left(\frac{\sigma}{1 - \sigma} \right)^{\frac{2}{3}} - p_{w1} \frac{\sigma}{1 - \sigma} \right] \qquad (4.102)$$

(first approximation)

$$P_w^{(0)} = \frac{\varrho_R}{\varrho} p_{wo} \qquad \text{(zero-th approximation)} \qquad (4.103)$$

190   4 How Light Scattering and Absorption Depend on the Content

Here, similarly to equation (4.50),

$$p_{wo} = \frac{z_{w1}}{z_{wR1}} \approx \frac{s}{s_R}$$

$$p_{w\frac{2}{3}} = z_{w\frac{5}{3}} - p_{wo}^{\frac{2}{3}} z_{wR\frac{5}{3}} \approx g - p_{wo}^{\frac{2}{3}} g_R$$

$$p_{w1} = z_{w2} - p_{wo} z_{wR2} \approx \frac{1}{K_b}(k - p_{wo} k_R) \tag{4.104}$$

Equation (4.102) has the new feature that both the second and the third terms in brackets are negative (not just the third as in eq. 4.87).

Next, the Kubelka-Munk functions for the test and reference pigments are written for equal pigment quantities $\hat{m}$, and the ratio $\hat{f}_w$ of the two is formed. In contrast to the previous derivation, an examination of the two perturbation terms shows that the $p_{w2/3}$ term must be kept because it relates to the attenuation coefficient for the scattering interaction of the white pigment. On the other hand, the intrinsic absorption of the white pigment (p_{w1}) can be neglected. Division of $P_w^{(1)}$ by $\hat{f}_w$ leads, finally, to the equations

$$P_w^{(1)} = \hat{f}_w \left[1 + a_w \left(\hat{f}_w^{-\frac{2}{3}} - 1\right)\right] \quad \text{(first approximation)} \tag{4.105}$$

where

$$a_w = z_{wR\frac{5}{3}} \left(\frac{\hat{m}}{m_{BP}} \cdot \frac{\varrho_{BP}}{\varrho_R}\right)^{\frac{2}{3}} \approx g_R \left(\frac{\hat{m}}{m_{BP}} \cdot \frac{\varrho_{BP}}{\varrho_R}\right)^{\frac{2}{3}}$$

$$\hat{f}_w = \frac{\hat{F}_{wR}}{\hat{F}_w} \cdot \frac{F_{BP} - \hat{F}_w}{F_{BP} - \hat{F}_{wR}}$$

$$P_w^{(0)} = \frac{\hat{F}_{wR}}{\hat{F}_w} \quad \text{(zero-th approximation)} \tag{4.106}$$

Equation (4.106) is another formula employed very frequently in pigment testing. It should be noted that the equation for the lightening power has the reference Kubelka-Munk function in the numerator and the test function in the denominator, in other words inverted to the relative tinting strength.

The zero-th approximation gives the lightening power in the Beer's law region; outside this region, the first approximation has to be used. The reason has to do with the relatively strong effect on the lightening power due to the scattering interaction. An example is shown in Figure 4-20. Equation (4.106) implies that the zero-th approximation is the ratio of the scattering coefficients per unit PMV for the test pigment and reference:

$$P_w^{(0)} = \frac{m_R}{m}\bigg|_{\Delta Y=0} \approx \frac{\sigma_R}{\sigma}\bigg|_{\Delta Y=0} = \frac{S}{S_R}\bigg|_{\sigma=\sigma_R} \approx \frac{S}{S_R}\bigg|_{m=m_R} \quad (4.107)$$

No account is taken of the scattering interaction. The lightening power is thus correctly represented by the ratio of scattering coefficients only in the Beer's law range, not at higher PVC. For this reason, the lightening power and the relative scattering power are identical only in the lower range; the difference between them grows larger and larger outside this range.

The first approximation requires a determination of g_R. An appropriate technique is the dilution method, since g can be obtained directly with equation (4.52). Chapter 8 will describe a method by which the lightening power of white pigments can be determined in a rational manner combining the adding pigment and dilution methods.

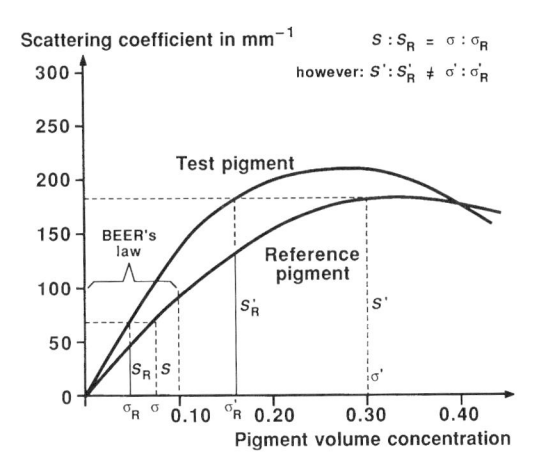

Fig. 4–20. Significance of lightening power.

4.7.2 Color Differences after Matching

The "color difference after color reduction" is the color difference still present after tinting strength matching; its analog after lightening power matching is called the "shade difference after lightening power matching." These are obtained by colorimetry using the CIELAB system; they can be broken down into lightness difference, chroma difference, and hue difference components in the first case, into shade strength and nature of shade in the second (see Section 2.3.3).

The color difference between the reduced test and reference pigment are calculated from the CIE tristimulus values of both. The X, Y, Z of the reduced test pigment are measured directly when the tinting strength and lightening power are determined (or the reflectometer values R'_x, R'_y, R'_z, when the tristimulus method is employed); X_R, Y_R, Z_R for the reference pigment at matching, on the other hand, have to be calculated. The PMV-dependent Kubelka-Munk functions of the CIE tristimulus values, equations (4.82) and (4.48), are available for this purpose; all that is necessary is to substitute the PMV of the reference at matching, $\psi_R = m_R/\mu_R$, in the formulas, and $m_R = \hat{m} P$ because the mass m_R of reference pigment at matching is very easily found with fundamental equation (4.75).

In the equations that follow, the needed relations will be listed for direct use in tinting strength/lightening power programs. The term with subscript 5/3 is dropped in the case of tinting strength, the term with subscript 2 in that of lightening power, because they are negligible. Thus the *color difference after color reduction* (tinting strength) is given by

$$\begin{aligned} F^{(2)}_{Xa,R} - F_{WP} &= \bar{z}_{XR_1}\,\psi_R\,(1 - \bar{z}_{XR_2}\,\psi_R) \\ F^{(2)}_{Ya,R} - F_{WP} &= \bar{z}_{YR_1}\,\psi_R\,(1 - \bar{z}_{YR_2}\,\psi_R) \\ F^{(2)}_{Za,R} - F_{WP} &= \bar{z}_{ZR_1}\,\psi_R\,(1 - \bar{z}_{ZR_2}\,\psi_R) \end{aligned} \qquad (4.108)$$

while the *shade difference after color reduction* (lightening power) is

$$\begin{aligned} \frac{1}{F^{(2)}_{w,Xa,R}} - \frac{1}{F_{BP}} &= z_{w,XR_1}\,\psi_R\,(1 - z_{w,XR\frac{5}{3}}\,\psi_R^{\frac{2}{3}}) \\ \frac{1}{F^{(2)}_{w,Ya,R}} - \frac{1}{F_{BP}} &= z_{w,YR_1}\,\psi_R\,(1 - z_{w,YR\frac{5}{3}}\,\psi_R^{\frac{2}{3}}) \\ \frac{1}{F^{(2)}_{w,Za,R}} - \frac{1}{F_{BP}} &= z_{w,ZR_1}\,\psi_R\,(1 - z_{w,ZR\frac{5}{3}}\,\psi_R^{\frac{2}{3}}) \end{aligned} \qquad (4.109)$$

Here

$$\psi_R = \frac{m_R}{\mu_R} = \frac{\hat{m}}{\mu_R}\,P \qquad (4.110)$$

with

$$\mu_R = m_{WP}\,\frac{\varrho_{WP}}{\varrho_R}$$

(and also P_w in place of P). When the determination is done by the tristimulus method, the needed coefficients z_{Ri} for the reduced reference pigment are determined once for all with the help of two mixtures containing different quantities of pigment. If the first approximation for the Kubelka-Munk functions is sufficient, one can set $F_{WP} = 1/F_{BP} = 0$ in the above equation and set the brackets on the right equal to 1.

The Kubelka-Munk functions of the CIE tristimulus values immediately imply the (simply) corrected CIE tristimulus values of the reduced reference, in accordance with equation (3.116) or (4.61) (the subscript R is dropped in the equations that follow):

$$\boxed{\begin{aligned}\frac{X^{++}}{X_n} &= 1 + F_X - \sqrt{F_X(F_X + 2)} \\[1em] \frac{Y^{++}}{100} &= 1 + F_Y - \sqrt{F_Y(F_Y + 2)} \\[1em] \frac{Z^{++}}{Z_n} &= 1 + F_Z - \sqrt{F_Z(F_Z + 2)}\end{aligned}} \qquad (4.111)$$

and hence the uncorrected CIE tristimulus values of the reduced reference pigment, in accordance with equation (3.10):

$$\boxed{\begin{aligned}\frac{X}{X_n} &= 0.04 + \frac{0.384 \dfrac{X^{++}}{X_n}}{1 - 0.6 \dfrac{X^{++}}{X_n}} \\[1.5em] \frac{Y}{100} &= 0.04 + \frac{0.384 \dfrac{Y^{++}}{100}}{1 - 0.6 \dfrac{Y^{++}}{100}} \\[1.5em] \frac{Z}{Z_n} &= 0.04 + \frac{0.384 \dfrac{Z^{++}}{Z_n}}{1 - 0.6 \dfrac{Z^{++}}{Z_n}}\end{aligned}} \qquad (4.112)$$

Now the CIE tristimulus values X, Y, Z of the reduced test and reference pigment are known, and with the scheme of Section 2.3.3 one can calculate ΔE_{ab}^* along with the breakdowns of color or shade difference after matching.

Both quantities are strongly affected by particle size, especially in pigments with high light-scattering power. The reason for this behavior will be explained in Chapter 5. By virtue of this relationship, the color and shade differences after matching can well be used in production control, where they serve as sensitive indicators of any departure from constancy of particle size. Because the only concern is with deviations, simplified results are also adequate here; for example, the difference between the CIE tristimulus values X and Z can be cited, or this difference related to the CIE tristimulus value Y, $(Z - X)/Y$.

4.7.3 Change in Tinting Strength

Section 4.1.3 treated the steps in the dispersing of pigments: wetting, disintegration, and stabilization. Now the disintegration process needs to be examined more closely, since it is directly related to the final tinting strength of the pigment. When pigments are dispersed in binders, it can be observed as a general rule that the relative tinting strength increases as the dispersing effort continues. Therefore the relative tinting strength is an excellent test instrument for describing the dispersing process ("tinting strength development") and the final dispersed condition.

The mathematical instrumentarium of reaction kinetics, well-known from physical chemistry, can be employed for the quantitative description of dispersing over time, since dispersing is a transformation of substances – of larger particles to smaller ones. For the sake of simplicity, just two classes of particles are considered, the particles to be dispersed (larger) and the particles already dispersed (smaller). It is still amazing that the process of dispersing can be handled by such a simple model and that the results are in such good agreement with observation.

First, some conventions: σ_∞ is the total PVC of particles in the mill base; this does not vary during the dispersing process. σ is the PVC of fine particles at time t, so that $\sigma_\infty - \sigma$ is the PVC of coarse particles. If the PVC is referred to the total PVC, the relative fraction of fine particles is given by $v = \sigma/\sigma_\infty$ and that of coarse particles by $1 - v$.

We now investigate the following cases:

– Size reduction of the coarse particles is due to the grinding action of the large particles themselves

– It results from grinding on the grinding medium and the walls of the grinding vessel

– It is the consequence of grinding by already dispersed (small) particles

Case 1: Size reduction by large particles themselves

The increase in the fraction of small particles dv over time must be proportional to the squared fraction of coarse particles $(1 - v)^2$. For a selected coarse particle is reduced in size faster the more coarse particles it encounters (simple propor-

4.7 Special Problems

tionality); because the same thing happens with all the coarse particles, however, there must be another proportionality to the fraction of coarse particles (double proportionality):

$$dv = \varkappa_c (1 - v)^2 \, dt \tag{4.113}$$

where $\varkappa_c$ is the rate of size reduction of coarse particles by coarse particles (units min^{-1} or h^{-1}). With the initial condition that the initial concentration (at $t = 0$) of small particles is zero ($v(0) = 0$), the integral of the differential equation yields

$$\frac{1}{v} = \frac{1}{\varkappa_c t} + 1 \tag{4.114}$$

as in the case of a second-order reaction. The rate of dispersion $\varkappa_c$ can also be expressed in terms of the half-time $t_{0.5}$:

$$t_{0.5} = \frac{1}{\varkappa_c} \quad \text{(units: min or h)} \tag{4.115}$$

Now the relative tinting strength can be written in accordance with equations (4.97) and (4.98), with $\overline{k}$ for the colorimetric $\overline{y}(\lambda)$ integrals of the absorption coefficients per unit PVC. The absorption coefficient $\overline{k}_\infty$ (corresponding to the "final tinting strength") is taken for the reduced reference pigment. In this way, the pigment is compared with its own final dispersion; as a consequence, there is a single pigment density $\varrho = \varrho_R$. If $P_o = \overline{k}_o/\overline{k}_\infty$ is the initial tinting strength, then

$$P = \frac{\overline{k}}{\overline{k}_\infty} = \frac{(\sigma_\infty - \sigma)\overline{k}_o + \sigma\overline{k}_\infty}{\sigma_\infty \overline{k}_\infty} = P_o + v(1 - P_o) \tag{4.116}$$

Rewriting and substitution of equation (4.114) leads to

$$\frac{1}{P - P_o} = \frac{1}{1 - P_o} + \frac{1}{(1 - P_o)\varkappa_c} \cdot \frac{1}{t} \tag{4.117}$$

If now one sets $P = \hat{F}/\hat{F}_\infty$ in accordance with equation (4.94), in other words making the relative tinting strength equal to the ratio of the Kubelka-Munk functions for equal pigment masses $\hat{m}$, and if one further neglects the (usually very small) initial tinting strength P_o, one gets

$$\boxed{\frac{1}{\hat{F}} = \frac{1}{\hat{F}_\infty} + \frac{1}{\hat{F}_\infty \varkappa_c} \cdot \frac{1}{t}} \tag{4.118}$$

This is the well-known PIGENOT equation, which states that a straight line results if the reciprocals of the Kubelka-Munk functions for the reduced pigments are plotted versus reciprocal time over the dispersion process. As equation (4.117) shows, the linearity can be improved if the initial function $\hat{F}_o$ is subtracted from $\hat{F}$ and $\hat{F}_\infty$ at the outset.

The PIGENOT equation holds more-or-less well for many dispersing operations, indicating that the disintegration of coarse particles on themselves is the dominant process in dispersion. The prerequisites for this are satisfied especially in mixtures for dispersion where the pigment concentration is high or the degree of filling of the grinding vessel is high.

Curve 1 in Figure 4-21 presents one dispersing curve for this form of size reduction. In the final phase, the last little bit of reduction comes about very slowly because of the declining quantity of coarse particles.

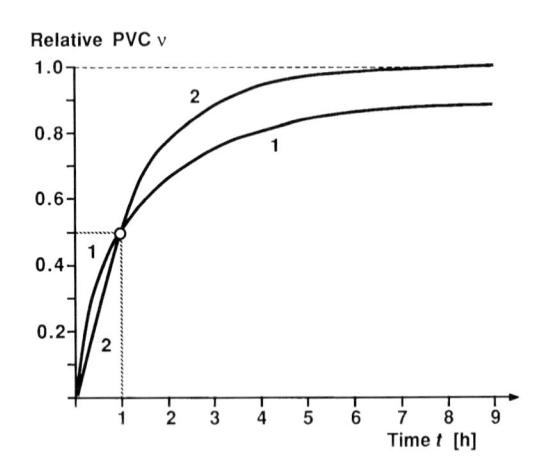

Fig. 4–21. Dispersing curves for the size reduction of coarse particles on themselves (1) and on the grinding medium and grinding vessel walls (2). The half-time is the same for both curves, $t_{0.5} = 1$ h.

Case 2: Size reduction by grinding medium and vessel walls

This process corresponds to a first-order reaction, so that the increment of fine particles is simply proportional to the fractional concentration of coarse particles:

$$dv = \varkappa_w (1 - v) \, dt \tag{4.119}$$

where $\varkappa_w$ is the rate of size reduction of coarse particles by grinding medium and wall, with units of min^{-1} or h^{-1}. With the same initial condition as above, the integral comes out to

$$v = 1 - e^{-\varkappa_w t} \tag{4.120}$$

The half-time is

$$t_{0.5} = \frac{\ln 2}{\varkappa_w} = \frac{0.693}{\varkappa_w} \qquad (4.121)$$

with units min or h. If $v(t)$ is substituted in tinting strength equation (4.116), the result after simplifying and taking logarithms is

$$\ln(1-P) = \ln(1-P_o) - \varkappa_w t \qquad (4.122)$$

Thus a plot on semilog paper, with $1 - P$ on the logarithmic scale versus t on the linear scale, will give a straight line.

This disintegration mode dominates when the mill base has a low pigment concentration (and thus also in the final phase) or the filling of grinding medium is high. Figure 4-21 also shows a dispersion curve for this type of disintegration (curve 2). The rate constant was selected so that the same half-time $t_{0.5}$ applies to curves 1 and 2. Obviously, this disintegration process is markedly more efficient than size reduction on the particles themselves, at least in the final phase.

Case 3: All forms of disintegration

In addition to the two modes with which we have already become acquainted, another disintegration process is possible: size reduction of coarse particles on fine ones. While this form will play a rather minor role, it must be allowed for in the global formula. It is not worthwhile, however, to examine it in isolation as we did the other two modes, for it contributes slightly to only the middle portion of the dispersion process. The reason is that there are no small particles at the beginning of the process, while there are no large ones at the end, so that conditions for $t = 0$ and $t = \infty$ cannot be introduced.

The increase in the content of small particles for this type of disintegration must be expressed as an interaction term between small and large particles, which is thus proportional to $v(1 - v)$. The differential equation for all disintegration processes is now

$$dv = [\varkappa_c (1-v)^2 + \varkappa_w (1-v) + \varkappa_f v(1-v)]\, dt \qquad (4.123)$$

where $\varkappa_f$ is the rate of size reduction of coarse particles by small particles (units min^{-1} or h^{-1}). With the usual initial condition, the integral is

$$\boxed{\frac{1}{v} = 1 + \frac{A}{e^{Bt} - 1}} \qquad (4.124)$$

where $A = (\varkappa_w + \varkappa_f)/(\varkappa_c + \varkappa_w)$, $B = \varkappa_w + \varkappa_f$, and the half-time is

4 How Light Scattering and Absorption Depend on the Content

$$t_{0.5} = \frac{1}{B} \ln(1+A) \approx \frac{A}{B}\left(1 - \frac{A}{2}\right) \approx \frac{1}{\varkappa_c + \frac{3}{2}\varkappa_w + \frac{1}{2}\varkappa_f} \qquad (4.125)$$

The approximation being obtained with $A \ll 1$. The constants A and B can be determined by iteration based on two measurements. A plot versus time does not directly give a straight line as in the previous cases, but for the initial values of v or $\hat{F}$ (i.e., $t \ll t_{0.5}$) the smallness of the exponent in equation (4.124) means that the Pigenot equation (4.118) holds to a good approximation. Figure 4-22 shows how v varies with t for rate constants $\varkappa_c = 1$, $\varkappa_w = 0.1$, and $\varkappa_f = 0.01$ h^{-1}.

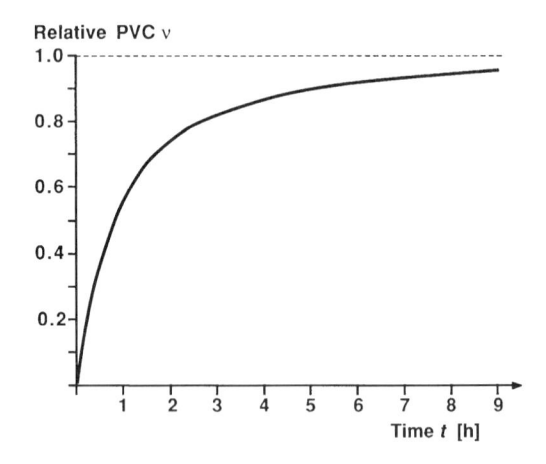

Fig. 4–22. A dispersing curve for all types of disintegration.

According to equation (4.116), the time derivative of the tinting strength at the onset of dispersing:

$$\dot{P}(0) = \dot{v}(0)(1 - P_o) = (\varkappa_c + \varkappa_w)(1 - P_o)$$

$$\text{with } \dot{P}(0) = \left.\frac{dP}{dt}\right|_{t=0} \text{ and } \dot{v}(0) = \left.\frac{dv}{dt}\right|_{t=0}$$

where the condition $\dot{v}(0) = \varkappa_c + \varkappa_w$, easily derivable from equation (4.123), has been used. We now define

$$\Psi = \frac{\dot{P}(0)}{P_o \dot{v}(0)} = \frac{1}{P_o} - 1 = \frac{\hat{F}_\infty}{\hat{F}_o} - 1 = \frac{1}{\hat{F}_o}(\hat{F}_\infty - \hat{F}_o) \qquad (4.126)$$

with the term "potential of strength"* for Ψ, because this quantity represents the difference between the initial and potential final tinting strengths, divided by the initial strength; on the other hand, Ψ does not say anything about how rapidly the final tinting strength is attained. The potential of strength is the rate of tinting strength development at the beginning of dispersion, divided by to the rate of dispersion; as the quotient of two quantities with the same dimensions, it has itself the dimension 1. Because the potential of strength is a relative quantity, its numerical value changes little from one type of dispersing equipment or dispersing process to another.

The half-time $t_{0.5}$ defined above is a valuable tool when using the dispersing performance to characterize dispersing equipment or when comparing pigments in terms of their dispersing resistance. The examples presented in Figure 4-23, based on the v versus t curves of Figure 4-22, clearly show that the potential of strength of a pigment does not necessarily have anything to do with the dispersing resistance. Both pigments have the same increase in strength $\Psi = 1$

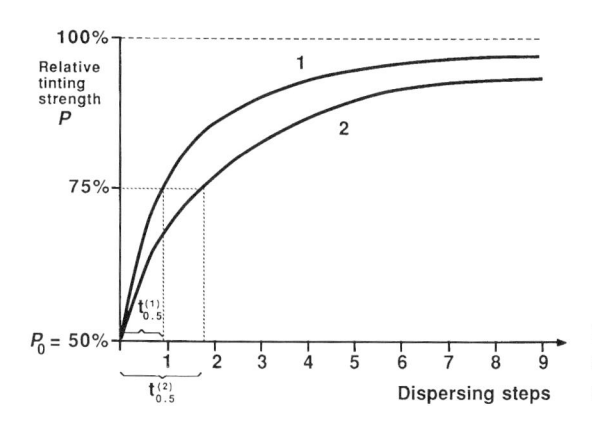

Fig. 4–23. Significance of dispersing resistance and potential of tinting strength.

(because the initial tinting strength P_o is the same for both), but the half-times differ by a factor 2; in other words, pigment 2 is twice as "hard" as pigment 1.

In accordance with the definitions given, the rate constants $\varkappa_c$ and $\varkappa_f$ also depend on the total PVC (singly proportional to σ_∞), while $\varkappa_w$ is independent of σ_∞. In general, the constants are functions of the mill base (i.e., the properties of the pigment and vehicle) and the dispersing equipment (e.g., degree of mill base filling, volume fraction and size of balls in the case of ball mills). Only when all other experimental conditions remain unchanged is it possible to assess the dispersion properties of pigments and of dispersing equipments by comparing the dispersing rate constants or half-times.

---

* In ISO 8781-1, the "*increase in strength*" IS is 100 times this value.

4.8 List of Symbols Used in Formulas

A, B	constants
a^*, b^*, L^*	CIELAB coordinates
$B_i = \Delta T_{Si}$	deviations from surfaces of constant standard color depth with i = 1/3, 1/9, 1/25
β	pigment mass concentration, PMC (g cm^{-3})
c	coefficients of the PVC-dependent Kubelka-Munk function for constant volume method

with subscripts:
- b black pigment
- w white pigment
- 1 linear term
- 5/3 term in 5/3 power
- 2 quadratic term

$\bar{c}$	c for colored pigments with same subscripts as c, $\bar{F}$
D	particle diameter, mm
ΔC^*_{ab}	chroma difference (CIELAB)
ΔE^*_{ab}	color difference (CIELAB)
ΔH^*_{ab}	hue difference (CIELAB)
ΔL^*	lightness difference (CIELAB)
ΔT_{Si}	see B_i
ε	small quantity ($\varepsilon \ll 1$)
η	attenuation factor of scattering interaction
$\hat{f}$	quantity formed from Kubelka-Munk functions

with subscript:
- w white pigment

$F = K/S$	Kubelka-Munk function

without subscript: test (sample)

with subscripts:
- a   adding pigment method
- A   initial value
- b   black pigment in white paste
- B   black system (constant volume method)
- BP black paste (adding pigment method)
- c   constant volume method
- d   dilution method
- PU full shade
- R   reference
- W   white system (constant volume method)
- w   white pigment in mixture with black paste
- WP white paste (adding pigment method)

4.8 List of Symbols Used in Formulas

$\overline{F}$	Kubelka-Munk function of CIE tristimulus values of colored pigments
	with subscripts:
	X, Y, Z CIE tristimulus values
	see also F
$\hat{F}$	Kubelka-Munk function for constant pigment mass $\hat{m}$
	with subscripts:
	0 initial
	∞ final
	see also F
$F^{(1)}, F^{(2)}$	first and second approximations to Kubelka-Munk function
	with same subscripts as F
$g = \gamma s$	attenuation term of scattering interaction
γ	coefficient in attenuation term, mm
h	film thickness, mm
K	absorption coefficient for diffuse irradiation, mm^{-1}
	with subscripts:
	F colored or black pigment
	see also c
k	absorption coefficient referred to PVC, mm^{-1}
k_β	absorption coefficient of dye for diffuse irradiation, referred to mass concentration, m^2 g^{-1}
k'	absorption coefficient for directional irradiation, mm^{-1}
k''_β	absorption coefficient of dye for directional irradiation, referred to mass concentration, m^2 g^{-1}
$\varkappa$	reaction rate constant, min^{-1} or h^{-1}
	with subscripts:
	c size reduction on coarse particles
	f size reduction on fine particles
	w size reduction on wall and grinding body
m	mass, g
	with subscripts:
	M medium
	see also F
μ	pigment mass equivalent to the medium volume (PVM), g
	with same subscripts as F
μ'	pigment mass equivalent to the medium volume of the mass portion (PMM)
$\nu = \sigma/\sigma_\infty$	relative volume fraction of fine particles during dispersing process
P	relative tinting strength
	with subscripts:
	d dye
	w lightening power
	0 initial
$P^{(0)}, P^{(1)}$	zero-th and first approximations to relative tinting strength
	with same subscripts as P

4 How Light Scattering and Absorption Depend on the Content

p	area fraction of individual particle (for scattering interaction)
p_i	coefficients of relative tinting strength (i = 0, 2/3, 1)
	with subscripts:
	0 absolute term
	2/3 term in 2/3 power
	1 linear term
	see also F, $\hat{F}$, c
Φ	coloring power, mm^{-1} = m^2 L^{-1}
	with same subscripts as F
φ	pigment volume portion
φ	hue angle in CIE chromaticity diagram
Ψ	potential of strength
ψ	pigment/medium volume ratio (PMV)
	with same subscripts as F
Q	any color quality
$q_D = \sigma/\sigma_A$	dilution ratio by volume
R	diameter of closest-packed spheres, mm
R'	reflectometer value
	with subscripts x, y, z for the tristimulus method
ϱ	density, g cm^{-3}
	with subscripts:
	D dispersion
	see also F, m
$\varrho(\lambda)$	spectral reflectance
	with same subscripts as F
$\varrho^*(\lambda)$	spectral reflectance after Saunderson correction
	with same subscripts as F
S	scattering coefficient, mm^{-1}
	with same subscripts as K
s	scattering coefficient referred to PVC, mm^{-1}
	with same subscripts as c
s	measure of saturation in the CIE system
σ	pigment volume concentration (PVC)
	with subscripts:
	ob black pigment in black paste
	ow white pigment in white paste
	U achromatic pigment in mixture
	UA achromatic pigment in mixture, initial value
	UP achromatic paste in mixture
	U0 achromatic pigment in achromatic paste
	see also F
T^*	transmission factor
T^*	color depth
t	time, min or h
	with subscript:
	0.5 half-time

V	volume, cm³ or L
	with same subscripts as F, m
V	CIELAB color locus
v	coefficient of PVC-dependent Kubelka-Munk function for dilution method
	with same subscripts as c
$\bar{v}$	v for colored pigments
	with same subscripts as $\bar{c}, \bar{F}$
w	pigment mass portion (PMP)
X, Y, Z	CIE tristimulus values
	with subscripts:
	n for CIE standard illuminant n
	see also F
X^{**}, Y^{**}, Z^{**}	
	CIE tristimulus values of Saunderson corrected reflection spectrum
	with same subscripts as F
X^{++}, Y^{++}, Z^{++}	
	CIE tristimulus values after Saunderson correction
	with same subscripts as F
x, y	CIE chromaticity coordinates
	with subscript:
	n CIE standard illuminant n
$\bar{x}(\lambda), \bar{y}(\lambda), \bar{z}(\lambda)$	
	CIE color matching functions
	with subscript:
	N normalized to CIE standard illuminant N
z_i	coefficients of PMV-dependent Kubelka-Munk function for adding pigment method
	with the same subscripts as c
$\bar{z}$	z for colored pigments
	with same subscripts as $\bar{c}, \bar{F}$

4.9 Summary

Particle size is one of the most important parameters of a pigment; there is no application-related property that is not affected by it. This chapter reports on the needed definitions (primary particle, agglomerate, aggregate) and the particle-size parameters. The colorimetric properties of pigments are, in practice, always tested after incorporation into a vehicle system. The preparation of the dispersion (the dispersing process) must be done with particular care. Quantitative description of concentration-related phenomena demands clear definitions of measures for the dye and pigment concentrations. The following quantities are derived from the relationships in mixed-phase systems: pigment mass portion (PMP), pigment mass concentration (PMC), pigment volume concentration (PVC), pigment mass equivalent to the medium volume (PVM), pigment mass equivalent to the medium volume of the mass portion (PMM), and pigment/medium volume ratio (PMV).

The absorption and scattering coefficients of Chapter 3 are not true constants of a material, since they also depend on the concentration of coloring material. The *absorption* of dyes and pigments is a linear function of concentration (Beer's law). There remains a real material constant, the "absorption coefficient referred to concentration" (or per unit concentration). If several coloring materials are present, the individual coefficients can generally be summed.

In the case of *scattering*, the relationship is linear only in the initial region (up to a PVC of ca. 10%). At higher concentrations, the scattering falls increasingly behind the linear function because of the scattering interaction between particles as they come closer and closer together. This effect is included in an earlier published equation having a term in the 2/3 power of the concentration. Experimental tests of this formula have shown good agreement with the actual behavior of the scattering curves for white pigments.

The laws for mixtures of pigments with achromatic pastes are systematized by the method used to prepare the mixture. The three most important methods in practice are

– Constant volume method. The PVC of the achromatic pigment in the paste is held constant. It is used chiefly in determining scattering and absorption coefficients

– Adding pigment method. The pigment is added, in measured quantities by weight, to a constant weight of paste. This is the most important of the three methods because it forms the basis for determining the tinting strength of pigments

– Dilution method. After a mixture is prepared, it is diluted by adding more vehicle. This method plays some role in the rationalization of test procedures

Each of these preparation methods is characterized by its own conservation law.

The concentration-dependent Kubelka-Munk functions for pigment/achromatic paste mixtures are obtained in the following way: the equations for the three standard mixing methods are derived from a generally valid formula. Application of the proper conservation law in each case yields the equations for black pigments in white paste and white pigments in black paste; separate equations are obtained for the constant volume, adding pigment, and dilution methods. The pigment/medium volume ratio is the appropriate concentration measure for the adding pigment method. The procedure is also used for colored pigments whose spectral Kubelka-Munk functions are to be derived. Then it is shown that similar approximate formulas can also be obtained for broad-band spectral distributions, and thus for the CIE tristimulus values. These are of great practical importance when tristimulus instruments are employed. In summary, the equations are all similar except for having different coefficients. Typical for these relations is that the Kubelka-Munk functions are linear in the concentration to a first approximation, while in the second approximation there are perturbation terms in the 5/3 and 2 powers respectively.

The tinting strength describes one of the most important optical qualities of coloring materials. It is a measure of the ability to impart color to other substances through *absorption.* All testing methods give the *relative* tinting strength, that is, the absorbing power relative to another pigment or dye. In the case of dyes, the relative tinting strength is also logically defined as a relative absorption; for pigments, on the other hand, the definition is plainly a measure of effectiveness, stating the mass ratio of test and reference pigments such that mixtures of the pigments with white paste appear alike. The lightening power (the tinting strength of white pigments) is based on *scattering* and is determined in black pastes. – The "standard experiment" of tinting strength determination thus consists in matching one pigment to the other in reduction by varying the mass of one pigment added. In matching, which means causing the mixtures to look alike or have the same color, the matching criterion plays a key role. An indispensable step in rationalizing computer-aided methods is to establish the tinting strength criterion. Any such criterion can be thought of as a set of equipotential surfaces in color space, which are intersected by the "tinting characteristic" (a curve connecting the color loci of mixtures obtained by adding larger and larger amounts of colored pigment to a white paste). Among the tinting strength criteria in use, two are of special significance: the lightness L^* (or CIE tristimulus value Y) and the color depth. Two color matching studies reported in the literature are consulted in order to establish what tinting strength criteria are preferred by colorists. In one of the studies (on inorganic pigments only), the question to be answered was this: which of the added pigment quantities produces the best match? A comparison with colorimetric data gave the unambiguous answer that the CIELAB lightness L^* (or CIE tristimulus value Y) is the crucial tinting strength criterion. In the second study (primarily on organic pigments), the one sample was to be chosen that best agreed in color depth with the reference pigment and, simultaneously, with the other samples already selected, right around the color circle. This study resulted in improved formulas for calculating color depth values. Because the inorganic pigments in the first case were nearly all of a dark and not highly saturated full shade, while in the second case the pigments used were more highly saturated, current practice is as follows: the lightness criterion is used for inorganic pigments, the color depth criterion for organic ones.

There follows a complete derivation of the tinting strength equation expressed as a dependence on pigment concentration for the lightness matching case. As a result of the development, it is found that Y matching is virtually identical with matching based on the Saunderson-corrected Y values, and that this correction can be applied directly to the CIE tristimulus values. Finally, it is shown that the coloring power defined in Chapter 3 can be regarded as an absolute tinting strength. Because the color depth formulas taken over from the textile field were not accurate enough, the studies were utilized in improving these formulas as applied to pigments. To rationalize the determination of tinting strength based on color depth matching, the required mass of the test pigment must be determined by an iterative method; the principle of spectral evaluation is well-suited to this purpose.

In the case of lightening power, the white-pigment analog of the relative tinting strength, the Kubelka-Munk function again serves as a tool for rationalization. The needed equations for the lightening power as a function of PVC are derived in a similar way to the tinting strength equations. The zero-th approximation of this equation provides a good description of the behavior in the Beer's law region; at higher pigment concentrations, the first approximation with the perturbation term must be employed. The color difference that remains after matching is called the "color difference after color reduction" in the case of tinting strength and the "shade difference after lightening power matching" in the case of lightening power. The CIELAB coordinates are accessible from the CIE tristimulus values (calculated for the pigment mass at matching), and from these in turn the color differences between sample and reference can be obtained. For white pigments, the shade difference after lightening power matching is a particularly suitable tool for the quantitative characterization of the pigment particle size. In the dispersing process, the disintegration of pigment agglomerates is crucial to the quality of the resulting dispersion. It is shown how a quantitative description of the dispersing process can be arrived at with the equations of reaction kinetics. For example:

- The size reduction of large particles on large particles is a second-order reaction and can immediately be described by the well-known PIGENOT equation, which is suitable for linearizing the measured values.

- Size reduction on the vessel wall and grinding medium is described by a first-order reaction equation.

- Size reduction by all mechanisms includes disintegration on small particles already produced.

In all three cases, the half-time can be used to characterize the rate of dispersion. The attainable tinting strength is described by an easily determined value called the potential of strength.

4.10 Historical and Bibliographical Notes

Even before KÄMPF's thorough work in 1965 [1], there was a large specialist literature dealing with the effect of particle size on pigment properties (see the bibliography in [1]). The particle-size concepts defined in Section 4.1.1 correspond to the classification of pigment particles proposed in 1963 by HONIGMANN and STABENOW [2], which has gained international adoption and has been set forth in DIN 53206 Part 1. As the existence of several "mean diameters" and other size measures shows, special care is needed when using the term "particle size" [1], [2], [3], [4]. KEIFER [5] and KALUZA [6] have worked on problems of flocculation.

Next to the production of the individual components, the most important operation in the manufacture of lacquers and paints is the dispersing process, and numerous publications have dealt with this topic (see bibliography in KALUZA [7]). Work by von PIGENOT [8] on the tinting strength development as a second-order reaction dates from the early 1960s; this line has been followed up by many other authors [9–13], who also considered the first-order reaction.

The linear dependence of absorption on dye concentration was discovered by Beer (1825–63) [14], whose work came long after Lambert's law was derived (see Section 3.7). The application of Beer's law to light scattering by pigments did not happen until our century; it was contained, for example, in the first edition of JUDD's well-known book [15]. Perturbation terms for the intrinsic scattering of the white or colored pigment being tested, in the form of quadratic terms [16], were initially used for formulation and rematching problems [17]; the first publication of the $\sigma^{2/3}$ perturbation term was in 1967 [18]. The quasi-linear concentration dependence of the Kubelka-Munk function (Chapter 4) is valid over a much broader concentration range than is a logarithmic dependence as suggested by von PIGENOT [8].

It is no longer possible to tell when the consensus was reached that the relative tinting strength of pigments should be determined as the ratio of portions by weight added to a white substrate. An early indication is found in a 1932 paper [19]. For the Kubelka-Munk function of reduced black pigments, the famous work of the discoverers (see Section 3.7) already gives a linear dependence on the black pigment concentration (first approximation; see Section 4.5.2). The ratio of the test and reference Kubelka-Munk functions for equal pigment masses, which is easily derived from their remark, is identical to the zero-th approximation for the tinting strength of black pigments, derived in Chapter 4. For colored pigments, the tinting strength criterion is important. The color matching studies discussed in Chapter 4 were performed by GALL [20] and KEIFER/VÖLZ [21]. Gall's publication derives the new color depth equations based on RAABE and KOCH's earlier equations for dyed textiles [22]. A table of $a(\varphi)$ factors, needed for computer calculations of color depth shades, appeared later [23]. A rational determination of the tinting strength of colored pigments, based on the principle of spectral evaluation (Section 3.6) and thus suitable for any arbitrary tinting strength criterion, was also developed by GALL [24]. SCHMELZER and co-workers have also proposed doing the matching for $B \neq 0$ [25].

The methods formerly used for lightening power have today been largely forgotten, for the most part with good reason (e.g., the REYNOLDS method [26]). This does not apply to the procedure of GRASSMANN and CLAUSEN [27], which uses the fact that the functions $\log F(\varrho)$ and $F(1/m)$ have linear shapes over wide range (see Sections 3.3.1 and 4.5.3 respectively), so that a straight calibration line is possible. This means that a graphical technique [28] can be used; despite being restricted to the Beer's law region, this technique produced good results in many laboratories before the introduction of computers. Another advance is KEIFER's graphical method [29], in which the level of pigmentation is varied and the relative scattering power and shade can be determined simultaneously with the lightening power; this method has been incorporated into standards. A rationalized method for determining the lightening power, suitable for use in present-day laboratories, is described in Chapter 8 [30]. It is based on the equations of Section 4.7.1 for P_w as a function of m. A comprehensive summary of solved and unsolved problems from Chapters 1–4 should also be mentioned [31]. On the determination of dye concentrations, see [32].

References

[1] G. Kämpf, *farbe + lack* **71** (1965) 353.
[2] B. Honigmann, J. Stabenow, VI. *Fatipec*-Congress Wiesbaden 1962, Congress-B., 89, Verlag Chemie Weinheim.
[3] K. Leschonski, Particle Size Analysis and Characterization ov a Classification Process, *Ullmann's Encyclopedia of Industrial Chemistry*, Vol. B2, 5th edit. 1988, VCH Verlagsgesellschaft Weinheim.
[4] T. Allen, *Particle Size measurement*, 3. Ed. 1981, *Chapman & Hall*, London/New York.
[5] S. Keifer, *farbe + lack* **79** (1973) 1161.
[6] U. Kaluza, *farbe + lack* **80** (1974) 404.
[7] U. Kaluza, Physikalische Grundlagen der Pigmentverarbeitung für Lacke und Druckfarben, Edition Lack und Chemie E. Moeller, Filderstadt 1980.
[8] D. v. Pigenot, VII *Fatipec*-Congress Vichy 1964, Congress B. 249, Verlag Chemie Weinheim.
[9] O. J. Schmitz. R. Kroker, P. Pluhar, *farbe + lack* **79** (1973) 733.
[10] U. Zorll, *farbe + lack* **80** (1974) 17.
[11] A. Klaeren, H. G. Völz, *farbe + lack* **81** (1975) 709.
[12] L. Gall, U. Kaluza, *defazet* **29** (1975) 102.
[13] U. Kaluza, XII. *Fatipec*-Congress G.-Partenkirchen 1974, Congress B. 233, Verlag Chemie Weinheim.
[14] A. Beer, *Grundriß des photometrischen Kalküls*, Braunschweig 1854.
[15] D. B. Judd, *Color in business, science and industry*, 1952, *J. Wiley & Sons*, New York.
[16] F. Vial, XII. *Fatipec*-Congr. G.-Partenkirchen 1974, Congr. B. 159.
[17] L. Gall, II. *AIC*-Congress York 1973, Congr. B. 153, *A. Hilger*, London.
[18] H. G. Völz, *Ver. Bunsen-Ges. phys. Chem.* **71** (1967) 326.
[19] Anonym, *Off. Digest Fed. Paint & Varnish* Prod. *Clubs* (1932) 1029.
[20] L. Gall, I. *AIC*-Congress Stockholm 1969, Congr. B. Vol. I, F 6.
[21] S. Keifer, H. G. Völz, *farbe + lack* **83** (1977) 180 und 278.
[22] P. Rabe, O. Koch, *Melliand Textilber.* **38** (1957) 173.
[23] L. Gall, K. Friedrichsen, *Defazet* **28** (1974) 158.
[24] L. Gall, IX. *Fatipec*-Congress Brüssel 1969, Congr. B. Sec. I, S. 34, Verlag Chim. d. peintures.

[25] G. Geißler, H. Schmelzer, W. Müller-Kaul, *farbe + lack* **84** (1978) 139.
[26] H. A. Gardner, G. G. Sward, *Paint testing manual, 12. Ed., 1962, S. 53*, Gardner Lab. Inc., Bethseda (Maryland).
[27] W. Graßmann, H. Clausen, *Deutsche Farben-Z.* **7** (1953) 211.
[28] W. Jaenicke, *Z. Elektrochem.* **60** (1956) 163.
[29] S. Keifer, *farbe + lack* **82** (1976) 811.
[30] H. G. Völz, *farbe + lack* **88** (1982) 356.
[31] H. G. Völz, *Die Farbe* **25** (1976) 147.
[32] G. Boot, H. Zollinger, K. McLaren, W. G. Sharples, Alan Westwell, Dyes, General Survey, *Ullmann's Encyclopedia of Industrial Chemistry*, Vol. 9, 5th edit., 1987, VCH Verlagsgesellschaft Weinheim, p. 73.

5 How Light Scattering and Absorption Depend on the Physics of the Pigment Particle (Corpuscular Theory)

5.1 Introduction

5.1.1 Description and Significance of the Corpuscular Theory

The term "corpuscular theory" refers to model concepts that form a bridge between the Kubelka-Munk coefficients and two physical-optical constants of the pigment particle, the refractive index and the absorption index. The development is centered on the theory of Mie, which links the physical constants, including particle size, with the scattering and absorption properties of the particle. With this theory, we complete the progression discussed in Chapter 1, finally getting from color back to the physics.

Chapter 3 briefly outlined what differentiates the Mie theory from the phenomenological theory, which also makes statements about scattering and absorption. Following up what was stated there, one can say in simple language that while the phenomenological theory knows about material properties but not about particles, the Mie theory knows about the particle – but just one particle. The step from here to the ensemble and on to the material constants will involve other models of corpuscular theory.

It is obvious why the corpuscular theory is of no importance for dye testing, since there are no discrete particles in dye solutions. In fact, however, it is possible to get along without this theory in pigment testing too, at least as far as direct application is concerned. If it is used here anyway – if it must be used – the reason lies in the unique insights permitted by Mie calculations of pigments. These are important for understanding circumstances, properties, and processes having to do with the optics of pigments. Only with its help can we understand why white pigments scatter differently or why otherwise identical colored pigments can exhibit different hues if they differ in particle size.

This causal link between color properties and particle size means that methods utilizing the Mie theory are also interesting for the determination of particle sizes and particle-size distributions. Such methods, however, are not colorimetric but purely spectroscopic; they are based on the spectrum alone and therefore were not supposed to be discussed in this book. On the other hand, the scope of these methods has grown so broad that a description of them could easily fill another book. All the same, the context of this chapter demands a short exposition of particle-size distributions, since the distributions are needed in order to establish the link between the Mie theory and the scattering and absorption coefficients. We will limit ourselves to explaining the basic terms, also leaving out the methods of determination, which can be found discussed in the extensive specialist literature.

Similarly, determination methods for the physical-optical constants can be dealt with only in a cursory and decidedly nonexhaustive way – so that there will be an opportunity to take up the proper subject, colorimetric test methods for coloring materials, and handle it comprehensively.

5.1.2 Particle-Size Distribution

The discussion in this section bears a direct relationship to the descriptions and definitions of Section 4.1.1; we will continue to use the standardized terminology.* The particle-size distribution comes from the statistical analysis of the particle sizes in an ensemble (population). Pigments are manufactured on an industrial high-tonnage basis and thus are never monodisperse. The distribution of a particle-size parameter (chiefly the equivalent diameter) can be described in terms of a particle-size distribution curve. These curves generally depend strongly on the type of dispersing medium, the degree of dispersing, and the method of determination. The wetting power and polarity of the medium, the dispersing work done to disintegrate agglomerates, the tendency to reagglomeration, and other factors play significant roles. For these reasons, the particle-size distribution in most cases must be determined in a different way for each application.

Statistical frequencies of the particle-size parameters defined in Section 4.1.1 (particle diameter, equivalent diameter, particle surface area, particle volume, particle mass and also the number of particles) are plotted versus the parameter of interest or some other one. A preferred type of plot for pigments is the "volume/diameter distribution," the function $v(D)$ obtained when the volume frequency is plotted against the diameter. The reason therefor has already been discussed: the fact that optical pigment properties such as scattering and absorption are functions of the volume.

The following definitions can be stated:

– Density of distribution: the relative frequency of the particle-size parameter in the particle ensemble as a function of this or some other parameter. Especially prominent among particle-size parameters is the equivalent diameter D, but any other measure can be used in its place. Distribution density functions must always be normalized; that is, the total population must yield 100 %

---

* See "Particle-Size Analysis" in Appendix 1.

5.1 Introduction

- Cumulative distribution: the normalized frequency (i.e., the frequency referred to the whole ensemble) summed up to a stated value of the particle-size parameter

- Size fraction and class: the size fraction is the particle ensemble between two fixed values of the selected particle-size parameter that bound the class

- Mean: of the many possible central measures for the particle-size parameter, some are preferred in practice, such as the arithmetic mean, the mode, and the median of the distribution

- Breadth of distribution: a measure characterizing the dispersivity of the particle size in polydisperse populations

Empirical particle-size distributions can be represented by the following means used in practice:

- Tables

- Diagrams, such as histograms (bar charts) or closed curves

- Analytic functions obtained by approximation techniques

As a rule, a diagrammatic representation is recommended as the way to gain an overview and to make a quick comparison between pigments. The frequencies can be plotted by class and represented as areas in a histogram or bar chart, as shown in Figures 5-1 (distribution density) and 5-2 (cumulative distribution). The horizontal axis for the plot of distributions is the particle-size parameter, commonly the equivalent diameter. Because the parameter can be chosen freely, however, the frequencies for the number of particles, particle surface area, or particle volume (mass) can also be graphed versus a different particle-size parameter selected as the independent variable. In general, these curves are not identical for a given pigment.

Special distributions are defined in special standards:*

- Logarithmic normal (lognormal) distribution
- Power-law distribution
- RRSB distribution

The lognormal distribution is particularly important for pigments, at least when it can be generated by a nucleation and grain growth process. While nuclei ordinarily have a normal distribution on creation, the growth rate of a particle

---

* See "Particle-Size Analysis" in Appendix 1.

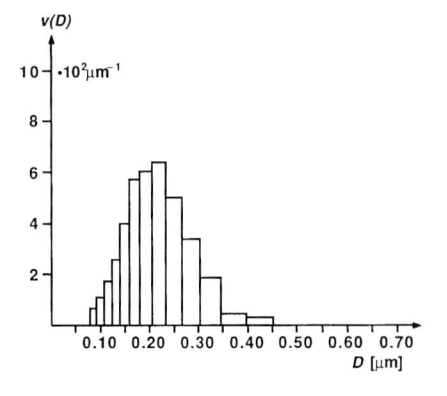

Fig. 5–1. Histogram for volume distribution density of a TiO₂ pigment.

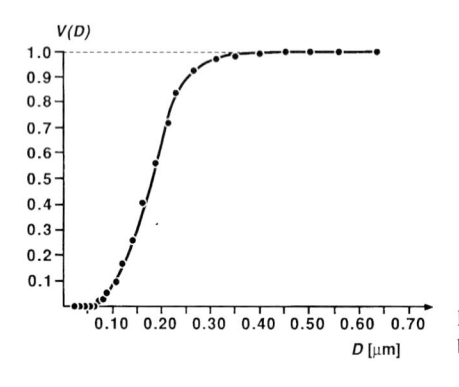

Fig. 5–2. Cumulative curve for the volume distribution of a TiO₂ pigment.

is usually proportional to its surface area. Integration of this growth rule gives a linear time dependence for the log of the particle diameter; substitution in the normal distribution automatically results in the lognormal distribution. A distribution states the probability of a particle being found between values x and $x + dx$ of the particle-size parameter. The equations needed are therefore completely identical to those derived in Section 2.4.1 for the statistical analysis of color coordinates. For example, the series of equations beginning with (2.17) yield the following for the volume/diameter distribution:

$$v(D) = A\, e^{-\frac{1}{2}\delta s^2} \quad \text{with} \quad \delta s = \frac{x - \bar{x}}{\sigma} \tag{5.1}$$

where now $x = \ln D$. The arithmetic mean is $\bar{x} = \ln \overline{D}_z$, which is the logarithm of the median of the distribution. Because the sum of the logarithms between 0 and 1 is exactly as great as the sum between 1 and ∞, this value divides the area under the distribution function exactly in half. The normalization factor in front of the exponential function changes to $A' = A/D$ because $d(\ln D) =$

dD/D. A German standard (DIN 66144) gives a lognormal coordinate grid. If the particle-size distribution function obeys the lognormal law, the cumulative frequency curve graphed on this grid yields a straight line. For details of the mathematical treatment, reference must be made to the specialist literature. As examples of the form of logarithmic normal distribution functions, which (for the reasons cited) can often be used with success to describe pigment distributions, Figure 5-3 shows several curves for various median diameters $\overline{D}_{zv1}$; the breadth of the distribution is the same in each case.

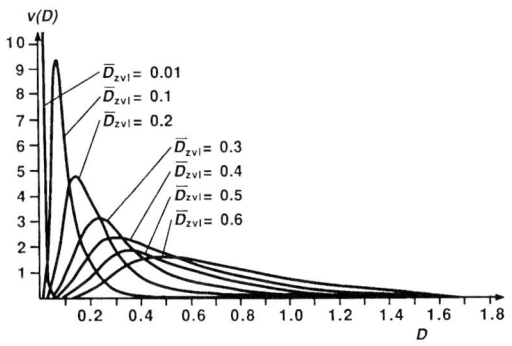

Fig. 5–3. Lognormal distributions. $\overline{D}_{zv1}$ is the median of the lognormal distribution; σ is a parameter characterizing the breadth of the distribution (σ = const = 1.6).

The type of *central measure*, as defined in the standards, will be selected case by case in accordance with the test method used or the type of central measure most pertinent to the pigment property of interest. Depending on the breadth of the distribution, a given ensemble or population can show markedly different values of these central measures.

In particle-size analysis (granulometry), "shape factors" are sometimes used. With these, equivalent diameters obtained by one determination method should be converted to "true" particle-size parameters. But the determination and use of shape factors is problematic; a better way to do the same thing is to use the statistical tool for the distribution of *anisometric* particles. While the frequency for isometric particles can be represented by a one-dimensional distribution curve, for anisometric particles it depends on at least two size parameters (e. g., the length L and the breadth B). Because of the dependence on two variables, a frequency surface $w = w(x, y)$ as shown in Figure 5-4 is obtained. To describe the distribution of, say, acicular pigments, the whole shape of the "hill" must be known, because only in this way is it possible to handle the correlation between L and B.

If the logarithms of L and B are denoted by x and y respectively, the applicable equations are those beginning with equation (2.24). For the same reasons as in the one-dimensional normal distribution case, the means are equal to the logarithms of the medians: $\bar{x} = \ln \overline{L}_z$ and $\bar{y} = \ln \overline{B}_z$; the normalization factor is then $B' = B/(\overline{L}_z \cdot \overline{B}_z)$. Now equation (2.27) with the quadratic form $\delta s^2 = 1$

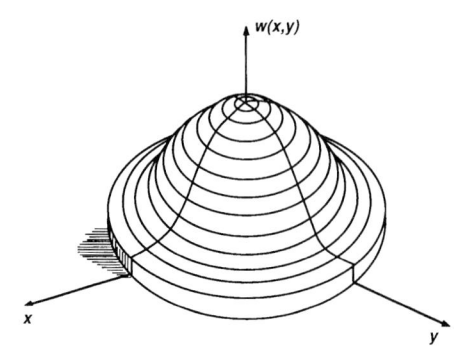

216   5 *How Light Scattering and Absorption Depend on the Physics*

Fig. 5-4. Two-dimensional normal distribution.

represents the "standard-deviation ellipse," which governs the exponents of the lognormal distribution. The family of such ellipses is the adequate form of visualizing the two-dimensional lognormal distribution; their coefficients yield the lengths of the ellipse axes, their slopes, and the eccentricity. At the same time, the coefficients determine the position of the median about which the ellipses of equal cumulative probability are grouped, as well as the maximum value of the probability density, which lies at the center of the constant-height ellipses (loci of equal probability density).

One way to represent such distributions graphically is the column chart (Fig. 5-5), here shown for the example of a yellow iron oxide pigment. The lognormal distribution is characterized by the concentric ellipses obtained when cumulative frequency is plotted to a logarithmic scale (Fig. 5-6); the constant probabilities on the curves increase from 10% to 90% in 10% steps. The eccentricity of the ellipses is a measure of the correlation between the length L and breadth B of the particles. If the eccentricity is 1, there is a rigorous dependence between L and B and the ellipses degenerate to straight lines. If, on the other hand, the eccentricity goes to 0, L and B are independent of one another and the ellipses become circles. The slope of the ellipse axes indicates how rapidly B increases with increasing L.

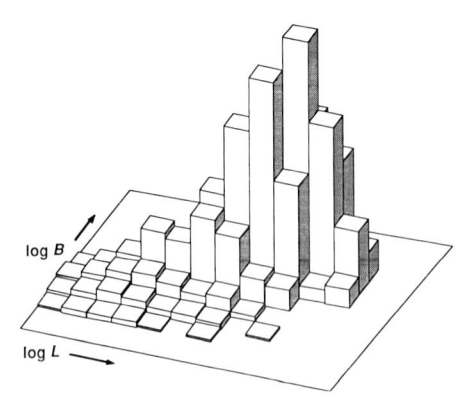

Fig. 5-5. Two-dimensional volume distribution density plotted as log L versus log B (acicular yellow iron oxide pigment).

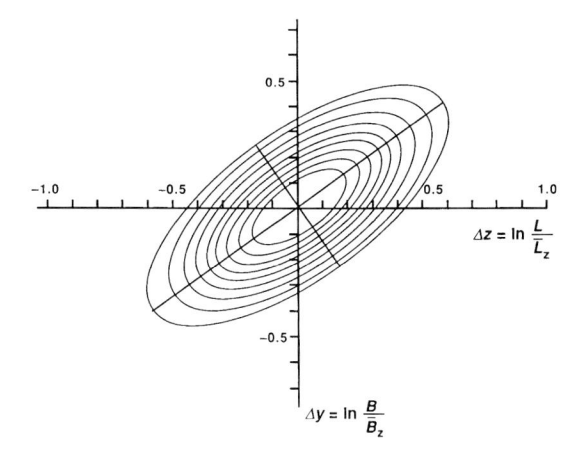

Fig. 5-6. Family of standard deviation ellipses for the two-dimensional log-normal distribution (yellow iron oxide pigment).

This type of mathematical analysis also prepares the way for cases involving more parameters, in which case higher-dimensional normal distributions are obtained.

The *methods of determination* (dispersoid analysis) can be discussed only briefly here. The following are applicable to pigments:

- Particle counting in electron micrographs
- Sedimentation analysis
- Light-scattering analysis
- Analytical techniques based on other physical volume effects such as electrical conductivity.

Particle counting is the most suitable of these methods for pigments. There are several reasons, a few of which are that counting can be done in the vehicle of interest; it can take place in the degree of dispersing that is under study; and concentration problems arise far less often than in other methods, some of which involve extremely high dilutions. The main advantages of counting are that particle shape is determined directly and that primary particles, agglomerates, and aggregates can be distinguished. Once again, further information can be found in the copious specialist literature.

5.1.3 The Optical Constants: Refractive Index and Absorption Index

In all the developments up to this point, light could be treated in a geometrical-optical fashion: as propagating continuously in rays. This model was even adequate for the phenomenological theory. The corpuscular theory, however, involves the interaction of light with matter and the constants of that matter, and so it is necessary to come back to the wave nature of light. Light is electromagnetic radiation, and electromagnetic theory (electrodynamics) is governed by Maxwell's well-known equations, which we need not write down here because they can be found in any textbook.

There are two differential equations for the electric and magnetic field vectors; their combination gives the universal wave equation for the electric field vector $\mathbf{E}(x, y, z, t)$:

$$\Delta \mathbf{E} = \frac{1}{v^2} \frac{\partial^2 \mathbf{E}}{\partial t^2}$$

$$\text{with } \Delta = \frac{\partial^2}{\partial x^2} + \frac{\partial^2}{\partial y^2} + \frac{\partial^2}{\partial z^2} \tag{5.2}$$

is the Laplacian and v is the propagation velocity, cm/s. An enormous variety of phenomena are described by this differential equation; for example, every radio and television program in the world is a solution! Here, one of the simplest cases is to be examined: that of a homogeneous plane wave in a dielectric, with the oscillation in the x-y plane and propagation in the z direction. The solution for the components of $\mathbf{E}$ is

$$E_x = A\, e^{i\omega(t - \frac{n}{c} z)}$$

$$E_y = B\, e^{i\omega(t - \frac{n}{c} z) + i\delta}$$

$$E_z = 0 \tag{5.3}$$

where $\omega = 2\pi v$ is the angular frequency, s^{-1}; v is the frequency, s^{-1}; c is the velocity of light in vacuum, cm/s; A and B are amplitude factors; δ is a phase constant (if there is a phase shift between E_x and E_y); and $n = c/v$ is the refractive index. These equations mark the first appearance in electromagnetic theory of a material constant, the refractive index n, that links the propagation velocity v in the dielectric and the propagation velocity c in vacuum. If damped waves are also allowed, n has to be replaced by the complex refractive index

$$n^* = n(1 - i\varkappa) \tag{5.4}$$

where $\varkappa$ is the absorption index. This damping constant, like the refractive index itself, is dimensionless. How n and $\varkappa$ can be further related to physical and molecular constants is not of interest here.

In connection with absorption and scattering problems involving pigments, however, the experimental determination of these quantities is important. It is necessary to look back at the wave equation in order to explain this point briefly. Suppose one writes this equation for a wavefront incident on the plane surface of a dielectric (x,y plane) at angles φ, ϑ. Elementary wave arguments

5.1 Introduction

lead to the following conditions that must be satisfied by the electric and magnetic field vectors at the interface: for the reflected wave (i = incident, r = reflected)

$$\vartheta_r = \vartheta_i \qquad \text{reflection law} \tag{5.5}$$

and for the transmitted wave (media 1 and 2)

$$\frac{\sin \vartheta_1}{\sin \vartheta_2} = \frac{n_2}{n_1} \qquad \text{refraction law (Snell's law)} \tag{5.6}$$

A boundary condition is needed if one is to answer the question what fraction of the incident radiation is reflected. The condition of continuous amplitudes at the interface is therefore imposed. As a result, the four famous *Fresnel equations* are obtained; here we use only the two that concern reflection:

$$r_\perp = \left[\frac{\sin(\vartheta_1 - \vartheta_2)}{\sin(\vartheta_1 + \vartheta_2)}\right]^2 \quad \text{and} \quad r_\parallel = \left[\frac{\tan(\vartheta_1 - \vartheta_2)}{\tan(\vartheta_1 + \vartheta_2)}\right]^2 \tag{5.7}$$

The reflection coefficient has been analyzed into a component in the plane of incidence ($r_\parallel$) and a component perpendicular to that plane ($r_\perp$). The angles ϑ_1 and ϑ_2 are linked by Snell's refraction law, equation (5.6). These equations were used earlier in Section 3.1.3. Thus, for normal incidence ($\vartheta_1 = 0$), r_o is found very easily (with the approximation $\sin \varepsilon \approx \varepsilon$ for small angles; $n_1 = 1$, $n_2 = n$):

$$r_o \approx \left(\frac{\vartheta_1 - \vartheta_2}{\vartheta_1 + \vartheta_2}\right)^2 = \left(\frac{\frac{\vartheta_1}{\vartheta_2} - 1}{\frac{\vartheta_1}{\vartheta_2} + 1}\right)^2 \approx \left(\frac{n-1}{n+1}\right)^2$$

The other coefficients are found by integrating over the half-space.

If the complex refractive indices are introduced for n_1 and n_2 in equation (5.6), two appropriate measurements have been done in equations (5.7), and n_1 and $\varkappa_1$ are known, then two equations are obtained from which the unknown optical constants $n_2(\lambda)$ and $\varkappa_2(\lambda)$ can be determined. This calculation used to be rather time-consuming, but in the computer age it has lost some of its terror. The reflection measurement should not be done at the air/specimen interface; instead, the specimen should be placed in contact with a material having a higher refractive index. The interfaces for this reflection measurement must be highly planar (i.e., as smooth as possible) and polished surfaces.

These brief remarks have pointed up all the difficulties that stand in the way of the direct determination of n and $\varkappa$ for pigments. The bounding surfaces of pigment particles are in the submicroscopic range, far too small for a reflection measurement – to say nothing of the fact that "good" surfaces (chemically homogeneous and identical with the pigment material) often cannot be obtained at all.

For these reasons, the $n(\lambda)$ and $\varkappa(\lambda)$ values used for pigments often have to be the values measured on single crystals of the same material. This substitution is not without its own problems, since it has been shown for many pigments that the particle surfaces can differ chemically from the compact material. For example, TiO_2 particles in their normal state have their surfaces covered with OH groups, and the same holds for other inorganic pigments. Organic pigments may also display chemical surface alterations resulting from the production process. As the particle size decreases, the area/volume ratio increases in inverse proportion to the diameter, so that the relative area of a particle measuring 0.1 μm is already a factor of 10 greater than that of a 1 μm particle. The chemical nature of the surface thus becomes rapidly more significant as the particle size decreases, with obvious impact on the optical constants. Thus no one up to now has been able to judge the extent to which constants measured on single crystals are also valid for pigments. One would do well to take a skeptical stance toward them and treat results obtained by them (e. g., in Mie calculations) as rough approximations at first.

Another source of trouble is that many pigments are optically anisotropic; that is, their n and $\varkappa$ values depend on direction. The optical constants referred to the crystal axes can differ widely, as Table 5-2 shows. Averaged constants may be characteristic of pigment populations, depending on the directional distribution of the crystallites.

Another technique for direct determinations on fairly compressible pigments is to press the powder into a compact tablet. Because of residual air included in

Table 5–1. Refractive indices and optimal particle sizes for some white pigments and fillers at a wavelength of $\lambda = 550$ nm.

Compound, structure	Mean refractive index in contact with Vacuum	Vehicle	Optimal particle size (in contact with vehicle) D_{opt}, μm
TiO_2, rutile	2.80	1.89	0.19
TiO_2, anatase	2.55	1.72	0.24
α-ZnS, sphalerite	2.37	1.60	0.29
ZrO_2, baddeleyite	2.17	1.47	0.37
ZnO, zincite	2.01	1.36	0.48
Basic lead carbonate	2.01	1.36	0.48
Basic lead sulfate	1.93	1.30	0.57
$BaSO_4$, barite	1.64	1.11	1.60

Table 5–2. Refractive indices of important inorganic colored pigments (after *Landolt-Börnstein*).

Mineral	Chemical formula	λ nm	n n_ω n_α	n_ε n_β	n_γ
Eskolaite	α-Cr$_2$O$_3$	671	2.5 (Note 1)	–	–
Greenockite	CdS	589	2.506	2.529	–
Goethite	α-FeOOH	589	2.275	2.409	2.415
Hematite	α-Fe$_2$O$_3$	686	2.988	2.759	–
Cobalt blue	CoAl$_2$O$_4$	blue red	1.74 > 1.78	– –	– –
Carbon (Note 2)	C	578	1.97	–	–
Crocoite	PbCrO$_4$	671	2.31	2.37	2.66
Magnetite	Fe$_3$O$_4$	589	2.42	–	–
Minium	Pb$_3$O$_4$	671	2.42	–	–
Ultramarine	(Note 3)		1.50	–	–

Note 1. Mean refractive index.
Note 2. Arc carbon.
Note 3. Chemical formula [AlSiO$_4$]$_6$(SO$_4$)(S, Cl)$_2$(Na, Ca)$_8$

the compact, mixing rules for the refractive index must be employed in evaluating the data. The surface chemical effects are thus included, at least as an average, and any anisotropy present is at least partly smoothed out.

Finally, the immersion method should be mentioned. When larger pigment particles are examined under the microscope, at least the real refractive index n can be determined from the point at which the bounding lines of the crystal vanish. Because of optical anisometry, even this method yields only rough approximations. In the case of the more strongly refracting pigments, there is also a lack of suitable highly refractive liquids or low-melting solids. A further hindrance is that the procedures are laborious and costly and, as a rule, give only the refractive index at a random wavelength or an average over the visible range of wavelengths.

In conclusion, it must be stated that exact values for the optical constants of pigments are hardly ever known, even though a knowledge of these values is essential for the development of new products and all analyses of optical performance require them.

5.2 Mie Theory

5.2.1 Integration of the Wave Equation

Our theoretical knowledge about the dependence of scattering and absorption on particle size and pigment refractive index is largely based on results derived from Mie theory. The complex of mathematical and physical knowledge that has been gained with this theory would occupy a whole book of its own. If an attempt is made here to trace the essential lines of thought in the theory, it will be done in quite brief form and under restrictive assumptions. For simplicity, the treatment is done with an eye to one objective, understanding the results of the theory as they apply to pigments.

The model is a dielectric sphere on which a plane wavefront is incident. Application of the wave equation reveals that the sphere is excited into oscillation; as a oscillating multipole, it then acts as the center from which new waves are emitted. These waves are superposed on one another. First, in wave equation (5.2) – including the Laplacian – one must separate the time function and rewrite the equation as a harmonic oscillation for the polar coordinates r, φ, ϑ suitable to the problem. The result is called the Helmholtz equation. Its solutions can be found in textbooks of theoretical physics. Any such solution can be written as a sum over products of three complex-valued functions; for example, the radial component of the electric field vector **E** is given by

$$E_r = \sum_{j=1}^{\infty} A_j(n, \alpha) \, \Theta_j(\vartheta) \, \Phi_j(\varphi)$$

Along with the refractive index n, this includes the typical dimensionless parameter

$$\alpha = \frac{D\pi n_o}{\lambda} \quad (5.8)$$

where D is the particle diameter and n_o is the refractive index (relative to vacuum) of the medium surrounding the particle. The complex-valued functions have the following meanings:

$A(n,\alpha)$: spherical Bessel function (n is the refractive index of the particle relative to the surrounding medium)

$\Theta(\vartheta)$: spherical functions (spherical Legendre polynomials)

$\Phi(\varphi)$: exponential function

5.2 Mie Theory

The summation takes care of the superposition of excited partial vibrations.

In order to get to intensities, one must form the absolute squares of **E**. Two expressions are obtained: $I_\|$ for the ray parallel to the incidence/observation plane and $I_\perp$ for the ray perpendicular to that plane. These are enough to describe unpolarized light. For specified sphere parameters n, α, these functions depend only on ϑ (by the axial symmetry of the system); they can also be taken from tables (see Section 5.3). If the complex refractive index $n^* = n(1 - i\varkappa)$ is formally introduced, the formulas once again include absorption.

The solutions of the Mie theory for small particles show that the entire particle oscillates symmetrically as a dipole, along and opposite to the propagation direction of the incident wave. In the case of larger particles, multipole oscillations also occur; what is more, the secondary waves do not oscillate in phase.

Figure 5-7 shows the first three partial oscillations as field lines on the surface of the sphere. The vibration plane of the light ray generating the waves is in the plane of the drawing, so that the direction of propagation is perpendicular to the drawing plane. This is a symmetry plane of the process, and so the front hemispheres shown must be complemented by the back hemispheres, on which the curves fall into congruent patterns.

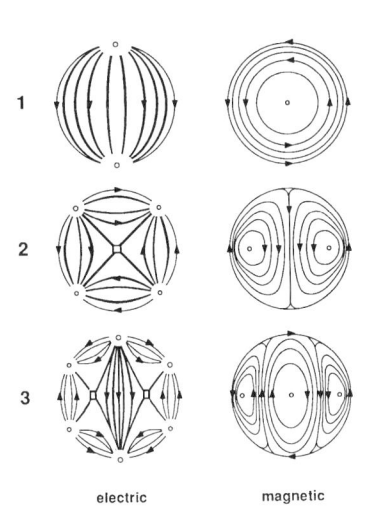

Fig. 5–7. The first three partial oscillation modes of a dielectric sphere after MIE.

Secondary waves outside the sphere are generated by the incident plane wave. For very small particles, the whole particle oscillates as a dipole; the excitation is symmetric, in and opposite to the propagation direction of the incident wave. Larger particles experience multipole oscillations; the secondary waves oscillate in unequal phases, and the superposition of the partial waves yields an increasingly complicated spatial distribution with a growing number of maxima and minima. There is also a decrease in the fraction scattered into the back half-space relative to the front half-space (see Fig. 5-8 for an example).

Similar calculations have been done for cubical, cylindrical, and ellipsoidal particles. These calculations are exceedingly complex, so that approximations are only used in practice. For the anisometric particles under consideration, one obtains a simplified scheme of two absorption and two scattering functions, one each perpendicular and parallel to the longitudinal axis of the particle.

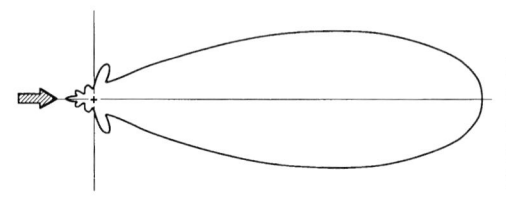

Fig. 5–8. Polar diagram for the emission distribution of a scattering sphere having $n = 1.7$ and $\alpha = 4.3$, after J. THIEMANN. (The center of the scattering particle is to be imagined at the cross+).

5.2.2 Absorption and Scattering of the Particle Ensemble

The solutions $I_{\parallel}$ and $I_{\perp}$ of the Mie theory will now be related to the Kubelka-Munk scattering and absorption coefficients. For it, with $I_{\parallel}$ and $I_{\perp}$, the fractions scattered into the space surrounding the sphere and absorbed in the sphere must be summed. This is done by integrating over the area 4π of the unit sphere* and dividing by the particle cross section $D^2\pi/4$. The integral over φ can be written out immediately ($I_{\parallel}$ and $I_{\perp}$ do not depend on φ):

$$Q = \frac{1}{\frac{D^2\pi}{4}} \cdot \frac{1}{4\pi} \int_{\varphi=0}^{2\pi} d\varphi \int_{\vartheta=0}^{\pi} f(\vartheta)\, (I_{\parallel} + I_{\perp})\, \sin \vartheta\, d\vartheta$$

$$= \frac{2}{D^2\pi} \int_{0}^{\pi} f(\vartheta)\, (I_{\parallel}(n^*, \alpha, \vartheta) + I_{\perp}(n^*, \alpha, \vartheta))\, \sin \vartheta\, d\vartheta$$

In this way, expressions not containing φ and ϑ are obtained: the scattering cross section $Q_S(n,\alpha)$ (real part) and the absorption cross section $Q_A(n^*,\alpha)$ (imaginary part). The function $f(\vartheta)$ takes care of the fact that the Mie theory (directional incidence of light) is not based on the same geometry as the Kubelka-Munk theory (diffuse incidence, plane-parallel layer). For absorption, $f_A = 2$ in agreement with Kubelka's assumption in Section 3.2.2; for scattering, according to MUDGETT and RICHARDS, $f_S = (3/8)\,(1 - \cos \vartheta)$. The quantities Q_A and Q_S are dimensionless; they represent the ratios of the optically effective cross sections to the geometrical cross section:

---

* In the case of scattering, integration over the back hemisphere, i.e., $0 < \theta < \pi/2$, is also recommended.

$$Q_A(n^*, \alpha) = \frac{q_{op}}{q_{geo}} = \frac{q_A}{\frac{D^2 \pi}{4}}$$

$$Q_S(n, \alpha) = \frac{q_S}{\frac{D^2 \pi}{4}}$$

i.e.
$$q_A(\lambda, n, \varkappa, D) = \frac{D^2 \pi}{4} Q_A(\lambda, n^*, \alpha)$$

$$q_S(\lambda, n, D) = \frac{D^2 \pi}{4} Q_S(\lambda, n, \alpha)$$

For practical use, the quantitative reference must be made to the pigment particle, for knowing how much one pigment particle scatters or absorbs is less important than knowing how much 1 L or 1 kg of pigment does. The single-particle quantities k_1, s_1 referred to volume are obtained by dividing by the particle volume $V = D^3\pi/6$ and using equation (5.8):

$$k_1(\lambda, n, \varkappa, D) = \frac{q_A}{V} = \frac{3}{2D} Q_A(\lambda, n^*, \alpha) = \frac{3\pi n_o}{2\lambda} \frac{Q_A(\lambda, n^*, \alpha)}{\alpha}$$

$$s_1(\lambda, n, D) = \frac{q_S}{V} = \frac{3}{2D} Q_S(\lambda, n, \alpha) = \frac{3\pi n_o}{2\lambda} \frac{Q_S(\lambda, n, \alpha)}{\alpha} \quad (5.9)$$

In the case of monodisperse particles, the quotient Q/α is thus proportional to the Kubelka-Munk coefficients.

The spectral Kubelka-Munk coefficients referred to PVC, on the other hand, are generally constants of the material for ensembles of polydisperse distributions. They are thus obtained by summing the contributions over the particle-size distribution $v(D)$:

$$k(\lambda, n, \varkappa) = \int_0^\infty v(D) k_1(\lambda, n, \varkappa, D) \, dD$$

$$s(\lambda, n) = \int_0^\infty v(D) s_1(\lambda, n, D) \, dD \quad (5.10)$$

In the case of distributions of *anisometric* particles $v(L, B)$, the angular distributions must also be known if one wishes to use a sum formula on the above pattern. If a random distribution (i.e., no preferred direction) is assumed, $v(L, B)$ can be written, similarly to the above case, provided the scattering and absorption cross sections parallel and perpendicular to the axis of the acicular or platy particle are known (see Section 5.2.1).

Two important special cases should be examined: those where the particles are very small and very large relative to the wavelength. Approximations will be obtained in both cases. For *absorption*, a fairly simple summary can be given: Thus, for particles smaller than a certain limiting size, absorption is independent of D; for large particles, absorption falls off as $1/D$. An explanation of this behavior will follow in Section 5.2.3.

The effect of particle size is somewhat more complicated for *scattering*. For large particles, scattering also decreases as $1/D$ (Fresnel reflection), since the sphere acts as a reflecting mirror. We have

$$s_1 = \frac{3}{2D} Q_S \quad \text{where} \quad Q_S = r_o = \left(\frac{n-1}{n+1}\right)^2$$

In the small-particle case, Rayleigh scattering is obtained. The small sphere is excited only to dipole oscillations. The result is

$$Q_S = \frac{8}{3} \alpha^4 M^2 \quad \text{where} \quad M = \frac{n^2 - 1}{n^2 + 2}$$

Here we are in the realm of molecular scattering, where the scattering cross section decreases as* $\alpha^4 \sim 1/\lambda^4$. This relation explains, say, why the sky is blue: The molecules of the atmosphere scatter violet light ($\lambda = 400$ nm) more strongly than red ($\lambda = 700$ nm) by a factor of $(700/400)^4 \approx 10$. The sky is blue, rather than violet, because the solar spectrum contains far less violet than blue and because the sensitivity of the human eye rapidly falls off toward zero as the wavelength becomes shorter.

Rayleigh scattering ceases to be important in the particle-size range of pigments, because Mie scattering dominates it by some orders of magnitude.

5.2.3 Scattering Behavior of Pigments

Using the behavioral rules outlined above, van de HULST gave an equation for nonabsorbing spheres and for $n = 1 \pm \varepsilon$:

$$Q_{\text{ext}} = 2 - \frac{4}{\beta} \sin \beta + \frac{4}{\beta^2} (1 - \cos \beta)$$

where $\beta = 2 \alpha | n - 1 |$ (5.11)

---

* In this text, $\sim$ denotes proportionality.

where the extinction cross section – the flux scattered away into all of space related to the geometrical cross section – is described as a function of α and n. This is one of the most important formulas in the whole field of Mie theory, for it describes the essential features of scattering. It is valid not just for n close to 1 but even for $n = 2$. This can be seen from the curves of Figure 5-9, which shows scattering curves for refractive indices $0.8 < n < 1.5$. The β scale at bottom applies to all five curves; there is a different α scale for each n.

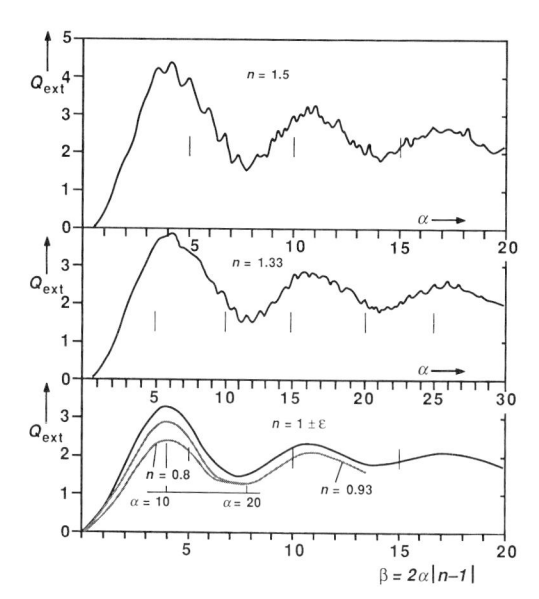

Fig. 5–9. Total scattering cross section Q_{ext} calculated by MIE theory for $n = 1.5$, 1.33, 0.93, and 0.8.

The most striking fact is the succession of maxima and minima. These extreme values result from interference between the refracted and transmitted light; maxima occur where the superposed waves reinforce, minima where they cancel. For small particles, the curves ought to begin as α^4 (Rayleigh scattering), but equation (5.11) does not properly reflect this fact. The ordinates Q_{ext} increase systematically for larger n because of grazing reflection. From the van de HULST calculated values, linear regression analysis yields the relation

$$Q_{max} = 2.716\, n + 0.3235 \tag{5.12}$$

The first minimum ("first Mie minimum") can still be clearly measured for isodisperse populations (e. g., latexes).

The existence of the principal maximum immediately implies that an optimal value α must exist. Hence it follows that

– There must be a particle size optimal with respect to lightening power

5 How Light Scattering and Absorption Depend on the Physics

- For a given particle size, there exists a wavelength at which the scattering takes on maximal values

Because what is wanted is a proportionality to the Kubelka-Munk PVC-referred coefficients s_1 (eq. 5.9), only scattering into the back half-space is to be counted. Semiquantitatively, this fraction can be set roughly proportional to Q_{ext} for a constant refractive index. Then, from the maximal value $\beta_{opt} \approx 4$, Figure 5-9, immediately gives

$$\beta_{opt} = 2\alpha(n-1) = 2\frac{D\pi n_o}{\lambda}(n-1) = 4 \quad (5.13)$$

If the refractive index is given, an expression can be obtained for the optimal α_{opt}; if λ is given, the optimal particle size can thus be found. In the case of TiO_2 in rutile form ($n = n_{TiO_2}/n_{medium} = 2.8/1.5 \approx 1.9$), for $\lambda = 550$ nm we obtain

$$D_{opt} = \frac{\beta_{opt}\lambda}{2\pi n_o(n-1)} = \frac{4 \cdot 0.55}{2\pi \cdot 1.5 \cdot 0.9} \approx 0.26 \text{ μm} \quad (5.14)$$

which is fairly close to the experimental value (see Table 5-1). Improved values are obtained if β_{opt} is taken not at the maximum of Q_{ext} but at the maximum of Q_{ext}/α as suggested by equation (5.9). The division shifts the optimum to somewhat smaller values, giving $\beta_{opt} = 2.93$, resulting in a D_{opt} of 0.19 μm for rutile. Figure 5-10 presents scattering curves of Q_{ext}/α for well-known white pigments, calculated with van de Hulst formula (5.11). As the refractive index decreases, the optimal particle size, typically, grows larger and less scattering takes place. Finally, when the refractive index approaches that of the vehicle, Q_{ext}/α decreases as $1/\alpha$, the behavior seen for Fresnel reflection, and vanishes. The optimal particle sizes for $\beta_{opt} = 2.93$ were calculated with equation (5.13) and are listed in Table 5-1.

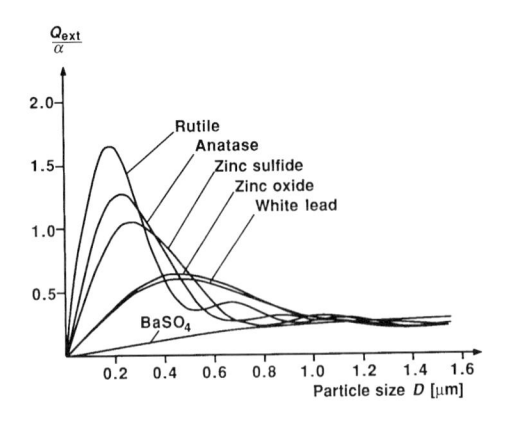

Fig. 5–10. Scattering curves (plots of Q_{ext}/α versus particle size) for some white pigments and fillers.

The wavelength dependence of scattering can be determined with equation (5.11); the curve for $n = 1 \pm \varepsilon$ (Fig. 5-9) can also be used in a semiquantitative analysis. The plot in terms of $\alpha \sim 1/\lambda$ must be read from right to left in order to trace what happens as the wavelength increases. The presence of the maximum indicates that particles smaller than D_{opt} preferentially scatter short-wavelength light while those larger than D_{opt} preferentially scatter long-wavelength light. This situation explains how white pigments in gray pastes can produce a colored shade through selective scattering: Finely-divided white pigments give bluish hues while larger ones give brownish hues.

5.2.4 Absorption Behavior of Pigments

Van de HULST also gives an equation for the absorption, but we need not discuss this because, fortunately, BROCKES has done systematic Mie calculations for fixed n and $\varkappa$. His curves (Fig. 5-11) are log-log plots of Q_A/α versus α for a range of n and $n\varkappa$. They imply the following:

- For very small particles, absorption is independent of α and thus of particle size. In other words, absorption will not increase on further size reduction. As the absorption index grows larger, absorption increases

- For very large particles, absorption is approximately the same for all refractive indices and absorption indices, falling off hyperbolically. The particle absorbs and reflects in accordance with its cross section, so that Q_A/α decreases as $1/D$

- The uppermost curve in Figure 5-11 is for pigments with high absorption index $\varkappa$ but low refractive index n, such as carbon black pigments. It shows that the optimal particle size is less than an identifiable limit

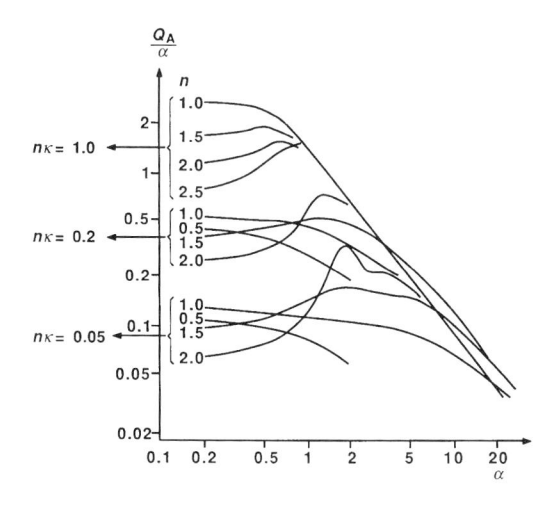

Fig. 5–11. Absorption curves of Q_A/α (log-log plot), after BROCKES.

– The lowermost curve is for pigments with small absorption index $\varkappa$ but high refractive index n, such as most inorganic pigments. This curve has distinct maxima, which should be thought of as higher and steeper than the log-log plot shows

One other important point should be mentioned because it is overlooked many times. At the positions $\varkappa(\lambda_o)$ of absorptions in the spectrum, the refractive index behaves in an anomalous way as described by the well-known dispersion formula

$$n - 1 \sim \frac{\varkappa(\lambda_o)}{1 - \left(\dfrac{\lambda_o}{\lambda}\right)^2}$$

Figure 5-12 presents an example of the refractive index discontinuity for real pigments that exhibit absorption peaks of finite breath. Because the scattering shows a direct dependence on n, it also has a discontinuity. The rule is that the refractive index drops abruptly on the rising edge of the absorption and rises abruptly on the falling edge. This behavior of the scattering characteristic may well lead to remarkable color effects.

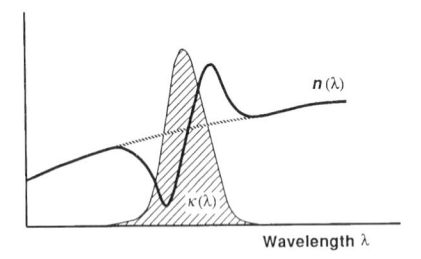

Fig. 5–12. Absorption and anomalous refractive index.

There are cases of pigments in which absorption and scattering are antagonistic and others in which they vary in the same sense. The red iron oxide pigments can serve as an example of the second type. For particles smaller than D_{opt}, equation (5.13) says that λ_{max} must also become shorter, if one would hold the maximum. This means that the position of maximal scattering shifts toward shorter wavelengths: The red takes on a yellow cast. The absorption maximum shifts at the same time, also to shorter wavelengths, and this effect strengthens the yellow cast. For particles larger than D_{opt}, yellow is taken away in both ways, and the pigment becomes redder. In pigment reduction with TiO_2, such coarse pigments are actually seen to take on a marked blue cast. Two factors can account for this effect: First, the absorption of the red pigment moves toward longer wavelengths, thus setting blue light free; second, the scattering increases suddenly because of the anomalous dispersion of the TiO_2 pigment, whose strong absorption ends near 400 nm.

The Mie theory has made it possible to predict the reflection spectrum of a pigment, and thus its color properties, given the complex refractive index, the concentration, and the particle-size distribution.

We have now reached the goal stated in Chapter 1 by tracing the amazing range of optical qualities seen in coloring materials – their technical color properties – back to measurable physical quantities.

5.3 List of Symbols Used in Formulas

$\alpha = D\pi n_o/\lambda$	parameter in the Mie theory
A, B	amplitude factors
$A(n, \alpha)$	spherical Bessel function
$\beta = 2\alpha\|n-1\|$	parameter in the van de Hulst equation
B	normalization factor for two-dimensional normal distribution
B	breadth of anisometric particle, μm
$\overline{B}_z$	median particle breadth, μm
c	velocity of light in vacuum, cm/s
δ	phase constant
δs^2	quadratic form of normal distributions
D	particle diameter, μm
$\overline{D}_z$	median particle diameter, μm
$\mathbf{E}$	electric field vector
I	radiance factor, W cm^{-2}
k	absorption coefficient per unit PVC, mm^{-1}
k_1	single-particle absorption coefficient per unit PVC, mm^{-1}
$\varkappa$	absorption index
λ	wavelength, nm
L	length of anisometric particle, μm
$\overline{L}_z$	median particle length, μm
M	abbreviated notation: $M = (n^2 - 1)/(n^2 + 2)$
n	(real) refractive index
	with subscript: 0 of the medium
$n^* = n(1 - i\varkappa)$	
	complex refractive index
$\nu = c/v$	frequency, s^{-1}
$\omega = 2\pi\nu$	angular frequency, s^{-1}
φ	meridian angle in spherical coordinates
$\Phi(\varphi)$	exponential function
q	optically effective cross section, cm^2
	with subscripts:
	A absorption cross section
	ext total scattering cross section (nonabsorbing sphere)
	S scattering cross section
Q	effective cross section with same superscripts as q

r	reflection coefficient
r	distance from origin in spherical coordinates
s	scattering coefficient per unit PVC, mm^{-1}
s_1	single-particle scattering coefficient per unit PVC, mm^{-1}
σ	standard deviation, a parameter of distribution breadth
ϑ	polar angle in spherical coordinates
$\Theta(\vartheta)$	spherical function
v	propagation velocity, cm/s
$v(D)$	volume distribution of diameter
x	general particle-size parameter

5.4 Summary

The corpuscular theory finally relates the material constants for absorption and scattering to the individual particle and its physical properties. At the center of the analysis is the M$_{\text{IE}}$ theory, which derives scattering and absorption properties from a few fundamental physical quantities. It is therefore important only for pigments, not for dyes, although even for pigments it yields insights that could not be received in any other way. The basic physical quantities are particle size, refractive index, and absorption index.

Because a pigment ensemble comprises fractions differing in particle size, the statistics – the particle-size distribution – must here be treated. The optical scattering and absorption properties are volume-dependent, so that the preferred way of representing these data is the distribution density of volume versus diameter: $v(D)$. The following concepts are discussed: distribution density, cumulative distribution, size fraction and class, measures of central tendency, and measures of distribution breadth. Representations include tables, diagrams, and analytic functions. When unlike pigments are to be compared, plots of distribution density and cumulative distribution are particularly useful. When distributions are to be described by analytic functions, the logarithmic normal (lognormal), power-law, and RRSB distributions can be used; of these, the lognormal is especially well-suited to pigment distributions. In this distribution, the central measure is the median. For describing anisometric (e. g., acicular or platy) particles, the tools of mathematical statistics are also available. The two-dimensional lognormal distribution of particle length L and breadth B also yields a graphical representation and makes it possible to calculate the parameters and central measures. The eccentricity of the calculated standard-deviation ellipse is a measure of the correlation between length and breadth. The principle can be extended when more than two particle-size parameters are needed. This chapter gave only a brief discussion of methods used to determine the particle-size distribution.

The phenomena treated here require for the first time that the wave nature of light must be considered. From the differential equation that describes oscillation processes (wave equation), the homogeneous plane-wave solution is de-

5.4 Summary

rived; it contains the refractive index n as a constant. For absorption, there is also a damping constant $\varkappa$; the complex refractive index n^* is formed by $\varkappa$ together with n. Reflection and refraction laws are effortlessly derived from the wave equation by postulating continuity at the interface between two media. An attempt to obtain the fractions reflected and transmitted leads to the Fresnel equations, with whose aid n and $\varkappa$ can be determined from reflection measurements. Because of the small size of pigment particles, unfortunately, such direct measurements are exceedingly difficult. In practice, it is necessary to employ physical constants measured on large single crystals instead, and ultimately this substitution is not satisfactory. What is more, the optical anisotropy of most pigments makes it hard to determine and use the constants. Reflection measurements to obtain the constants can be done on pressed pigment powders if the material is sufficiently compressible. The immersion method has sometimes been used to determine n. At present, there are just a few values of the optical constants directly measured on the pigment.

The MIE theory, which uses a model of a plane wave incident on a dielectric sphere, is discussed as it aids in understanding pigment phenomena. The theory of differential equations gives solutions containing $\alpha = D\pi n_o/\lambda$ as a typical parameter. These are also functions of the angles φ and ϑ in the three-dimensional coordinate system. After introducing the complex refractive index, one can split the solutions into scattering and absorption parts. The results show that the spherical particle acts as an emission center for new waves that interfere with one another. Even anisometric particles have been calculated with the MIE theory.

The procedure for linking the radiance factors in Mie theory to the Kubelka-Munk coefficients involves first integrating over the solid angle (giving scattering cross section and absorption cross section). The result is divided by the sphere volume and then summed over the particle-size distribution. The same technique can be used with anisometric particles if the angle-dependent cross section values are known. Rayleigh scattering on very small particles (not important for pigments) and Fresnel reflection on very large particles are discussed.

A formula stated by van de HULST is examined as a way of describing how the scattering cross section depends on the refractive index and the parameter α. The result displays maxima and minima, which can be attributed to interference between the partial oscillations. The principal maximum shows that there must be optimal particle sizes, or optimal wavelengths, for scattering. From the position of the principal maximum, the optimal particle sizes for well-known white pigments and fillers are derived along with the scattering curves. This procedure provides an explanation for the shade of white pigments in gray pastes.

The laws describing the absorption of inorganic and organic pigments are obtained from Mie calculations performed by BROCKES. Once again, maxima appear. Anomalous dispersion, seen at absorption edges, also has some effect. The results also account for the hue shift in iron oxide pigments when the particle size changes. The MIE theory makes it possible to calculate in advance the color properties of pigments.

5.5 Historical and Bibliographical Notes

The significance of particle size to the pigment properties has already been referred to, as have the various particle-size parameters; see [1], [2], [3], and [4] in Section 4.10. These prior publications also deal with the parameters and representations suitable for particle-size distributions. A special analysis of the logarithmic normal distribution appears in [1]. The mathematical treatment and representation of distributions of anisometric particles was first published just a few years ago [2].

The MIE theory [3], dating from 1908, is based on the principles of classical electrodynamics, which has been covered in textbooks [4], [5]. MIE devised this theory as a physical explanation of the colors displayed by gold sols having various particle sizes; it was a complete success. The application to very small particles is due to RAYLEIGH [6]. Manual Mie calculations used to be very tedious because of the summations that had to be done, but later workers [7, 8, 9] have performed calculations for a range of refractive indices; one much-used set of tables is due to LOWAN [10]. Nonspherical particles such as cubes and ellipsoids have also been calculated [11]. Although there were no calculations for refractive indices in the range of $1.6 < n < 1.9$, the very range of interest for pigments, an attempt was made at an early point to generalize the results into a single universal scattering function, with the object of predicting optimal particle sizes for the most important pigments. One such procedure is described by JAENICKE ([26] in Chapter 4). The approach taken in this book to determining the optimal particle size with respect to scattering was devised by van de HULST, whose monograph *Light Scattering by Small Particles* [12] is the best treatment of this area. KERKER's book on scattering [13] should also be cited, as should the book by KORTÜM ([19] in Chapter 3).

Computers have greatly simplified the process of doing Mie calculations; this capability has been heavily used, for example by DETTMAR [14] for higher refractive indices. DETTMAR also performed the calculations for iron oxide pigments cited by BROSSMANN [15] and those for the absorption of pigment particles, which were used by BROCKES [16] in the publication discussed in Section 5.2.4. The application of Mie calculations to practical problems has been reported by MAIKOWSKI [17], HAUSER and HONIGMANN [18], and HEDIGER and PAULI [19]. An interesting test is the one carried out by FELDER [20]: Spherical pigment particles made of a soft material (dyed PVA) were deformed into ellipsoidal particles by stretching the medium in which they were embedded. The results revealed how absorption changes on the transition from spheres to ellipsoids.

The literature on methods for finding the optical constants of pigments by the immersion method is rather sparse; the same can be said for measurements done on colored pigments (see, e.g., [21]). In methods based on pressed powder measurements, the presence of residual air in the compact means that refractive index mixing rules must be employed [22]. Given two reflection measurements, the Fresnel equations simultaneously yield n and $\varkappa$, as NASSENSTEIN has shown [23]. Unfortunately, this technique applies only to easily compressible pigments, a class that does not include inorganic pigments. Some work remains to be done in this direction.

References

[1] J. Aitchinson, J. A. C. Brown, *The Lognormal Distribution*, Cambridge University Press, London, 5. Reprint 1973.
[2] H. G. Völz, *farbe + lacke* **90** (1984) 642 und 752.
[3] G. Mie, *Ann. Physik* **25** (1908) 377.
[4] G. Joos, *Lehrbuch der Theoretischen Physik*, Geest & Portig, 7. edit., Leipzig 1950.
[5] J. Honerkamp, H. Römer, *Grundlagen der klassischen Theoretischen Physik*, Springer, New York 1986.
[6] J. W. Rayleigh, *Proc. Roy. Soc., Ser. A*, **90** (1914) 219.
[7] H. Blumer, *Z. Physik* **32** (1925); **38** (1926) 920; **39** (1926) 195.
[8] E. J. Meehan, H. Beattle, *J. Opt. Soc. Am.* **49** (1959) 735.
[9] R. D. Murley, *J. Phys. Chem.* **64** (1960) 161.
[10] A. N. Lowan, Tables of scattering functions for spherical particles, *Nat. Bur. Stand., Appl. Math. Ser.* 4 (1949), US Gov. Print. Off. Washington 1948.
[11] R. Gans, *Ann. Physik* **37** (1912) 881.
[12] H. C. van de Hulst, *Light scattering by small particles*, J. Wiley & Sons, London 1957.
[13] M. Kerker, *The scattering of light*, Academic Press, New York etc. 1969.
[14] H. J. Metz, H.-K. Dettmar, *Kolloid-Z.* **192** (1963) 107.
[15] R. Broßmann, *AIC* "Color 69" Stockholm, Congr. B. Vol. 2, 15.
[16] A. Brockes, *Optik* **21** (1964) 550.
[17] M. A. Maikowski, *Z. Elektrochem.* **71** (1967) 313.
[18] P. Hauser, B. Honigmann, *farbe + lack* **81** (1975) 1005.
[19] H. Hediger, H. Pauli, XIX. *Fatipec* Congress Aachen 1988, Congr. B. II, 503.
[20] B. Felder, *Helv. Chim. Act.* **49** (1966) 440.
[21] V. Koutnik, *Chem. Prumysl (Chem. Ind. CSSR)* **11** (1964) 604.
[22] W. Heller, *Gov. Res. Rep. US Opt. Commerce*, Nr. AD 608 337 (1964).
[23] H. Nassenstein, *Int. Color Congress*, Luzern 1965, Congr. B 12.

Part II
Test Methods

6 Measurement and Evaluation of Object Colors

6.1 Reflection and Transmission Measurement

6.1.1 Gloss Measurement and Assessment

When light falls on a body, part of it is generally reflected at the surface of the body, even before the absorption and scattering processes described in Chapter 1 can take place in the interior of the layer (Fig. 1-1). The fraction reflected at the surface, as described by the Fresnel equations, is a function of the refractive index n and the angle of incidence. For perpendicular irradiation, the reflected fraction is given by $r_0 = (n - 1)^2/(n + 1)^2$; for paint and lacquer films with refractive indices of $n \approx 1.5$, this formula gives $r_0 \approx 0.04$. The fraction increases very slowly with increasing angle of incidence and attains high values only at grazing incidence, that is, at incidence angles just below 90° (Fig. 6-1).

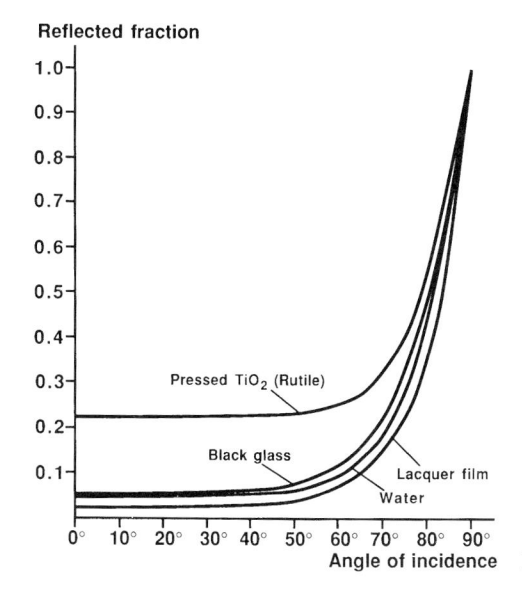

Fig. 6–1. Fresnel reflection at some surfaces.

If the fraction reflected at the surface is plotted as an indicatrix, that is, as a function of the observation angle ε_2 in polar coordinates, two extreme cases can be identified:

- Directional reflection only (Fig. 6-2)
- Diffuse reflection only (Fig. 6-3)

The first case corresponds to regular specular reflection; in the second case, numerous microscopically small irregularities at the surface give rise to backscattering independent of the angle. The indicatrix here obeys the Lambert cosine law: $r = \cos \vartheta$. Real paint and lacquer coatings commonly exhibit a mixture of the two extreme behaviors.

It should be emphasized that this reflected light is no longer available to take part in absorption and scattering processes in the coating. As a consequence, the effect of surface reflection must be properly considered in color evaluation. We will hear more on this point in Section 6.1.2.

Here, what is of interest about surface reflection is whether it is directional or contains directional components. One speaks of the "gloss" of a surface and distinguishes between "high gloss" and "mat gloss" (Fig. 6-4). These terms do not only refer to the purely physical phenomenon of surface reflection but rather to a visual experience. It follows that gloss has a physiological component plus a psychological one.

Studies of the way in which observers evaluate gloss have identified at least two types of observers:

- The "analyzing" observer, who considers the sharpness with which reflected objects (e. g., window cross) are imaged
- The "integrating" observer, who considers intensity contrasts of the reflected light; for example, the "Golden Helmet" (formerly attributed to REMBRANDT) gives the illusion of gloss without reflective surfaces, simply through the juxtaposition of light and dark.

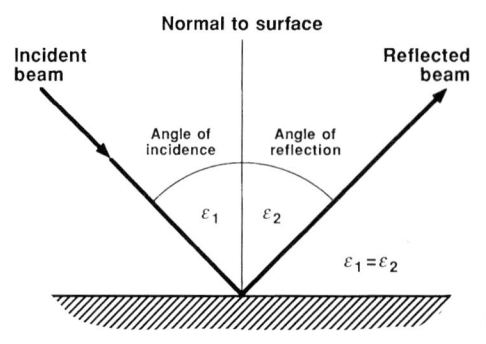

Fig. 6–2. Specular (directional) reflection.

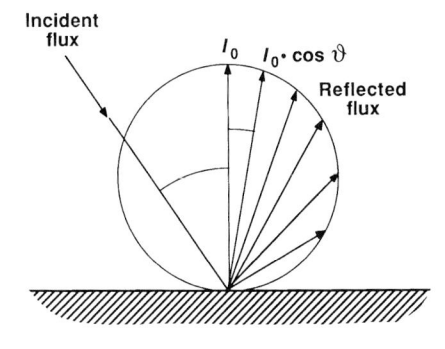

6.1 Reflection and Transmission Measurement 241

Fig. 6–3. Diffuse reflection.

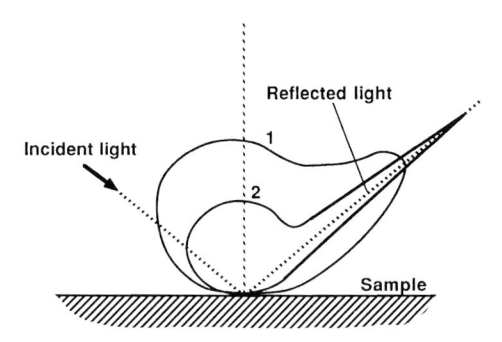

Fig. 6–4. Indicatrices for a nearly mat surface (1) and a high-gloss surface (2).

In visual gloss evaluation, as a rule, the evaluator includes varying proportions of both observers.

It is immediately clear that the observer behavior just described can scarcely be imitated by any gloss measurement technique with a purely physical basis. The situation is similar to that in colorimetry, though in that instance the dilemma could be resolved by postulating a standard observer and basing definitions on matching judgments. It is not entirely out of the question that a similar approach might work in gloss evaluation; it is, however, made difficult by the two modes of visual assessment and the accidental mixing of these modes.

In view of these circumstances, the gloss has to be determined as a function of observation angle in the great majority of cases. The preferred type of plot is the polar gloss indicatrix, as shown in Figure 6-4, but a bell-shaped curve in Cartesian coordinates is quite usual. Such a distribution curve is analogous to the spectral distribution in colorimetry (Chapter 2); the fact that no "standard gloss observer" exists here stands in the way of a more physiologically-oriented treatment of the measured values. The complete gloss curve is measured with a gonioreflectometer (goniophotometer), an instrument in which the angle of incidence can be set to any value and a probe (photocell) is placed on the observation side (Fig. 6-5). The probe is traversed across the cone of scattered light as the intensity distribution is measured and plotted. The resulting curve (indica-

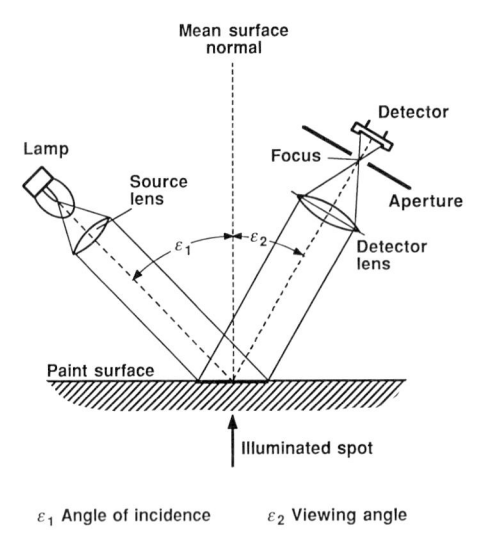

Fig. 6–5. Diagram of the reflectometer with ray paths.

trix) can often be described in terms of a few parameters, chiefly the maximum value and half-value width.

The proper functioning of a goniophotometer imposes certain requirements on construction, chiefly relating to the parallelism of the ray path and the aperture angle on the observing side. These essential details can be found in standards.* The same standards also describe simplified glossmeters in which definite aperture angles are set for three distinct incidence angles (20°, 60°, 85°). The viewing angle should be equal to the angle of incidence and must not differ from it by more than ±0.1°. The incidence angles are recommended as follows for best differentiation in various applications:

- 20° for high-gloss specimens
- 60° for specimens with moderate gloss
- 85° for mat specimens

Suitable primary standards are plane, polished plates of black glass having a refractive index of 1.567 for a wavelength of 589 nm. For each of the three incidence angles, the reflectometer reading for this standard is set equal to 100. If the refractive index differs from the stated value, the reflectometer reading for

---

* See "Gloss, Measurement" in Appendix 1.

6.1 Reflection and Transmission Measurement

each of the three angles can be calculated with the Fresnel equations (which are stated in DIN 67530).

Two gloss-related effects due to pigmentation should be mentioned here. First, it is often observed that the gloss initially decreases as the pigment concentration increases; at higher concentrations, particularly just before the critical pigment volume concentration is reached, the gloss may increase again. The cause of this phenomenon lies in the fact that the pigment particles become more and more ordered as the critical condition is approached; at the critical concentration they take on a close-packed configuration with binder filling the voids. This change is accompanied by smoothing of the surface, which was irregular at low concentrations because of individual pigment particles protruding from it. Second, when the mirror image of a very bright, small source of light in such a paint film is examined, a halo is seen around the image of the source ("gloss haze"). An incandescent filament thus appears as if viewed through a thin veil; that is, diffuse reflection is more intense in directions near the specular angle than far from it. Gloss haze is due to microscopic irregularities in the outermost layers of the surface, which may be present because, for example, inadequate pigment dispersing has left pigment agglomerates behind. The effect is easily measured with the goniophotometer by making the aperture angle as small as possible and traversing the measuring probe over a few degrees about the specular angle.

Gloss effects with metal-effect paints are also caused by pigments (metal-effect pigments). Because the optical phenomena on such coatings have to be characterized by color measurement, they will not be described until later (Section 6.2.3).

The term "gloss development" denotes the variation of the degree of gloss as a function of the dispersing time. As a rule, gloss development is a way of gaining information about the progress of the dispersion process (Section 4.1.2). The measure used, as set forth in ISO 8781–3, is the dispersing work required to attain a given reflectometer value. Alkyd resin systems* and alkyl/melamine resin systems** can be used as test media. These test methods, similarly to those for tinting strength development (Sections 4.7.3 and 8.2.4), makes it possible to compare pigments, mill base compositions dispersing equipment, or dispersing processes.

Preparation method: The test and reference pigments are dispersed in a suitable vehicle in accordance with suitable conditions (vehicle and conditions are to be agreed on). As the dispersing process continues, samples are taken from the batch and applied to a substrate; after drying or stoving, reflectometer values are measured.**

---

* See "Ease of Dispersion" in Appendix 1.
** See "Gloss Measurement" in Appendix 1.

Evaluation method: A diagram is prepared showing the reflectometer value (ordinate) versus the dispersing work (abscissa), that is, the dispersing time or stage. A curve is fitted to the points and, from the curve, the dispersing work required to attain a pre-established reflectometer value is read off.

6.1.2 Measurement and Evaluation Conditions

We now turn back to the discussion in Section 2.1.2, where it was shown what geometrical configurations are used in the measurement of reflectance, reflection factor, transmittance, and transmission factor. Now we will explain how these geometries can be realized in instruments; it will be seen that the geometrical conditions of irradiation and observation, the "measuring conditions," play a vital part.*

We begin with the 45/0 geometry, as illustrated in Figure 6-6. The source aperture angle should be no larger than 20°, the receptor aperture angle no larger than 10°. For specimens with surface texture, the reflection factor R_c depends on the orientation of the specimen relative to the direction of incidence; the predominant texture direction should be parallel to the plane of incidence. The figure also shows that the rays are not directional but conical (quasi-parallel).

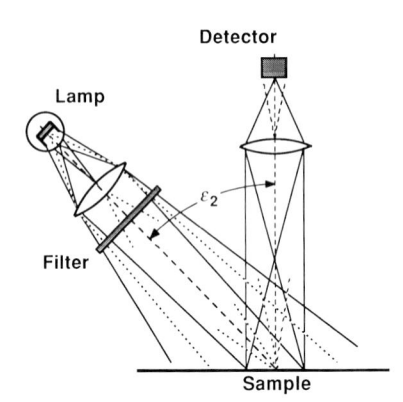

Fig. 6–6. The 45/0 geometry.

In geometries with diffuse irradiation or diffuse observation, an integrating sphere is used to create a hemispherical distribution of flux. Figure 6-7, for example, shows the use of the integrating sphere for the 8/d geometry (reflectance ϱ_c). If incidence on the specimen is diffuse (hemispherical), the setup of Figure 6-8 is employed, giving the d/8 geometry (reflection factor R_{dif}). Here

---
*See "Colorimetry" in Appendix 1.

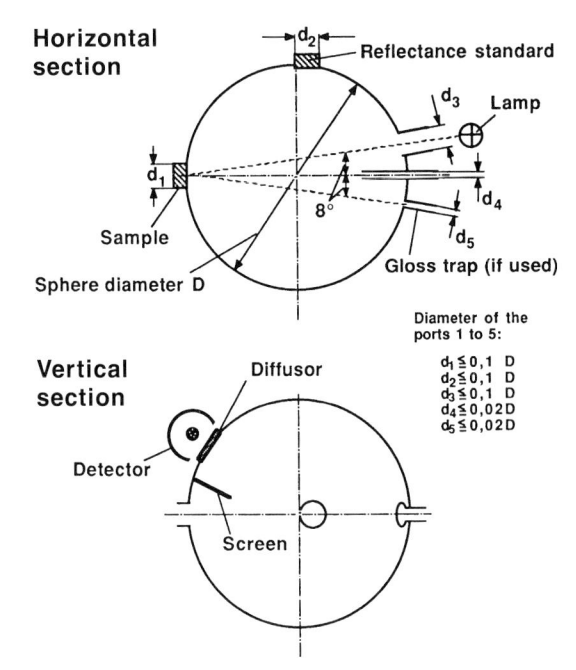

6.1 Reflection and Transmission Measurement   245

Fig. 6-7. Integrating sphere used in measurement of ϱ_c (8/d geometry).

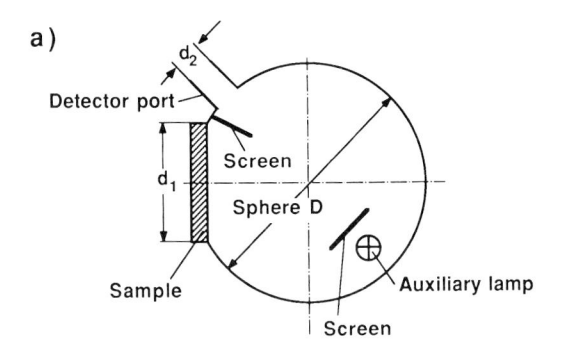

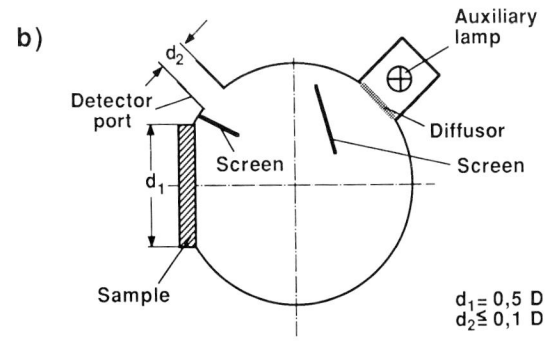

Diameter of the ports 1 to 5:
$d_1 \leq 0{,}1\ D$
$d_2 \leq 0{,}1\ D$
$d_3 \leq 0{,}1\ D$
$d_4 \leq 0{,}02\ D$
$d_5 \leq 0{,}02\ D$

Fig. 6-8. Integrating sphere used to create diffuse irradiation (a and b show alternative configurations).

$d_1 = 0{,}5\ D$
$d_2 \leq 0{,}1\ D$

the specimen and standard are measured in a comparative method; they are placed at the respective ports in the sphere either simultaneously or one at a time. Geometries involving the integrating sphere have this peculiarity: If the directionally reflected fraction of the radiation is absorbed or allowed to escape from the sphere ("gloss exclusion" by a "gloss trap"), what is measured is solely the scattered reflection without directional components (Fig. 6-7). Geometries with gloss exclusion therefore give values similar to those for the 45/0 geometry, in which, similarly, the directional components on the observing side are not measured. Whenever necessary to maintain this distinction, the subscript letters "in" (inclusive, *with* surface reflection) or "ex" (exclusive, *without* surface reflection) will be added to the symbols for the reflection values.

The decision whether or not to measure the directional components of the reflected radiation is extremely important for the measurement of nonluminous perceived colors. As a consequence, the geometries just described can also be classified by whether the surface reflection is included or excluded:*

- *Method A:* Geometry 8/d (reflectance $\varrho_{in,c}$) or d/8 (reflection factor $R_{in,dif}$), with no gloss trap in either case. The surface reflection is included in the measurement. Evaluation: Before calculating the color difference in accordance with ISO 7724/3, a correction for the surface reflection ($r_o = 0.04$; see preceding section) must be subtracted from the measured values.
- *Method B:* Geometry 45/0 (reflection factor R_c), resp. 8/d or 0/d both with gloss exclusion (reflectance $\varrho_{ex,c}$), and their inverses. The surface reflection is not measured. Evaluation: The color difference is calculated in accordance with ISO 7724/3, directly from the measured values (i.e., without the subtractive correction for surface reflection).

In Method A, the color differences determined between test and reference specimen are those having to do with differences in coloring materials. The results are generally not affected by differences in specimen surface quality. Accordingly, Method A is particularly suitable for pigment testing. The only exception is in the full-shade determination of black pigments (see Section 6.2.2). If the gloss changes, the total surface reflection, which consists of a directional and a diffuse component, remains constant when Method A is employed for the measurement. Because the total surface reflection is included in the measurement and a correction must be made by subtracting the surface reflection from the measured value, color differences obtained in this way are strictly due to differences or changes in the interior of the system. If the measured values were used directly (i.e., without subtracting the surface reflection) for calculating color differences, the results would be too small because of the convexity of the cube-root curves for X^*, Y^*, and Z^* (see Chapter 2). This holds particularly for color-saturated or dark specimens. Essentially, the effect of calculating out the surface reflection is that color differences are roughly the

---

* For standards see "Color Differences" in Appendix 1.

same as seen by an observer looking in a "gloss-free" manner at high-gloss specimens, that is, if the surface reflection were "reflected out" at the time of observation. In the case of mat specimens, the color differences are equal to what an observer would see if diffuse surface reflection were eliminated, that is, if a coat of lacquer binding medium were applied to the specimen to render it glossy and the specimen were then examined in a gloss-free manner. Colorimetric evaluation by this method, then, has a direct visual analog only for high-gloss specimens.

In Method B, the color differences found between test and reference specimen are those that stem from the coloring materials as well as differences in surface quality. Accordingly, specimen color differences come out approximately the same as in visual color matching with gloss exclusion. If a specimen loses its gloss, as in the course of a durability test, the diffuse fraction in the surface reflection becomes larger, so that Method B yields a larger measured value. The method therefore gives integral color differences that include differences in coloring materials together with surface gloss differences. The method is therefore not well-suited to routine pigment testing, because comparisons between pigments must be free of coating surface effects. On the other hand, in the case of durability tests, a change in gloss is to be determined by a separate gloss measurement. Nonetheless, Method B can yield marked departures from visual assessment, frequently because of the heavy angular dependence of the surface reflection for semi-mat specimens or because of the gloss haze for glossy specimens. These deviations are especially apparent when the 45/0 geometry is not exactly observed for visual matching of specimens with gloss haze; often, the illumination employed for matching is not directional or the angle of incidence deviates from 45°.

Following this discussion, it should be emphasized that similar to identical results can be obtained with these two methods only on high-gloss specimens. In all other cases, the measuring method must be selected on the basis of what information is desired: Is the difference in the interior of two specimens (difference due chiefly to pigments or dyes) wanted, or the total difference (including surface effects)? Differences in gloss will thus result in unequal color differences, not just unequal between Methods A and B but from one instrument to another (depending on the accuracy of the measuring geometry in the instruments).

Special problems in colorimetry arise when coloring materials must be examined in media whose surfaces are not smooth or homogeneous but display more-or-less regular textures and variable degrees of gloss. Again, Method A or B can be employed, but the 0° geometries should be preferred so that the azimuthal dependence will be minimized in the measurement. It is desirable to use circularly symmetrical irradiation, for example circular/0 or d/0.

Agreements between the interested parties must always take account of the following points:

– Whenever possible, the same model of instrument should be used, but in any case the same measuring geometry. If possible, a check should be made in

order to verify that the selected geometry yields color coordinates that are compatible with visual evaluation.
– An attempt should be made to create color reference standards whose surfaces match as closely as possible the surfaces of the color specimens to be tested.
– Acceptability can be determined only on the basis of color sample collections that have been visually evaluated by experienced colorists. The collection of color samples must be large enough that measurements lead to statistically reliable limits on the acceptable color differences (see section 6.5).

6.2 Practical Evaluation of Color Differences

6.2.1 Preparation of Specimens

Preparation of the dispersion must be preceded by proper sampling. This is done with a suitable sampling device such as a scoop or thief for solids and appropriate equipment for liquids of high and low viscosity.*

Good dispersion is essential for testing pigments in vehicles. Various standards** afford guidance in the selection of test media and dispersing methods. The standards describe media and techniques suitable for testing the dispersion properties of pigments, but these are also valid for colorimetric testing:

ISO 8780
    Part 1, Introduction
– Dispersing equipment:
    Part 2, Oscillatory shaking machine
    Part 3, High-Speed Impeller Mill
    Part 4, Bead Mill
    Part 5, Automatic Muller
    Part 6, Triple-Roll Mill
DIN 53238 (shortly as ISO 8780)
– Binders:
    Part 30, Alkyd Resin System, Low Viscous, Drying by air oxidation
    Part 31, Alkyd-/Melamine Resin System, Low Viscous, Stoving Type
    Part 32, Alkyd-/Melamine Resin System 2, Low Viscous, Stoving Type

There are problems in the testing of pigments in latex vehicles, because reproducible measurements are virtually impossible in these media. If such tests are unavoidable, special care must be taken in the choice of vehicle and the preparation of specimens.

---
* For standards see "Sampling" in Appendix 1.
** See "Ease of Dispersion" in Appendix 1.

A second group of standards covers the testing of pigments in plastics. These include a detailed description of how to prepare basic mixtures and specimens:

DIN 55773 Parts 1 and 2: PVC Pastes (Plastisols)
DIN 55774 Parts 1 and 2: PVC-U (Unplasticized PVC)
DIN 55775 Parts 1 and 2: PVC-P (Plasticized PVC)

In solvent-based systems containing more than one pigment (and frequently also in full-shade systems), flocculation can give rise to color changes ("rub-out effect"), especially when mechanical force is employed during application and drying. This behavior can be utilized to indicate a tendency to unwanted flocculation in the **Rub-out test:**

Apply liquid system to cardboard or sheet metal; after stated time under exhaust hood, rub part of surface with a finger until the color difference between rubbed and unrubbed surfaces has reached largest possible value and remains constant, as indicated by resistance to rubbing. From color measurements on rubbed and unrubbed surfaces, determine color difference* (see Section 6.2.3).

In the remainder of Part II, specimen preparation techniques used with a number of test methods are described in more detail.

6.2.2 Color Measuring

All color measurements start with the reflection values for the color sample. In principle, any instrument used for color measurement includes the following main components:

– Illumination device
– Monochromator (or wide-band filter)
– Sample holder
– Radiation-measuring device

A common arrangement has the monochromator and sample holder interchangeable to give two distinct configurations (Fig. 6-9). In the chain lamp-specimen-chromator-detector, the specimen is irradiated with polychromatic light; in the other case, irradiation is monochromatic.

The light source may be a halogen lamp, a xenon arc lamp, or a xenon flash lamp. Heating of the specimen and resulting changes of color coordinates must be anticipated with the xenon arc lamp.

---

* See "Color Differences" in Appendix 1.

6 Measurement and Evaluation of Object Colors

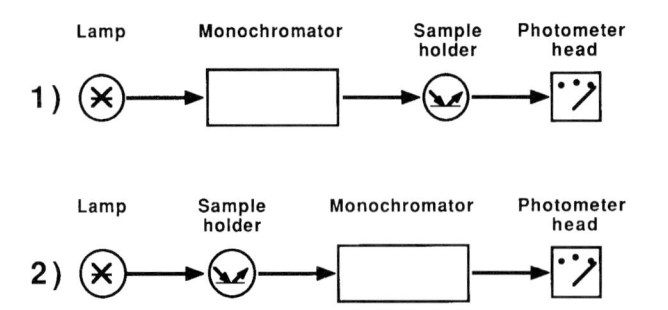

Fig. 6-9. Paths followed by light in colorimeters with (1) monochromatic and (2) polychromatic irradiation.

The instrument is calibrated with a black standard (hollow light trap) for the zero and a $BaSO_4$ white standard* (see Section 2.1.2). For practical tests, working standards calibrated to $BaSO_4$ (secondary standard) are usually employed as white standards for the ideal matte white surface.

In Chapter 2, it was shown that color measurement can be reduced to determining x, y, and Y; this means, according to the equations presented there, the measurement of the CIE tristimulus values X, Y, Z.* The product $\varrho \cdot S$ or $R \cdot S$ for reflection measurements, or $\tau \cdot S$ or $T \cdot S$ for transmission measurements, must be multiplied by the CIE color matching functions $\bar{x}, \bar{y}, \bar{z}$ (see Table 7-4) in order to obtain the three curves shown in the last row of Figure 6-10; X, Y, and Z are given by the areas under these curves. The areas, in turn, are equal to the integrals in equation (2.3), and because these admit of two different interpretations, there are two distinct types of measuring instruments:

- Tristimulus method (tristimulus colorimeters)
  The transmission curves of optical filters together with the lamp spectrum and the detector sensitivity correspond to the $S \cdot \bar{x}$, etc., so that the measured intensities yield the CIE tristimulus values after a simple calculation
- Spectral method (spectrophotometers)
  The reflection curve is measured with a spectrograph. Built-in analog computers or internal or external digital computers multiply by the $S \cdot \bar{x}$, $S \cdot \bar{y}$, $S \cdot \bar{z}$, then integrate.

Both types are used in colorimetry; the manufacturers have developed photometers, and recent years have seen improvements in them. A *tristimulus colorimeter* has, instead of the monochromator, three wide-band filters whose transmission curves combine with the source spectrum and the photometer charac-

---

* See "Colorimetry" in Appendix 1.

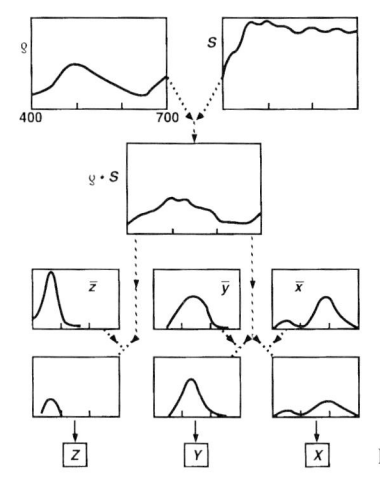

Fig. 6-10. Schematic representation of color measurement.

teristic to give the reflectometer values* R'_x, R'_y, R'_z. The CIE tristimulus values are obtained directly from these values as

$$X/X_n = 0.798 \frac{R'_x}{100} + 0.202 \frac{R'_z}{100}$$

$$Y = R'_y$$

$$Z/Z_n = R'_z/100 \tag{6.1}$$

where X_n and Z_n are the CIE tristimulus values of the CIE standard illuminant. (Because $\bar{x}$ has two maxima, X is made up of R'_x and R'_z.)

In *spectrophotometers*, the monochromator uses optics (entrance and exit slits, prisms, interference filters, optical gratings) to split up the visible spectrum into bands 10–20 nm wide. Gratings and filters have completely displaced prisms in modern instruments, because of their higher transparency, constancy of optical dispersion, and lower manufacturing cost. Interference filters take two distinct forms: a single filter with fixed wavelength and a graded filter continuously traversing the spectrum. The bell-shaped transmission curves of interference filters, however, exhibit relatively large half value widths (ca. 20 nm), so that deviations relative to gratings can occur in spectra having steep slopes.

The sample holder is matched to the desired geometry, usually 0/d, 8/d, or 45/0.

---

* According to the standards, reflectometer values are symbolized by R' (to distinguish them from reflection factors R); see "Photometric Properties of Materials" in Appendix 1.

6 Measurement and Evaluation of Object Colors

The light-measuring device, finally, can be either a photomultiplier (for low irradiation intensities) or a silicon cell (for higher intensities). Silicon elements process the input signals in quasi-parallel fashion, so that instruments so equipped can offer measuring cycles shorter than 3 s.

In the past – usually on grounds of cost and short cycle times – tristimulus instruments dominated in the fields of production control and other applications as well. The situation has been changing since new technologies have made it possible to build fast spectral instruments at competitive prices. Another factor here is that tristimulus photometers cannot tell metameric color samples apart, because these instruments do not recognize metamerism.

Table 6–1. Sample computation based on color measurement. Pigment: test
(Figures in italics represent handwritten entries.)

CIE color notation
a) Spectral method

b) Tristimulus method

λ (nm)	$\varrho(\lambda)$	$\bar{x}_{C/2}(\lambda)$	$\varrho \cdot \bar{x}$	$\bar{y}_{C/2}(\lambda)$	$\varrho \cdot \bar{y}$	$\bar{z}_{C/2}(\lambda)$	$\varrho \cdot \bar{z}$
400	0.0126	0.044	*0.0006*	0	*0*	0.187	*0.0024*
420	0.0125	2.926	*0.0366*	0.085	*0.0011*	14.064	*0.1758*
440	0.0119	7.680	*0.0914*	0.513	*0.0061*	38.643	*0.4599*
460	0.0119	6.633	*0.0789*	1.383	*0.0165*	38.087	*0.4532*
480	0.0122	2.345	*0.0286*	3.210	*0.0392*	19.464	*0.2375*
500	0.0125	0.069	*0.0009*	6.884	*0.0861*	5.725	*0.0716*
520	0.0137	1.193	*0.0163*	12.882	*0.1765*	1.450	*0.0199*
540	0.0176	5.588	*0.0983*	18.268	*0.3215*	0.365	*0.0064*
560	0.0357	11.751	*0.4195*	19.606	*0.6999*	0.074	*0.0026*
580	0.0967	16.801	*1.6247*	15.989	*1.5461*	0.026	*0.0025*
600	0.1726	17.896	*3.0888*	10.684	*1.8441*	0.012	*0.0021*
620	0.2072	14.031	*2.9072*	6.264	*1.2979*	0.003	*0.0006*
640	0.2226	7.437	*1.6555*	2.897	*0.6449*	0	*0*
660	0.2403	2.728	*0.6555*	1.003	*0.2410*	0	*0*
680	0.2675	0.749	*0.2004*	0.271	*0.0725*	0	*0*
700	0.3010	0.175	*0.0527*	0.063	*0.0190*	0	*0*

Σ: *10.96*   *7.01*   *1.44*

CIE tristimulus values   (X)   (Y)   (Z)

Reflectometer values (C/2°)

$R'_x = 13.71$
$R'_y = 7.01$
$R'_z = 1.22$

$\boxed{X} = 0.782 \cdot R'_y + 0.198 \cdot R'_x$

$10.721 + 0.242 = 10.96$

$\boxed{Y} = R'_y$

7.01

$\boxed{Z} = 1.182 \cdot R'_z$

1.44

CIE tristimulus values

CIE chromaticity coordinates

$\boxed{x} = \dfrac{X}{X+Y+Z}$ $\boxed{y} = \dfrac{Y}{X+Y+Z}$

0.5649 0.3613

6.2 Practical Evaluation of Color Differences

Table 6–2. Sample computation of color differences.
(Figures in italics represent handwritten entries.)

CIELAB color coordinates			Pigment: test		
$X^* = \sqrt[3]{X/98.07}$		*0.4817*	$(L^*) = 116 \cdot Y^* - 16$		*31.83*
$Y^* = \sqrt[3]{Y/100}$		*0.4123*	$(a^*) = 500 \cdot (X^* - Y^*)$		*34.70*
$Z^* = \sqrt[3]{Z/118.22}$		*0.2296*	$(b^*) = 200 \cdot (Y^* - Z^*)$		*36.54*

Note: If radicand
$r \leq 0.008856$ then:
$X^* = 7.787 \cdot r + 0.138$
(similarly for Y^* and Z^*)

CIELAB color differences

	Reference (R)	Test (T)		$\Delta = T - R$	
L^*	30.71	*31.38*	(ΔL^*)	*1.12*	Lightness difference
a^*	35.45	*34.70*	(Δa^*)	*− 0.75*	
b^*	35.62	*36.54*	(Δb^*)	*0.92*	
	$C_{ab}^* = \sqrt{a^{*2} + b^{*2}}$				
C_{ab}^*	50.25	*50.39*	(ΔC_{ab}^*)	*0.14*	Chroma difference

$(\Delta E_{ab}^*) = \sqrt{\Delta L^{*2} + \Delta a^{*2} \Delta b^{*2}}$	*1.63*	Color difference	
$\|\Delta H_{ab}^*\| = \sqrt{\Delta E_{ab}^{*2} - \Delta L^{*2} - \Delta C_{ab}^{*2}}$	*1.18*		
$a_R^* \cdot b_T^* - a_T^* \cdot b_R^*$	*+59.33*	$\left.\begin{array}{l}\text{Sign } (\geq 0) = +1 \\ \text{Sign } (< 0) = -1\end{array}\right\}$ *+*	
$(\Delta H_{ab}^*) = \text{Sign} \|\Delta H_{ab}^*\|$	*+ 1.18*	Hue difference	

After color measurements have been performed on the test and reference specimens, the X, Y, Z are transformed into the CIELAB system so that the color differences can be calculated for the full-shade determination.

Practical (manual, so to speak) evaluations of colorimetric data for both tristimulus and spectral methods are illustrated in Tables 6-1 and 6-2. The values of the CIE color matching functions used here, from Stearns' 20 nm table (see Chapter 3), are not found in any standard. The example shows very clearly the burden of computation that used to be necessary and has now been taken off the colorimetrist's back by the computer age.

6.2.3 Full Shade

A standardized test of dry pigment powder is possible for white pigments.* In the procedure, the CIE tristimulus value Y can be determined as a way of characterizing the lightness of white pigment and filler powders in pressed form. A suitable powder press has to be used to prepare the tablets. The measured Y value depends on a number of factors (rate of pressing, degree of compaction, surface texture, etc.), so that specimen preparation conditions must be clearly agreed on. Naturally, the CIE tristimulus value Y of the powder has no power to indicate the lightness of the pigment in media; it is suitable – at most – as a supplementary receiving inspection method for pigments.

Thus the colorimetric values obtained on such tablets are not in any way typical, for the colors of colored, white, and black pigments must be evaluated in suitable media (binders, vehicles). This means that final paint and lacquer coatings as well as plastics can be evaluated by these methods. Full-shade testing is described in separate standards** for black, white, and colored pigments. These also define the terms "full shade" and "mass tone," which were formerly used inconsistently. A *full-shade system* is a pigment/vehicle dispersion with a *single* pigment, and the color of such a system in an optically hiding layer is called the *full shade*. In contrast, the *mass tone* is a full-shade system in a non-hiding layer. Thus no mass tone is unambiguously defined unless information is provided on the preparation, composition, application, and substrate; many mass tones can be made with the same full-shade system, differing in film thickness and substrate properties.

A special role in pigment testing is played by the test media used, first because of their actual effect and second because the use of suitable media makes it possible not only to avoid certain problems (inadequate dispersion, flocculation, etc.; see Section 4.1.2) but also to achieve certain economies. Chief among the latter are time savings in the testing procedure; one time-saving step is the use of nondrying pastes as vehicles. In addition to the time saving, the measured values also gain in significance, because full-shade mea-

---

* See "Lightness" and "Reflection Factor" in Appendix 1.
** See "Color in Full-Shade Systems" in Appendix 1.

surements on pigmented pastes exhibit much smaller preparation errors than those on dried vehicles. Other proposals for test systems and their preparation can be found in Section 6.2.1.

A good vehicle for *nondrying pastes* (DIN 55983) is an alkyd resin, Alkydal® L 64 (Bayer AG). This binder combines the advantages of easy dispersion with the suppression of any tendency for the pigment to flocculate.

Preparation method: Mix 97% by weight Alkydal with 3% by weight Aerosil® 200 (Degussa), taking steps to prevent dust losses. Pass the mixture through a triple-roll mill twice. Weigh out 3 g of this mixture and a quantity of pigment such as to give a PVC of 10–20%. Put the Aerosil paste in the middle of the lower plate of an automatic muller, scatter the pigment on the test medium; work together with a spatula without applying pressure. Next, distribute the paste in dots or a circle about 35 mm away from the center of the lower plate. A useful marking technique is to place beneath the lower glass plate a ring-shaped paper pattern with an inside diameter of 40 mm and an outside diameter of ca. 100 mm. Add a 25 kg weight to the upper plate. Wipe as much as possible of the paste from the spatula onto the upper plate. Mull the paste in two stages of 25 revolutions each. Transfer the finished pigmented paste to a special paste film holder made of metal or polyamide (see Fig. 6-11) and draw down with the doctor blade included.

Among *drying systems* (DIN 55983), air-drying systems are preferred to stove-drying ones because the preparation errors are smaller in the first case. Especially good choices are Alkydal F 48 (Bayer AG) and URALAC® 6008 (Scado GmbH), both in 50% solution. URALAC does not require an additional dryer and dries in 1–2 h (see also DIN 53238 Part 30); Alkydal has to have a dryer added.

Preparation method: Weigh out 5 g of the test medium and the quantity of pigment to give a PVC of 10–20%. Using an automatic muller, proceed as before. Apply hiding coats of the finished reference and test pigment mixtures to a suitable substrate (glass plate, metal sheet, contrast card, lacquered or aluminum-lined cardboard), employing the same procedure in each case; the use of a Bird applicator is desirable. Coatings are hiding if a coating applied on a black/white contrast substrate exhibits a color difference $\Delta E_{ab}^* < 0.2$ after drying.

Measurement method: Using a suitable colorimeter, measure test and reference preparations in such a way that the measurement conditions for Method A (i.e., use of integrating sphere with no gloss trap, correction by subtracting surface reflection; see Section 6.1.2) are fulfilled. Depending on the configuration of the measuring apparatus, either the reflectance ϱ or the reflection factor R is obtained (in tristimulus instruments, the reflectometer values R'_x, R'_y, R'_z are obtained). The corrections for surface reflection must be subtracted from each such quantity. Evaluation is done with the CIELAB formulas, equations (2.5) to (2.14).

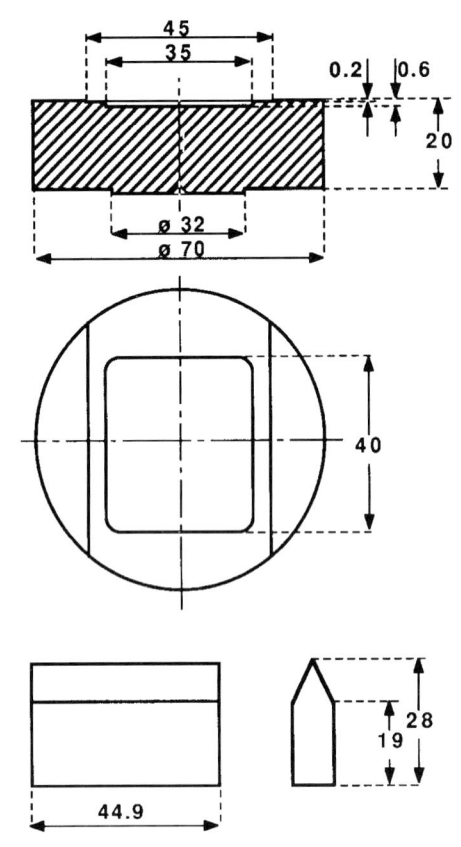

Fig. 6–11. Paste film holder with doctor blade, corresponding to DIN 55983 (dimensions in mm).

Figure 6-12 shows a computer printout illustrating the calculations for a full-shade measurement on yellow iron oxide pigments.

Special aspects in the evaluation for *white pigments* have already been discussed (Section 2.3.3). In this instance, it is preferable to state as the color designation the "hue of near-white specimens"* and its strength Δs (eq. 2.15). Figure 6-13 presents an example for titanium dioxide pigments.

Similar considerations apply for *black pigments* and gray coatings: The recommended statement of the result is "hue of near-black (or gray) specimens" instead of color differences (Section 2.3.3). Another special point is the more stringent requirements on the colorimeter for the full-shade measurement of black pigments. These conditions, which are met by very few instruments today, relate to the display (five decimal places for reflection measurements, three for CIE tristimulus values), zero adjustment (calibration with a hollow light trap as special black standard), and reproducibility (standard deviation of CIE tristimu-

---

* See "Hue of Near-White Specimens" in Appendix 1.

6.2 Practical Evaluation of Color Differences

Program "FULL SHADE COLORED"
Full shade of colored pigments, according to ISO 787-25**********************
Date: ..-..-..   Pigment: Iron oxide yellow
Reference is A   Test is B
Binder: Alkydal L 64   PVC: 20%
Colorimeter, Standard observer: N.N., C/2°
Measuring conditions: d/8(incl.)
Input: ***

	R´(x)	R´(y)	R´(z)
Reference	17.00	10.57	4.95
Test	15.96	10.39	5.26

Output: ***

	a*	b*	L*	C*(ab)	h(ab)
Reference	34.6	38.3	30.8	51.7	47.9°
Test	30.6	33.4	30.4	45.3	47.5°

Color differences

Delta E*(ab)	Delta a*	Delta b*	Delta L*	Delta C*(ab)	Delta H*(ab)	Delta h(ab)
6.4	-4.0	-4.9	-.4	-6.3	-.3	-.4°

Fig. 6-12. Full shade, colored, yellow iron oxide (computer output).

Program "FULL SHADE WHITE"
Full shade of white pigments, according to ISO 787-25************
Relative hue of near white specimens
Date: ..-..-..   Pigment: Titanium dioxide
Reference is A   Test is B
Binder: Alkydal L 64   PVC: 20%
Colorimeter, Standard observer: N.N., C/2°
Measuring Conditions: d/8 (incl.)
Input: ***

	R´(x)	R´(y)	R´(z)
Reference	94.52	93.91	91.48
Test	94.54	94.18	92.55

Output: ***

Color difference		Hue	
Delta (E* (ab))	Delta (L*)	Type	Magnitude Delta (s)
.6	.1	blue	.6

Fig. 6-13. Full shade, white (relative hue of near-white specimens), titanium dioxide (computer output).

lus values less than 0.001). An example, based on a full-shade measurement of black iron oxide pigments, appears in Figure 6-14.

In the case of pigment-grade carbon blacks, it is also possible to define a "black value"* that gives an improved (logarithm-like) curve at low reflection values for this special case.

The full-shade determination of *dyes* in solution is accomplished with colorimeters capable of transmission measurements. The CIELAB evaluation is similar to that for nonluminous perceived colors.

Determinations of CIELAB color differences are described, as secondary procedures, in a number of other standards, for example under methods for determining the thermostability of plastics.**

```
Program "FULL SHADE BLACK"
Full shade of black pigments, according to ISO 787-25****************
Relative hue of near black specimens
Date: ..-..-..                          Pigment: Iron oxide black
Reference is A                          Test is B
Binder: Alkydal L 64                    PVC: 20%
Colorimeter, Standard observer: N.N., C/2°
Measuring Conditions: d/8 (excl.)
Input:   ************************************************************
              R´(x)       R´(y)        R´(z)

Reference     .491        .442         .471
Test          .369        .348         .429
Output:  ************************************************************
      Color difference                  Hue
Delta (E* (ab))   Delta (L*)     Type        Magnitude Delta (s)
    1.3             -.8        blue green            .9
```

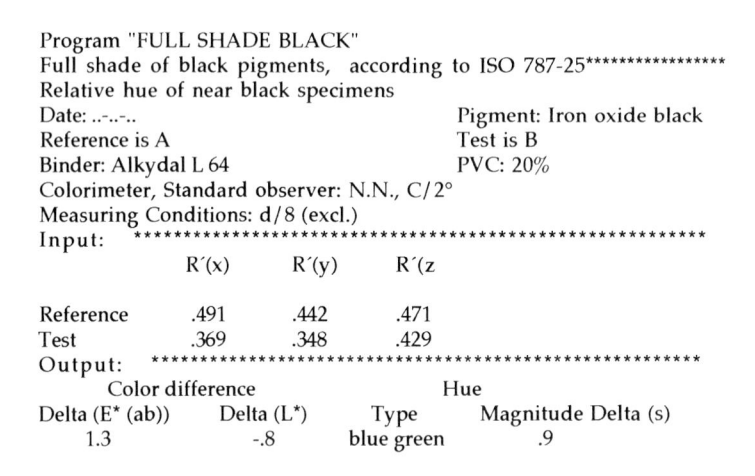

Fig. 6–14. Full shade, black, black iron oxide (computer output).

6.2.4 Special Problems

Metamerism (see Section 2.2.2) is commonly of secondary significance for pigment testing, though in practice it is much more important and can therefore become a factor in special testing situations. In formulation (as also in the pigmenting of lacquers and paints), one often cannot simulate the reflection spectrum of a given color, because (among other reasons) to do so would require having an unlimited number of pigments on hand. Instead, one has to be satis-

---

* See "Black Value" in Appendix 1.
** See "PVC, Plasticized", "PVC, Unplasticized", and "Thermoplastics" in Appendix 1.

fied with reproducing the three CIE tristimulus values, even though this means that the new color will match only for the precise illuminant used. Such "metamerically" matched colors will therefore show differences between the "daylight" color (under CIE standard illuminant C or D65) and the "incandescent light color" (under CIE standard illuminant A). In order to describe how much the color equality of two objects under a certain illuminant changes on a change of illuminant, the metamerism index* is employed. The color difference between the originally matching colors under the new illuminant is expressed in the perceptually-based CIELAB system. Figure 6-15 shows reflectance curves for two metameric specimens.

Even more important in pigment testing is the variety of metamerism called "geometric." While normal metamerism involves a change in color when the illuminant is changed, in geometric metamerism a color change is brought about when the irradiation or viewing geometry relative to the specimen is altered. A special form of geometric metamerism is encountered in the *Silking effect,* where the reflectance changes when the specimen is rotated about the surface normal. Several conditions must hold for the Silking effect to occur. The first condition is an anisometric pigment, one made up of acicular, bladed, or rod-shaped particles, whose scattering and absorption coefficients are different in the principal axis direction and perpendicular to this axis. What is more, the pigmented medium must be applied in such a way that the particles are oriented, that is, in a quasi-parallel arrangement. The Silking effect can be measured only in geometries involving directional fluxes (e. g., 45/0). In practice, the Silking effect can be rather a problem, for example producing the "picture-frame effect" when a painting job has been finished up by painting once more around the edge of a large area in order to equalize the coat.

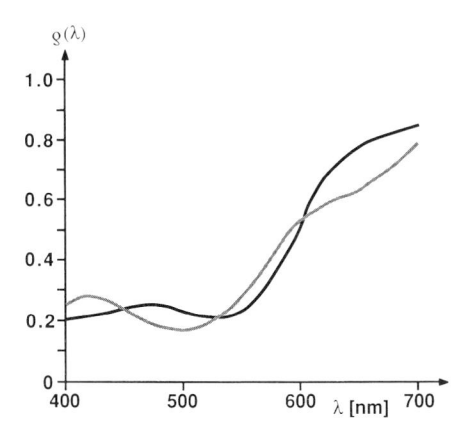

Fig. 6–15. Spectral reflectance for two yellow specimens having the same appearance in CIE standard illuminant D65 for the 10° observer.

---

* See "Metamerism Index" in Appendix 1.

Thermochromism is the property of coloring materials that show changes in reflection with changes in temperature. When performing colorimetric measurements on thermochromic specimens, it is necessary to wait for temperature equalization or, better, to use an irradiation source that does not heat up the specimen (monochromatic source or flash lamps). The following method is a suitable way of testing for the effect:

Measure specimen a number of times in succession without removing it from the measuring port. Determine color differences relative to first measurement and note the time up to which differences remain constant.

It should be noted that thermochromism can also be simulated by a change in the moisture content of the specimen.

Fluorescence occurs when absorbed radiation is reflected after a very short time (ca. 10^{-6} s) in the longer-wavelength visible. If the reflection is delayed for a longer time, the behavior is called phosphorescence. Both these phenomena come under the general heading of luminescence. Fluorescent coloring materials can be used in molecularly dissolved form or as fluorescent pigments. Often, fluorescent coloring materials in low concentrations are added to paints or inks to enhance their brilliance ("optical brighteners"); the attenuated fluorescence is not perceived as luminosity but as greater brightness or saturation. The exciting radiation can lie in a more-or-less wide range of wavelengths, between 350 and 550 nm, depending on the chemical nature of the colorant; similarly, the fluorescent radiation is generally not monochromatic.

Media and coatings that contain fluorescent coloring materials are, in fact, self-luminous, at least to the degree that the reflection is caused by them. Nonetheless, they are treated as nonluminous perceived colors, that is, referred to a mat white surface, so that reflectances $\varrho > 1$ can occur. Fluorescent object colors also differ from ordinary nonluminous perceived colors in that the reflectance and color stimulus depend on whether the incident light contains sufficient exciting radiation. For this reason, fluorescent specimens must be irradiated with the illuminant to which the measured values will be referred, that is, under which they will be visually matched. The content of fluorescent radiation is proportional to the intensity of the exciting radiation, and so if the incident radiation is deficient in the exciting range then there will be little fluorescent light; as a consequence, the measured color will no longer match with the visual results. This means that specimens must not be monochromatically irradiated for the spectral method; hence the only candidate configuration is number 2 in Figure 6-9. Figure 6-16 presents curves for a "right" and a "wrong" measurement; the absorptions can be seen clearly in the "right" curve.

Furthermore, mirrors and lenses in the instrument, and particularly the integrating sphere, will more-or-less strongly alter the spectral makeup of the incident radiation and, as implied by the discussion just presented, can again lead to changes in the color loci. It is therefore recommended that Method B of Section 6.1.2 be employed for these measurements, but only in the 45/0 geometry. Lamps with a daylight-like spectrum can be used to irradiate the specimen,

6.2 Practical Evaluation of Color Differences

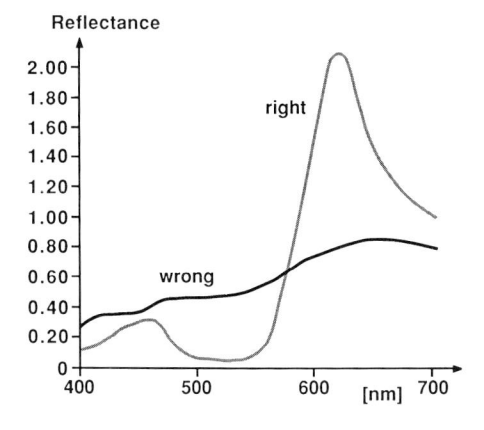

Fig. 6–16. Spectral reflectance for a fluorescent red color, measured in a "wrong" configuration (number 1 in Figure 6–9) and a "right" one (number 2 in Figure 6–9).

because daylight contains sufficient radiation in the ultraviolet near 400 nm. This is why high-pressure xenon lamps or xenon flash tubes are used in instruments. CIE standard illuminant C is not appropriate because it is deficient in this radiation; at present, illuminant D65 lacks a technical implementation that can be used for routine measurements.

Metal-effect paints, also called simply "metallics," are made with metal-effect pigments (Section 1.2). Their color properties are governed mainly by specular reflection on blade-shaped pigment particles, which are relatively large in diameter (up to 20 μm) and are predominantly oriented parallel to the paint surface. The fundamental physical processes of absorption and scattering observed on pigmented coatings are here joined by specular reflection. The interplay and interaction of these three processes give rise to the attractive aspect of a metal-effect paint. The directional distribution of the returned radiation depends heavily on the irradiation angle and the viewing angle, so that significant and abrupt changes of lightness and hue can occur (lightness flop, hue flop).

There are major difficulties associated with unsolved problems in color measurement on metallics. To determine the overall behavior, goniometric methods would have to be applied on both the irradiation and viewing sides of the system; this would lead to a multitude of plotted curves or measured values. Rationalization will not be possible until the colorimetric characterization of such paints can be accomplished with a limited number of angle settings, but no theoretical approach has yet been devised to this problem.

The situation is not much better with regard to instruments, plainly because of the lack of theoretical initiatives. One goniometric instrument that used to be available has now disappeared from the market, and devices that can be set to at least certain predetermined angles are slow in coming out. The present state of the art comprises instruments with conventional features, but these permit only the study of special aspects. Given these circumstances, for example, the only way to acquire new information about the conditions for hue and lightness flop is to analyze the scattered and reflected components separately to whatever extent this is possible.

Problems with *nacreous pigments* stem from the need to measure interference colors. This again involves an angle-dependent indicatrix, so that conventional colorimeters fall down. Only the beginnings of an approach to measurement exist now; development into routine lab methods will have to wait.

Generally accepted techniques for measurements on metal-effect and nacreous pigments are not yet in prospect. The methods gap is due to a lack of both theory and apparatus. Unfortunately, there is an intimate interaction between theory and hardware. This example shows that even colorimetric problems cannot be solved until the right instrument is available.

6.3 Test Errors

6.3.1 Calculation of the Standard Deviation Ellipsoid

A number of computer programs are needed for the calculations demonstrated in Section 2.4. These can be classified into four groups:

– Calculation of the coefficients of the ellipsoid
– Visualization of the resulting ellipsoid
– Calculation of test errors and significance
– Calculation of acceptability

It is desirable to program these calculations for a suitable computer.

The calculation of ellipsoid coefficients is preceded by measurements performed in a "reproducibility series" of at least 20 color samples. This term refers to a series of preparations made in the same way with the same pigment. The program takes the CIELAB coordinates (two decimal places) as input and calculates the means and the standard deviation ellipsoid.

First, the arithmetic means of the color coordinates a^*, b^*, and L^* are calculated. By substituting these means into the successive formulas, it calculates the barred values of the polar coordinates C^*_{ab} and h_{ab} (to two decimal places) with equations (2.8) and (2.9).

Next, the variances and covariances are calculated (equations 2.70 and 2.71) and arrayed to form the covariance matrix $\mathbb{V}$ (equation 2.72). The program now inverts the matrix (equation 2.73) to get $\mathbb{G}$. Appropriate software is often available for matrix inversion on the computer; an individual sample calculation is presented here. The formula is $g_{ik} = \hat{v}_{ik}/\det \mathbb{V}$, where $\det \mathbb{V}$ denotes the determinant of matrix $\mathbb{V}$, $\hat{v}_{ik}$ is the cofactor of element v_{ik}, obtained from matrix $\mathbb{V}$ by deleting the i-th row and the k-th column and taking the determinant of the matrix that remains, with the sign $(-1)^{i+k}$. The program requests the desired probability, divides the coefficients by the appropriate significance level δs^2_{cum} (Table 2-4), and rounds the g_{ik} to four significant figures. It outputs the means and the coefficients of the standard deviation ellipsoid for the desired probability.

The programming is not very complicated and can well be done on a medium-sized desktop computer. Figure 6-17 presents a typical printout. With the aid of such a program, the CIELAB coordinates for 20 preparations of a chromium oxide pigment have been transformed to the means and the coefficients g_{ik} of the standard deviation ellipsoid for a 70% confidence interval. A (very detailed) sample computation is contained in DIN 55600 Part 2.

In this instance, the 70% confidence interval means that 70% of all measured values lie inside the ellipsoid, or that there is a 70% probability of an individual measurement lying in the interior (see also Section 2.4.1).

```
Program "STANDARD DEV. ELL."
Standard deviation ellipsoid, according to DIN 55 600*******************
Date: ..-..-..              Pigment: Chromium oxide green
Binder: Alkydal F48         PVC: 10%
Dispersing equipment: Automatic Muller, 2*100 Rev.
Colorimeter (Measuring conditions, Standard observer): N.N.,d/8°,C/2°
Input:***********************************************************
Pigment is A
No.     a*       b*       L*
--------------------------------------
1     -19.09    16.08    45.48
2     -19.10    16.12    45.37
3     -19.10    16.01    45.39
4     -19.09    16.10    45.36
5     -19.02    16.06    45.36
6     -19.06    16.12    45.47
7     -19.12    16.16    45.36
8     -19.11    16.06    45.41
9     -19.10    16.09    45.47
10    -19.15    16.10    45.43
11    -19.08    16.15    45.43
12    -19.08    16.03    45.46
13    -19.13    16.20    45.36
14    -19.09    16.07    45.46
15    -19.09    16.14    45.40
16    -19.13    16.16    45.46
17    -19.11    16.12    45.47
18    -19.14    16.17    45.44
19.   -19.10    16.12    45.37
20    -19.14    16.11    45.39
Output:**********************************************************
Means:
    a* = -19.10     b* = 16.11     L* = 45.42
 C*(ab) = 24.99                   h(ab) = 139.85°
Ellipsoid (70%):
 g(1,1) = 424.7     g(2,2) = 152.3     g(3,3) = 139.5
 g(1,2) = 137.5     g(2,3) = 21.64     g(1,3) = 22.62
```

Fig. 6–17. Means and calculated standard deviation ellipsoid for 20 preparations of a chromium oxide pigment (computer output).

6.3.2 Visualization of the Ellipsoid

It has proved useful in practice to visualize the ellipsoid in the CIELAB color space. When comparing two series of color measurements, it is often not enough to state just the standard deviations associated with the measurements; detailed conclusions about a measurement or preparation method can be arrived at only when the location and orientation of the error ellipsoid in color space are accessible for comparison. Two programs are needed for this purpose:

– Calculation of principal axes
– Projection of the ellipsoid onto the principal planes of the CIELAB color space (plot program)

For the *calculation of principal axes*, the g_{ik} of the ellipsoid become the inputs. The program first verifies that the coefficients do pertain to an ellipsoid (and not to a different second-order surface); this is done with equation (2.39). If the result is affirmative, the secular equation (2.43) is then solved, for example by the trigonometric method. From its roots λ_a, λ_b, λ_c, the three principal axes of the ellipsoid are then obtained as vectors **a**, **b**, **c**, each expressed in terms of its Cartesian coordinates a^*, b^*, L^* (equations 2.45-2.48) and its three-dimensional polar coordinates r, φ, ϑ (equation 2.49).

An example in the form of a computer printout is given in Figure 6-18. The calculation relates to the total error ellipsoid, which will be described more fully in Section 6.3.3. The three principal axes are expressed as vectors **a**, **b**, **c**, each with its components. Listed next are the three-dimensional polar coordinates (one length plus polar angles; see Fig. 2-14). A quick comparison of the lengths of the three principal axes shows that one is very long ($|\mathbf{a}| = 0.1606$) and two are shorter. One of the short axes ($|\mathbf{b}| = 0.0762$) is roughly 3.5 times as long as the other ($|\mathbf{c}| = 0.02133$). The ellipsoid is thus shaped like a surfboard or a somewhat flattened cigar with an elliptical cross section; the long

```
Program "SEMIAXES"
Ellipsoid semiaxes, according to VÖLZ, Industrial Color Testing***************
Input:  Ellipsoid  coefficients***************************************************
g(1,1) =   228.5      g(2,2) =  345.9     g(3,3) =   1835
g(1,2) =   104.7      g(2,3) = - 730      g(1,3) = - 344.1
```

Output:	Semiaxes					
Vector	a*	b*	L*	Length	Phi	Theta
a	.0402	.1412	.0652	.1606	74.10°	66.08°
b	.0725	- .02294	.00493	.0762	- 17.55°	86.29°
c	- .00382	- .00788	.01945	.02133	-115.83°	24.25°

Fig. 6–18. Calculation of principal axes of a total error ellipsoid (computer output).

6.3 Test Errors 265

axis of the cigar is some eight times as long as the short axis and twice as long as the longer axis of the cross-sectional ellipse. The polar angles give the spatial orientation of the figure: The long axis (**a**) lies in the positive quadrant (a^* and b^* positive, φ positive) of the color plane and also in the positive octant of the CIELAB color space (L^* and ϑ also positive). It is nearer to the b^* axis than to the a^* axis of the three-dimensional coordinate system ($\varphi = 74.1°$ closer to $90°$ than to $0°$), and also somewhat nearer to the color plane than to the L^* axis of the coordinate system ($\vartheta = 66.1°$ closer to $90°$ than to $0°$).

Next we come to the *plot program for the ellipsoid projections*. After the g_{ik} are input, the coefficients of the three projected ellipses (in the a^*,b^*, a^*,L^*, and b^*,L^* planes, equation 2.52) are calculated along with the axes (equation 2.53) and the paired section lines (equation 2.57), and these results are output by the plotter.* Figure 6-19 presents an example, a plot of the total error ellipsoid from Section 6.3.3 and Figure 6-18. With some practice, one can gain an idea of the position and orientation in three-dimensional color space by examining the plot. Fig. 6-19 (left-hand side) shows the three projections, i.e. a vertical and two horizontal projections; these are superimposed to simplify the display (Fig. 6-19, right-hand side). The outermost points of the projections are

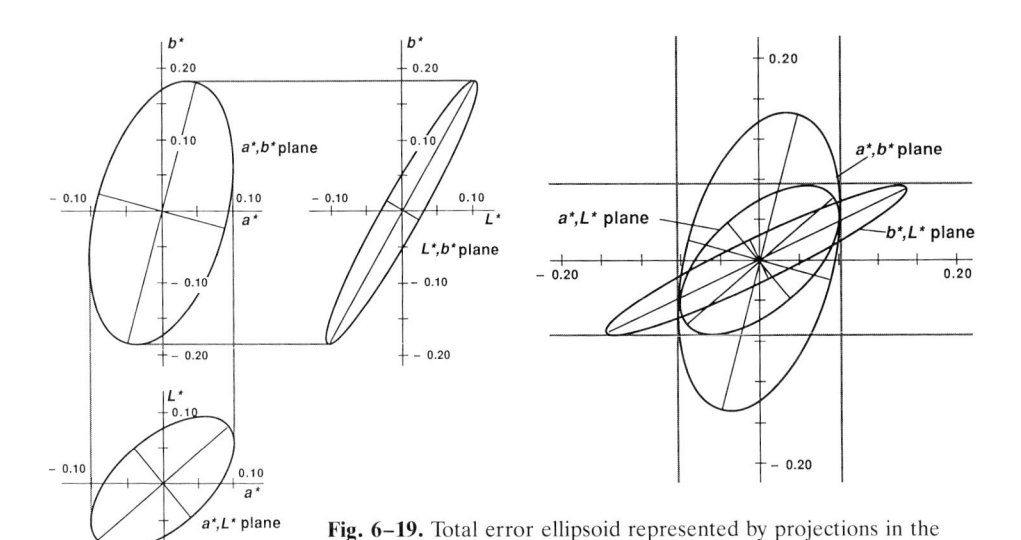

Fig. 6–19. Total error ellipsoid represented by projections in the principal planes of the CIELAB color space (from 20 preparations).

---

* In all plots, the horizontal axis is named first and the vertical axis second; thus the phrase "a^*,b^* plane" denotes the plane with coordinates a^* (abscissa) and b^* (ordinate).

6 Measurement and Evaluation of Object Colors

connected in pairs by section lines (dotted). If a color plotter is available, each of the three projections can be drawn in a different color.

As has already been shown, the ellipsoid represented here has the shape of a flattened cigar. The projections bear out the description just obtained: The b^*, L^* projection clearly shows the lengthwise orientation, and it together with the two others shows the flattening of the figure. It should be noted that the ellipse axes drawn are not identical with the principal axes of the ellipsoid but are merely their projections.

In order to get a still better idea of the three-dimensional shape of the figure, the ellipsoid can also be represented in its own principal-axis system. As equation (2.45) suggests, the three axis lengths, in the form of their reciprocal squares, are simply substituted in equation (2.44) and g_{12}, g_{23}, and g_{13} are simultaneously set equal to zero. The plot for the total error ellipsoid transformed in this way is illustrated in Figure 6-20. It confirms what has been said about the shape of the figure.

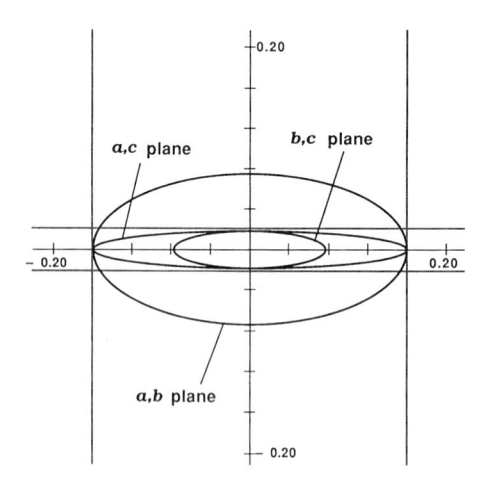

Fig. 6–20. Total error ellipsoid of Figure 6–19, each projection into its own principal-axis plane.

6.3.3 Total Error and Its Components

The total error of a colorimetric pigment test is made up of the error of preparation and the error of measurement. The covariance matrix of the three-dimensional normal distribution can be split up exactly as the variance of the one-dimensional distribution; equation (2.74) is then modified, becoming

$$\mathbb{V}\text{ (total)} = \mathbb{V}\text{ (prep.)} + \mathbb{V}\text{ (meas.)} \tag{6.2}$$

where $\mathbb{V} = \mathbb{G}^{-1}$. Because the error of measurement is commonly very small, preparation methods are crucial for the size and location of the total error ellipsoid. If the preparation error is too large, an attempt must be made to split it into several contributions so that error sources can be identified and eliminated. The preparation error can, for example, be analyzed with the scheme of equation (6.2), for example into manufacture error (grinding, milling), application error (coating, drawing down), and homogeneity error (equality over the entire specimen). The homogeneity error results from measurements performed at points that are statistically distributed over the entire specimen surface; the application error stems from measurements on drawdowns of 20 preparations. The manufacture error can be obtained as the difference

$$\mathbb{V}(\text{man.}) = \mathbb{V}(\text{prep.}) - \mathbb{V}(\text{hom.}) - \mathbb{V}(\text{appl.})$$

In the example under discussion, the color dispersions are prepared by the paste method of Section 6.2.2. The measurement of the tristimulus reflectometer values R'_x, R'_y, R'_z was carried out by Method A of Section 6.1.2 (8°/d) as follows:

Measure R'_x, R'_y, R'_z on one preparation in quick succession. Measure the next preparation after about 2 min, and so forth. After every five preparations, measure preparation No. 1 in order to recalibrate colorimeter to the values acquired on this preparation at the outset.

The differences between actual and nominal values in recalibration were always less than 0.05 reflectometer units.

The reproducibility series (dispersions of a pigment) used to define the total error ellipsoid was independent of random variations after 20 preparations. Confirmation of this claim is obtained by comparing Figure 6-19, the ellipsoid containing all 20 preparations, with the ellipsoids based on groups of 10 from the 20 preparations (Figures 6-21 and 6-22). These total error ellipsoids were determined for pastes (binder Alkydal® L 64) pigmented with a red iron oxide pigment (10% PVC).

In contrast to the total error, the error of measurement is supposed to show how much the measured values will differ when repeated measurements are done on one and the same preparation within a series; the cause of this error lies in short-term instabilities of the instrument.

Place on the instrument, 20 times, exactly the same portion of a selected preparation (e.g., No. 1 of the pastes prepared for the total error ellipsoid of Figure 6-19). After each placement (at time intervals of 3 min), measure the reflectometer values R'_x, R'_y, R'_z without recalibrating the instrument.

Figure 6-23 presents an ellipsoid for the error of measurement derived in this way. It has a somewhat different appearance from the total error ellipsoid of Figure 6-19, being more discus-shaped, flattened in the direction of the a^*, L^* diagonals. To reduce or eliminate the scatter of the measured values due to

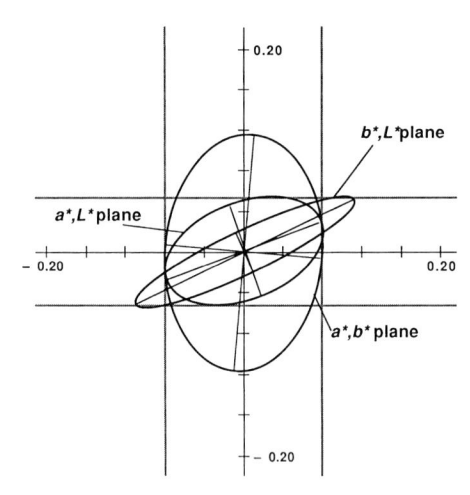

268   6 *Measurement and Evaluation of Object Colors*

Fig. 6–21. Total error ellipsoid of the first 10 preparations from Figure 6–19.

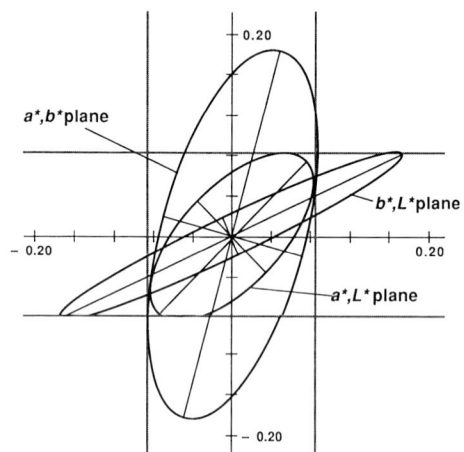

Fig. 6–22. Total error ellipsoid of the last 10 preparations from Figure 6–19.

long-term instabilities of the instrument, the same procedure as described for the total error must be employed:

> Maintain a time interval of 2 min between every two measurements; recalibrate colorimeter after every five measurements. In addition, perform each of the 20 measurements for this measurement error ellipsoid in the two-minute pause between every two of the 20 measurements for the total error ellipsoid.

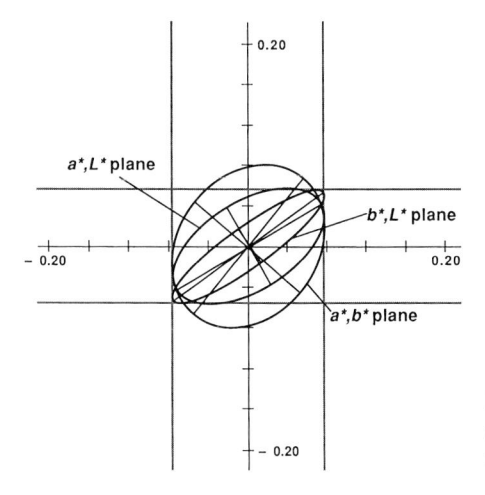

Fig. 6–23. Ellipsoid of measurement error, with no recalibration of instrument (measurement interval 3 min).

This methodology insures that the measurement error ellipsoid is contained inside the total error ellipsoid. The measurement error ellipsoid acquired in this way is shown in Figure 6-24. It is much smaller than the ellipsoid in Figure 6-23 and also differs in shape and orientation.

Certain fluctuations in the values resulting from specimen preparation can be eliminated or at least reduced if they are caused by

- Weighing errors
- Lack of constancy in dispersing
- Inhomogeneities due to flocculation or floating of pigments
- Surface defects produced when drawing down the pastes
- Temperature effects
- Moisture effects

A reduction in the preparation error can be achieved if the error components can be identified and the larger error contributions minimized through changes in the preparation process.

The g_{ik} values characterizing the preparation error are formed, as stated in equation (6.2), by writing the covariance matrices (total error minus measurement error) and inverting the resulting difference matrix. The relative contributions of preparation and measurement error to the total error can be determined by comparing the ellipsoids for the total error (Figure 6–19), the preparation error (Figure 6-25), and the measurement error (Figure 6-24). Note the differences in size, shape, and (in part) orientation of the ellipsoids in Figures 6-24 and 6-25. The preparation error ellipsoid is nearly as large as, and also quite similar in appearance to, the total error ellipsoid of Figure 6-19 (surfboard); thus it can be said that the preparation error is the dominant constituent of the total error in this example.

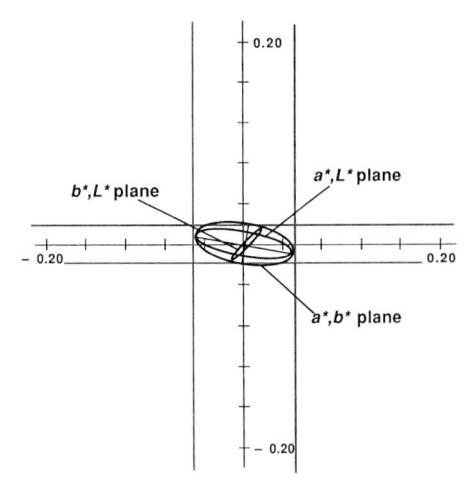

270   6 Measurement and Evaluation of Object Colors

Fig. 6–24. Measurement error ellipsoid contained in the total error ellipsoid of Figure 6–19. (Measurement interval 2 min, instrument recalibrated after every five measurements.)

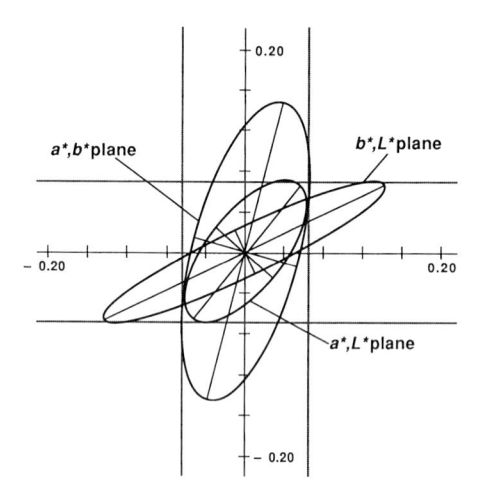

Fig. 6–25. Ellipsoid of preparation error, contained in the total error ellipsoid of Figure 6–19.

6.4 Significance

6.4.1 Calculation of Standard Deviations

Now, with the equations in Section 2.4.3, there is no further difficulty in calculating standard deviations. The standard deviations of the color coordinates will be calculated on the basis of a test error ellipsoid g_{ik} associated with these standard deviations. It is advisable once again to write a program using equations (2.60) to (2.67). A printout appears in Figure 6-26; the program was tested on the reference from the color difference test with red iron oxide pigments (see below).

When comparing standard deviations in order to evaluate errors, it should be kept in mind that the confidence intervals of the three-dimensional normal distribution here are systematically different from the confidence intervals of the ordinary one-dimensional distribution. In the 3D case, we consider the 70 % ellipsoid, assuming the associated standard deviation equal to σ, by analogy with the one-dimensional case. In the 1D case, ca. 95 % of the measured values fall between twice the positive standard deviation and twice the negative standard deviation ($\pm 2\sigma$), but the situation with the 3D distribution is different: As Table 2-4 shows, replacing the 70 % with the 95 % confidence interval changes the value obtained to $(7.81/3.665)^{1/2}\sigma \approx \pm 1.5 \cdot \sigma$. The other values of interest are found in a similar way; Table 6-3 offers a comparison.

The procedure for standard deviations of color differences is exactly the same. Another computer printout is given in Figure 6-27 (reference and test of a red iron oxide pigment).

Table 6–3. Comparison of confidence intervals for one- and three-dimensional normal distributions.

Confidence interval	1D	3D ($\sigma = \sigma(70\%)$)
70 %	$\pm 1.04\,\sigma \approx \pm\sigma$	$\pm 1\,\sigma \quad = \pm\sigma$
90 %	$\pm 1.64\,\sigma$	$\pm 1.31\,\sigma$
95 %	$\pm 1.96\,\sigma \approx \pm 2\,\sigma$	$\pm 1.46\,\sigma \approx \pm 1.5\,\sigma$
99 %	$\pm 2.58\,\sigma$	$\pm 1.76\,\sigma$
99.5 %	$\pm 2.81\,\sigma$	$\pm 1.87\,\sigma$
99.9 %	$\pm 3.09\,\sigma \approx \pm 3\,\sigma$	$\pm 2.11\,\sigma \approx \pm 2\,\sigma$

```
Program "ST. DEV. COORD."
Standard deviations of CIELAB coordinates, according to DIN 55 600
Input:*****************************************************
Pigment is Red iron oxide A
Medium: Alkydal L 64
Dispersing: Automatic muller
Colorimeter: Tristimulus Nr. 2
Means:    a* =  27.02        b* = 19.76        L*   =  39.40
C*(ab)       =  33.47                          h(ab) =  36.18°
Ellipsoid:
g(1,1) =  146.7      g(2,2) =  137.6     g(3,3) =   192.0
g(1,2) = - 50.49     g(2,3) = - 134.0    g(1,3) = - 30.52
Output:  *****************************************************

  Sigma(a*)    = .08    Sigma(b*) =  .09    Sigma(L*)    = .07
  Sigma(C*(ab)) = .10                        Sigma(h(ab)) = .18°
```

Fig. 6–26. Standard deviations of color coordinates (computer output).

```
Program "ST. DEV. DIFF."
Standard deviations of CIELAB color differences, according to DIN 55 600
Input:      ****************************************************************
Standard is Red iron oxide A              Test is Red iron oxide B
Medium: Alkydal L 64
Dispersing: Automatic muller
Colorimeter: Spectral Nr. 3
Means: Delta(a*) =  -.03       Delta(b*) = -1.20      Delta(L*)    =  -.71
       Delta(h(ab))  = -1.67°                         Delta(H*(ab)) = -.96
       Delta(C*(ab)) = - .71                          Delta(E*(ab)) = 1.39
Ellipsoid: g(1,1)  =  135.7    g(2,2) =   99.50       g(3,3)  =  484.7
           g(1,2)  =   36.37   g(2,3) = -165.1        g(1.3)  = -190.0
Output:     ****************************************************************
Sigma(Delta(a*))        =  .09
Sigma(Delta(b*))        =  .10
Sigma(Delta(L*))        =  .05
Sigma(Delta(h(ab)))     =  .14°
Sigma(Delta(H*(ab)))    =  .08
Sigma(Delta(C*(ab)))    =  .15
Sigma(Delta(E*(ab)))    =  .14
```

Fig. 6–27. Standard deviations of color differences (computer output).

In the case of color differences, it is commonly not enough to state the standard deviations alone, especially when these are of the same order of magnitude as the measured values. It is important to ask whether a color difference is significant, in other words, to perform a test of significance. This will take account of the number of measurements from which the mean color differences were obtained.

6.4.2 Test of Significance

In Section 6.3.3 it was shown that 20 preparations are ordinarily both necessary and sufficient for the calculation of standard deviations. The same applies to the test of significance. Once again, it is useful to write a computer program based on the information and formulas in Section 2.4.4. Such a program can be generated by combining our sample programs STANDARD DEV. ELL. for color differences (Figure 6-17) and ST. DEV. DIFF. (Figure 6-27).

The procedure begins with values evaluated by the CIELAB method (Section 2.3.3), that is, test and reference are measured one right after the other. The covariances to be formed will depend on the sequence of measurements, since they are supposed to describe relationships between two distinct measured quantities. Thus the results will include not only colorimetric relationships (e.g., a^* and b^* are calculated from the same CIE tristimulus value Y, and in addition the CIE color matching functions from which the X, Y, Z and ultimately the a^*, b^* are formed may overlap considerably) but also common differences in the method of preparing the color samples as well as short-term fluctuations of the colorimeters. It is therefore necessary to leave the measurements in the order in which they were acquired, forming the differences between two successive measurements of test and reference. If, on the other hand, an earlier ellipsoid is used again for the reference and it is desired to dispense with the separate determination of the g_{ik} for the color-difference ellipsoid, equation (2.64) can be employed again because there is no correlation between the measurements on the reference and on the test.

First, as in Section 6.3.1, the means of the color measures $\overline{a_R^*}$, $\overline{b_R^*}$, $\overline{L_R^*}$, $\overline{a_T^*}$, $\overline{b_T^*}$, $\overline{L_T^*}$ and of the differences $\overline{\Delta a^*}$, $\overline{\Delta b^*}$, $\overline{\Delta L^*}$ are formed. Because it is always the case that $\overline{\Delta a^*} = \overline{a_T^*} - \overline{a_R^*}$ and so on, it is enough to obtain two of each triplet of means (of a_R^*, a_T^*, Δa^*, etc.) by averaging, then simply use these relations to find the third. In series measurements against the same reference, the quantities $\overline{a_R^*}$, $\overline{b_R^*}$, $\overline{L_R^*}$ are already in hand and all that is necessary is to form the means of the Δa^*, Δb^*, ΔL^*. This can be done when the covariance matrix is calculated, since the same sums are used in both cases.

```
Program "SIGNI", Page 3
Significance of CIELAB color differences, according to DIN 55 600 ******
Output:  Significance  test  *********************************************
Number of measurements: n = 20
Conf. int.     Color differences (abs.)      Crit.value  Color difference is
-------------------------------------------------------------------------

   70%         Delta(a*)       =  .03        >  .02      significant
   95%         Delta(a*)       =  .03        ≤  .03      non-signif.
   99%         Delta(a*)       =  .03        ≤  .03      non-signif.

   70%         Delta(b*)       = 1.20        >  .02      significant
   95%         Delta(b*)       = 1.20        >  .03      significant
   99%         Delta(b*)       = 1.20        >  .04      significant

   70%         Delta(L*)       =  .71        >  .01      significant
   95%         Delta(L*)       =  .71        >  .01      significant
   99%         Delta(L*)       =  .71        >  .02      significant

   70%         Delta(C*(ab))   =  .71        >  .03      significant
   95%         Delta(C*(ab))   =  .71        >  .05      significant
   99%         Delta(C*(ab))   =  .71        >  .06      significant

   70%         Delta(h(ab))    = 1.67°       >  .03°     significant
   95%         Delta(h(ab))    = 1.67°       >  .05°     significant
   99%         Delta(h(ab))    = 1.67°       >  .06°     significant

   95%         Delta(H*(ab))   =  .96        >  .03      significant
   95%         Delta(H*(ab))   =  .96        >  .04      significant
   99%         Delta(H*(ab))   =  .96        >  .05      significant

   70%         Delta(E*(ab))   = 1.39        >  .03      significant
   95%         Delta(E*(ab))   = 1.39        >  .05      significant
   99%         Delta(E*(ab))   = 1.39        >  .05      significant
```

Fig. 6–28. Test of significance (n = 20; computer output).

6 Measurement and Evaluation of Object Colors

It should be pointed out again that only statistical fluctuations, in the sense of stochastically independent events, may enter into the variance and covariance calculation in the determination of standard deviations or significances. As a result, systematic errors, which lie outside the bounds of probability theory, must be carefully eliminated by appropriate means. Systematic errors include heating of the specimen in the instrument during the measurement, miscalibration of the instrument, and other such constant and regular deviations.

The standard deviation ellipsoid generated in the test of significance described here is valid for

– A given reference pigment with color locus a_R^*, b_R^*, L_R^*
– A given binder
– A given dispersing method
– A given type of colorimeter

```
Program "SIGNI", Page 3
Significance of CIELAB color differences, according to DIN 55 600*******
Output:  Significance test  *********************************************
Number of measurements: n = 1
Conf. int.     Color differences (abs.)     Crit.value  Color difference is
---------------------------------------------------------------------------
   70%         Delta(a*)      =  .05          ≤ .09     non-signif.
   95%         Delta(a*)      =  .05          ≤ .13     non-signif.
   99%         Delta(a*)      =  .05          ≤ .15     non-signif.

   70%         Delta(b*)      = 1.12          > .10     significant
   95%         Delta(b*)      = 1.12          > .15     significant
   99%         Delta(b*)      = 1.12          > .18     significant

   70%         Delta(L*)      =  .66          > .05     significant
   95%         Delta(L*)      =  .66          > .07     significant
   99%         Delta(L*)      =  .66          > .08     significant

   70%         Delta(C*(ab))  =  .60          > .15     significant
   95%         Delta(C*(ab))  =  .60          > .22     significant
   99%         Delta(C*(ab))  =  .60          > .26     significant

   70%         Delta(h(ab))   = 1.64°         > .14°    significant
   95%         Delta(h(ab))   = 1.64°         > .21°    significant
   99%         Delta(h(ab))   = 1.64°         > .25°    significant

   95%         Delta(H*(ab))  =  .95          > .08     significant
   95%         Delta(H*(ab))  =  .95          > .12     significant
   99%         Delta(H*(ab))  =  .95          > .15     significant

   70%         Delta(E*(ab))  = 1.30          > .14     significant
   95%         Delta(E*(ab))  = 1.30          > .20     significant
   99%         Delta(E*(ab))  = 1.30          > .24     significant
```

Fig. 6–29. Test of significance (n = 1, computer output).

All these items must therefore be noted every time a dispersion ellipsoid is calculated. If such an ellipsoid already exists, the procedure for determining the significance can also be applied to other color differences under the same conditions (i.e., same reference, same vehicle, same dispersing method, same instrument type).

Figure 6-28 is a significance test listing for a red iron oxide pigment, based on measurements of 20 specimens (n = 20). It is worth mentioning here that Δa^* is so small for the higher confidence levels (95 % and higher) that it is no longer significant, while all the other color differences are still significant. Another example, this one for a single color measurement (n = 1) with an already existing dispersion ellipsoid, appears in Figure 6-29. The analysis is based on one measurement of the 20 in the same example. It should be noted that the critical values are higher, but the only additional rejection is for Δa^* and confidence level 70 %.

6.5 Acceptability

6.5.1 Specimen Preparation

Growing interest in colorimetric production control, not just for pigments but also for textile dyes, has slowed down the process of creating test methods based on proposals made in the literature. In any such method, two aspects must be given special attention:

– The preparation of color samples needed for matching
– The clear statement of task for colorists who take part in matching

The determination of acceptability calls for a collection of samples. When testing the acceptability of pigments, full-shade systems (air drying systems) as discussed in Section 6.2.3, with side-by-side drawdowns of test and reference, can be prepared. It is recommended that color samples be created with systematic departures all around the color locus of the reference pigment. Such a group of samples might be prepared by adding to the reference small amounts of suitable pigments in the directions of the four colors red, yellow, green, and blue, while seeking to remain in the a^*,b^* plane, that is, maintaining $\Delta L^* = 0$ (see Figure 6-30). This requirement comes about because lightness differences are weighted much more heavily in visual evaluation than are CIELAB chroma or hue differences of equal magnitude. Experience in matching tests shows that a lightness difference as small as $|\Delta L^*| > 0.3$ between reference and test will cause the color differences – even if chroma and hue are equal – to be deemed unacceptable. This point again calls attention to a weakness of the CIELAB system: Lightness differences are underestimated. In addition to the

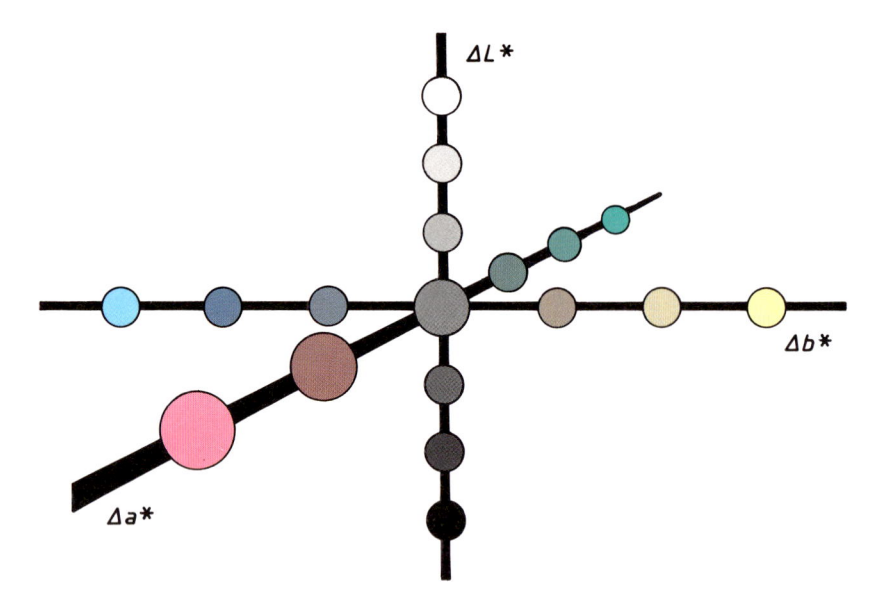

276   6 Measurement and Evaluation of Object Colors

Fig. 6–30. Color samples deviating systematically from a gray reference (the spacing between samples is greatly exaggerated for clarity).

samples with pure hue-direction deviations, further samples with deviations in the lighter/darker direction should also be manufactured.

The available set of samples must, of course, be sufficiently large. The number of samples to be matched can grow quite large if every one is laid next to every other one in the form of side-by-side drawdowns (like test and reference). An example: Suppose there are three difference steps in the four color directions and in each light/dark direction; then the total number of color samples including the reference is 3·6 + 1 = 19 (see Figure 6-30). If every sample is laid next to every other, (19·18)/2 = 171 side-by-side drawdowns must be matched. The number of "double" drawdowns increases rapidly with the number of deviating steps away from the reference. Adding just one more step in each of the six directions raises the total number of double drawdowns to (25·24)/2 = 300. Numbers of this order are also required because the orientation of the acceptability ellipsoid is initially unknown. It is therefore useful to do preliminary matches by colorists of a few typical drawdowns in order to determine rough acceptability limits. The difference steps selected on the basis of these preliminary trials should then extend out to twice the acceptability limits, so that the desired ellipsoid will definitely lie inside the region spanned by the samples.

The trick of laying every color sample next to every other one not only increases the number of double drawdowns to be matched, but also permits a closer approach to the desired normal distribution. In any case, not too much importance should be attached to departures from the normal distribution that result from this procedure. If the requirement of normally distributed measurements were taken too strictly, there would be virtually no possibility of applying statistical methods to practical problems, for intact normal distributions are quite rare in practice. Textbooks state that it is sufficient for the random sample to have a single-peaked distribution, and this will always be the case here because the reference color locus lies at the center of the ellipsoid.

6.5.2 Matching and Color Measurement

Well-defined conditions for color matching have already been implied by the requirement of a clear task statement for the colorists. Where the resultant acceptability ellipsoid is intended for later use in production control, say of pigments or paints, the colorists should be selected from those who are daily involved in batch inspection, so that they can take advantage of their experience in this form of matching. The difference between acceptability* and perceptibility of a color difference is that perceptibility characterizes the visual threshold for small color differences, while acceptability represents the range of color differences that manufacturer, purchaser, user, and consumer are prepared to accept. Acceptability must, further, be in accord with the variations resulting from the methods used to produce the coloring materials and media. If acceptability limits are set so narrow that they are smaller than the fluctuations due to process variations, it is no longer possible to make a product in conformity with the standard. If, on the other hand, they are set too wide, there is a danger that they will be rejected by consumers as excessive. All these relationships and insights become part of the experienced colorist's background – and thus the task in hand is to embody this experience in the acceptability ellipsoid, so that the computer can be supplied with the information needed for colorimetric production control.

The following points are crucial for color matching proper, and must be treated with special care**:

– The illuminant under which colors are matched
– The geometric conditions of color matching
– The physiological conditions of color matching

The *illuminant* should be chosen from among the well-known CIE standard illuminants (Section 6.1.1). For artificial average daylight, corresponding to illuminant D65, the standard gives a method for determining the quality of the

---

* The use of the term "tolerance" in place of "acceptability" is sometimes noted. This usage is not recommended, because tolerances correspond just as well to perceptibility.
** See "Color Matching" in Appendix 1.

lamp spectrum with the help of test color pairs. A factor working against the use of illuminant D65, however, is that no satisfactory technical implementation has yet been devised; CIE standard illuminant C is better realizable. The lamp service time must be monitored (elapsed time indicator).

The *geometric* conditions have to do with the direction of incidence and the viewing direction (e. g., 0° incidence and 45° viewing), the aperture angle of the irradiation, and the position and size of the specimen.

Physiological conditions include the visual capabilities of the viewer (ability to distinguish colors, color-normality), the luminance level, the state of adaptation of the eye, the surroundings (e. g., gray, CIE tristimulus value $Y = 25$), and the field of view.

The conditions listed here must be fixed exactly; they should correspond to the situation previously existing in coloristic-visual production control. It is helpful to have a suitable matching room (light box) painted gray on the inside, in which appropriate light sources and a pivoting specimen stage with angular scale are mounted.

In matching, all the double drawdowns are shuffled at random like a deck of cards, then matched in a single pass. The color samples are arranged by the colorist in two groups:

– Conforming with the standard ("Yes"), that is, accepted
– Not conforming with the standard ("No"), that is, not accepted

In reports of matching trials of this type, it is sometimes stated that even experienced colorists suffer from a "test" psychosis and feel themselves in a stress situation as a result. The consequence is usually an excessively strict set of decisions, that is, too many rejections. To counter this effect, it may be useful to allow two passes for matching the samples. In the first pass, the task is to classify as "color difference perceptible" or "not perceptible." In the second pass, the actual task is set: to classify as "color difference acceptable" or "not acceptable." In this way, the colorist is enabled to differentiate better between color differences that he can just see (though this question is not asked) and color differences that he does perceive but just regards as conforming. These two passes can be combined into one if the colorist is asked to form three groups:

– No difference perceptible: accepted
– Conforming with the standard despite difference: accepted
– Not conforming with the standard, difference too great: not accepted

The first and second groups together form the "Yes" group; the third is the "No" group. The number j of "Yes" evaluations is tabulated for each of the numbered samples.

Naturally, special care is necessary in color measurement as well, because any error made in this operation can lead to incorrect decisions later on in colorimetric production control. The requirement must be that the color measurements take place within a short time of matching by the colorist, because samples may be subject to color changes due to aging phenomena. It is desirable to sandwich the matching task between two color measurements, whose results should be averaged. Every individual measurement is tabulated under the same sample number as in matching.

6.5.3 Evaluation and Example

Calculation of the ellipsoid coefficients with the likelihood function of equation (2.78) requires an appropriate computer program. Finding the maximum amounts to determining the zeroes of the partial derivatives of ln M. All methods for finding parameters by iteration, such as Newton's method (there is no difficulty in forming the second derivative), as well as other numerical methods, are candidates for this operation. Once the initial values for the g_{ik} and p are specified, the Δa^*, Δb^*, ΔL^* associated with specimen number i are input to equation (2.34) together with the number j_i of "Yes" judgments:

$$\delta s^2 = g_{11} \Delta a^{*2} + g_{22} \Delta b^{*2} + g_{33} \Delta L^{*2} +$$
$$+ 2 g_{12} \Delta a^* \Delta b^* + 2 g_{23} \Delta b^* \Delta L^* +$$
$$+ 2 g_{13} \Delta a^* \Delta L^*$$

Aside from $p = 0$, the initial values for the iterative procedure should be as close to the expected values as possible given the approximate knowledge of the acceptability ellipsoid. This will prevent convergence problems. If little is known about the expected ellipsoid, it is also possible to begin with values $g_{11} = g_{22} = g_{33} = 1$ and $g_{12} = g_{23} = g_{13} = 0$. A calculation can take from minutes to hours, depending on the number of double drawdowns, the number of colorists, the type of computer, and the iterative method chosen.

As discussed in Section 2.4.5, this procedure yields the coefficients g_{ik} of the acceptability ellipsoid plus a "false alarm rate" $p = 1 - q$. This rate points to the reproducibility of the color-matching experiment; if one works carefully, it can be held as small as desired. If there are problems with convergence, different initial values for the g_{ik} should be tried.

The resulting acceptability ellipsoid can, for example, form the basis for computer-aided colorimetric production control. The following approach seems practicable: The 50% and 75% ellipsoids are calculated by dividing the g_{ik} by the δs^2_{den} factors from Table 2-4. Two concentric ellipsoids result; these divide the color space into three regions. If a color locus is inside the inner ellipsoid, it is "conforming to the reference"; if it is outside the outer ellipsoid, it is "not conforming." But if the color locus is in the region between the two ellipsoids, it can be identified as "tentatively conforming."

The confidence limits of an acceptability test must not be confused with the confidence limits of the tolerances for mass-produced articles in production control (e.g., length of nails, cross-sectional area of thermometer capillaries).

6 Measurement and Evaluation of Object Colors

The 50% confidence level here means that half of colorists will accept this specimen; the 75% level means that three-fourths of colorists will accept the specimen. Accordingly, 50% can be viewed as the limit of acceptability, since here there are equal numbers of "Yes" and "No" judgments. A specimen that more than half of colorists reject is thus rightly deemed "not accepted." What is more, the ellipsoids represent the result of psychophysical experiments (color matching by colorists), and so human shortcomings (errors, blunders, and prejudices) enter into them and "blur" the acceptability limits. Ultimately, the uncertainty due to the significance, that is, the effect of the (much smaller) test error ellipsoid, must also be taken into account. These are the reasons for the recommended confidence limits; the "tentative" region defined by them provided a good buffer between the two effects (see also Section 6.5.4).

Figure 6-31 presents an example of two such ellipsoids. The drawing was produced by a plot program as discussed in Section 2.4.2. Table 6-4 lists the ellipsoid coefficients. The use of a computer to determine conformity or nonconformity can then be implemented roughly as illustrated in the following examples.

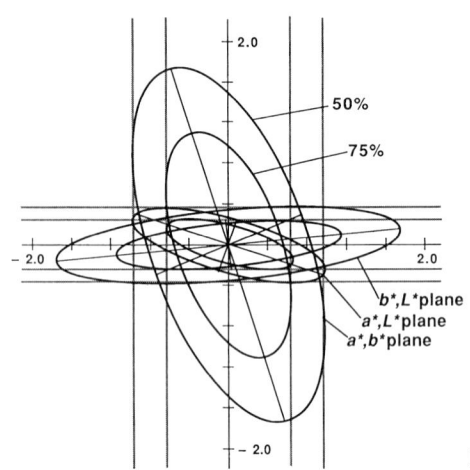

Fig. 6–31. Acceptability ellipsoids (projections).

Suppose there are three specimens, which are to be tested for conformity to the standard. Their color differences relative to a reference are given in Table 6-5. Using the g_{ik}, the program calculates the squared distances from the reference $\delta s_{50\%}^2$ and $\delta s_{75\%}^2$. These are entered in the right-hand columns of Table 6-5. The values indicate that

Specimen 1 is conforming because $\delta s_{75\%}^2 < 1$

Specimen 2 is tentatively conforming because $\delta s_{50\%}^2 < 1$ and $\delta s_{75\%}^2 > 1$

Specimen 3 is not conforming because $\delta s_{50\%}^2 > 1$

Table 6–4. 50 % and 75 % acceptability ellipsoids for a pigment.

Ellipsoid	g_{11}	g_{22}	g_{33}	g_{12}	g_{23}	g_{13}
50 %	2.063	0.520	9.83	0.558	0.0845	2.069
75 %	4.97	1.253	23.69	1.345	0.2037	4.99

Table 6–5. Inspection of three specimens for conformity to the standard.

Specimen	Δa^*	Δb^*	ΔL^*	$\delta s^2_{50\%}$	$\delta s^2_{75\%}$	Judgment
1	0.2	0.3	0.1	0.38	0.92	Yes
2	0.2	0.3	0.2	0.76	1.84	Tentative
3	0.3	0.4	0.2	1.05	2.54	No

In this way, the acceptability can be calculated by the computer and output as a "Yes," "No," or "Tentative" on the printer.

The form of the ellipsoids in Figure 6-31 once again makes it clear how much the lightness differences are underestimated by the CIELAB system. Strictly speaking, such a statement can be arrived at only with perceptibility ellipsoids, but in fact there is little doubt in specialist circles that acceptability is not entirely independent of perceptibility. Perceptibility ellipsoids in an intact color-difference system would have to be spheres in the color space; acceptability ellipsoids should be expected to display a greater similarity to a sphere. Practical trials have shown again and again that this is not the case. Accordingly, in no case known up to the present can the acceptability be characterized by citing a ΔE^*_{ab} value alone (corresponding to the radius of this sphere). Often the agreed-on limits include not only ΔE^*_{ab} but also the components ΔC^*_{ab} and ΔH^*_{ab}. One must, however, keep in mind that this expedient is nothing more than that, because such limits do not exactly divide acceptable from non-acceptable color differences. Figure 6-32 illustrates this statement clearly. If limits $[\Delta C^*_{ab}]$ and $[\Delta H^*_{ab}]$ were taken, the color locus P would be wrongly accepted; in reality, it should be rejected because it lies outside the ellipsoid. Thus it has to be stressed that a statistically correct division into acceptable and unacceptable can be accomplished only with the acceptability ellipsoid.

In the field of standards development, the first efforts are now being made to deal with the acceptability problem. In DIN 6175, however, one should perhaps see nothing more than a preliminary attempt to establish the first solid methods. But the prescribed use of ΔE^*_{ab} alone is in no way adequate, as has been stated already. This value should at least be refined by breaking it down into components, even if doing so does not make this technique capable of replacing the acceptability ellipsoid method. If sufficient experimental data are

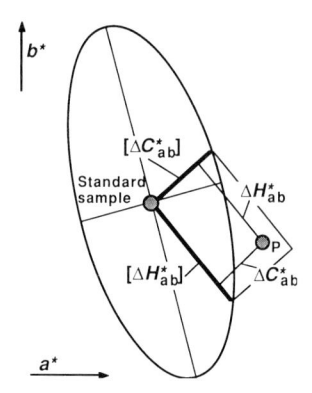

Fig. 6–32. Acceptability ellipsoid. $[\Delta C_{ab}^*]$ and $[\Delta H_{ab}^*]$ are agreed-on limits. The color locus P is wrongly judged conforming to the standard.

available, it is conceivable that the ellipsoid coefficients g_{ik} can be made accessible to a "running" calculation, that is, treated as functions (linear wherever possible) of the color locus.

In a further step, the "Color" subcommittee of the German Standards Committee (DIN) is preparing a report titled "Methods for Agreement on Color Tolerances." Its report, which should be out soon, should help remedy the information gap that now exists among manufacturers and users in this field.

A trend that should not be emulated is the practice of some software vendors who deal with the acceptability of textile dyes but do not disclose the color-difference formulas or the test criteria used. One might well ask whether this secrecy is an attempt to mask the inadequate scientific basis for the algorithms! Even if it is not, the result is very negative because it prevents discussion of the systems, possible recommendations for improving the color-difference formulas, and – ultimately – advances in what is today the most important specialty area of our field.

6.5.4 Acceptability and Test Errors

It has been shown that the acceptability ellipsoid is a useful criterion for the approval of color formulations, since it decides what color differences are tolerated. The test error ellipsoid (total error ellipsoid, see Section 6.3.3), in contrast, tells only how well a color laboratory can perform the tests. Nonetheless, test errors are of some importance in acceptability testing, because from what has been said it should be clear that the acceptability ellipsoid can be meaningfully applied only when it is much larger than the total test error ellipsoid. The term "test error ellipsoid" here denotes the 2σ ellipsoid, whose confidence interval roughly corresponds to the 3σ interval in the one-dimensional case (see Table 6-3).

If one uses the method described in Section 6.5.3, with 50% and 75% acceptability ellipsoids, this problem does not arise. The reason is that the color

6.5 Acceptability

measurements used in finding the acceptability ellipsoids already contained errors. It is true that the errors average out on the whole, but they do contribute to increasing the standard deviation, and thus the colorist's uncertainty. If, accordingly, the region between the 50% and 75% ellipsoids is regarded as a "tentative" region, this "gray area" takes care of the test error in the great majority of cases. As a quantitative treatment shows, the error ellipsoid in this case can have linear dimensions as large as 1/5 of those of the 50% acceptability ellipsoid. If this is so, then the exact size and shape of the total error ellipsoid no longer play any decisive role, because the figure vanishes in the "tentative" region. This 1:5 requirement is more than satisfied in many practical cases because test errors are commonly much smaller.

The question now arises what has to happen if the linear dimensions of the error ellipsoid are larger than 20% of the acceptability ellipsoid. First, the color lab must make an effort to reduce the total error below this threshold. The measures by which this can be accomplished are described in Section 6.3.3.

It is desirable to state the 1:5 limit in terms of the coefficients of the two ellipsoids. There is a linear factor of five between the two ellipsoids if the traces of the two ellipsoid matrices differ by a factor of $5^2 = 25$:

$$sp\,(\mathbb{G}_T) > 25\,sp\,(\mathbb{G}_A)$$

or, written out in full:

$$g_{11}^T + g_{22}^T + g_{33}^T > 25\,(g_{11}^A + g_{22}^A + g_{33}^A)$$

where $\mathbb{G}_A$ is the acceptability ellipsoid and $\mathbb{G}_T$ is the test error ellipsoid. If this limit cannot be maintained, it will be necessary to combine the test error ellipsoid with the acceptability ellipsoid. Two cases must be distinguished:

– A color difference that is just acceptable can be rejected as a result of test error, because it appears to be shifted outside the acceptability ellipsoid
– A color difference that is just unacceptable can be accepted as a result of test error, because it appears to be shifted inside the acceptability ellipsoid

Only the second of these cases is dangerous, since it gives the appearance of acceptability when there is none. For this reason, the best way to deal with excessively large test error is to diminish the acceptability ellipsoid by the test error ellipsoid. This must be done through the covariance matrices, as shown in Section 6.3.3. We then have

$$\mathbb{G}_A^{-1} - \mathbb{G}_T^{-1} = \mathbb{G}_{A(corr)}^{-1}, \text{ that is,}$$
$$\mathbb{V}_A - \mathbb{V}_T = \mathbb{V}_{A(corr)}.$$

6 Measurement and Evaluation of Object Colors

This relation is illustrated in an obvious way by Figure 6-33. After subtraction, the matrices must be re-inverted so that the corrected acceptability ellipsoid $\mathbb{G}_{A(corr)}$ can be obtained. The correction must be carried our with both the 50% and 75% acceptability ellipoids, using the same error ellipsoid.

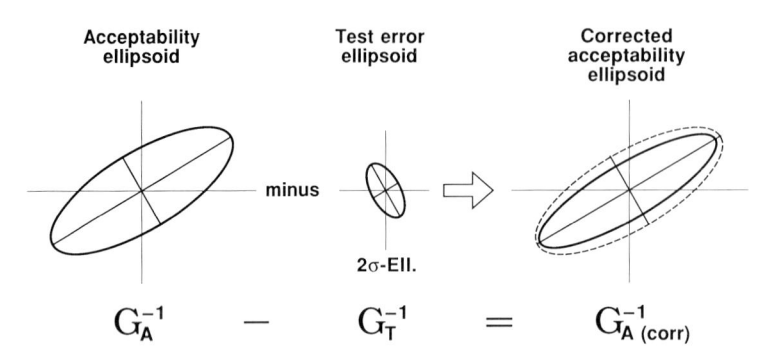

Fig 6-33. Schematic representation of the integration of the test error ellipsoid into the acceptability ellipsoid.

7 Determination of Hiding Power and Transparency

7.1 Measurement of Film Thickness

7.1.1 Selection of Method

Any determination of hiding power, transparency, or color strength requires a knowledge of the spectral scattering and absorption coefficients; in other words, these must generally be determined first. For this purpose, paint or plastic layers are prepared and used in reflection measurements with suitable photometers. This information is enough for the calculation of the *relative* scattering and absorption coefficients (relative to a reference pigment). For the *absolute* quantities, which are needed for the determination of hiding power and transparency, reflection measurements alone are not sufficient: the thickness of the layer must also be determined, as the equations in Section 3.3.3 show.

The significance of film-thickness measurement in the determination of scattering, absorbing, and hiding power and transparency is immediately evident from an error analysis: If one forms the differential of $\ln S$ in equation (3.124), one sees that the percent error of the scattering coefficient $dS/S = -dh/h$ is absolutely equal to the percent error of the film thickness. Films used in such determinations often have to be so thin that the substrate has a marked effect on the reflectance; with highly pigmented media, films only 25–35 µm thick may well occur. If the error in the thickness measurement is 3 µm (a level that is not attained by all methods), the relative error will be ca. 10%, and this will propagate in linear fashion to the scattering or absorption coefficient. The fact that the film-thickness measurement alone has such a heavy influence on the final result is not tolerable, however, because smaller errors would be necessary even if the only requirement imposed were that of conformity to a standard.

What follows is a survey of methods commonly used to determine film thickness. These are also listed in standards*, along with brief descriptions and error ranges. We follow this classification, adding the pneumatic method and also creating higher-level groupings.

---

* See "Film Thickness" in Appendix 1.

- Mechanical methods
  Dial-gauge or micrometer methods
  Pneumatic method
- Optical methods
  Wedge cut method
  Profile measuring method (only for transparent films with known refractive index; not nondestructive)
  Microscopic measurement of film thickness by cross section method (not nondestructive)
- Chemical methods
  Gravimetric method
  Titrimetric method (not nondestructive)
- Electrical and magnetic methods (require metallic substrates)
  Coulometric method
  Capacitive method
  Eddy-current method
  Magnetic method
- Methods based on absorption or scattering (not applicable to all pigments or all substrates)
  Beta-ray backscatter method
  X-ray fluorescence method

The selection of methods suitable for the determination of the absorbing, scattering, and hiding power must be based on the following considerations:

- Range: 0–2000 μm
- Accuracy: ± 0.5 μm (better still, ± 0.2 μm)
- Noncontact measurement (no contact with the film)
- Nondestructive measurement
- Measurement on small areas, under 4 mm^2
- Measurement in same film areas where photometric measurements are performed
- Measurement independent of vehicle and pigment, pigment volume concentration, and film surface quality
- Method suitable for opaque black/white glass plates between 2 and 8 mm thick as substrates
- Measurement of mean film thickness (i.e., for films with no smooth surface) possible if necessary
- Partly or fully automated measurement with printout of results

These requirements imply right away that most of the methods listed above are not candidates. A short list includes just four methods: dial-gauge/micrometer, wedge-cut, gravimetric, and pneumatic. All other techniques are unsuitable, chiefly because they are not nondestructive or do not permit the use of glass plates as substrates. Nor do the other methods satisfy the accuracy requirements. The first three of the short list methods, indeed, do not entirely fulfill some of the requirements (in particular, errors of measurement are too large; what is more, the dial-gauge/micrometer and wedge cut methods also require the destruction of the film, if only in small areas), as will be discussed in depth.

7.1.2 Gravimetric, Wedge Cut, and Dial-Gauge/Micrometer Methods

The basic procedure in the *gravimetric methods* is to determine the mass of the film per unit area by weighing, in an analytical balance, a known area of the substrate with and without the film and taking the mass difference. Since the density of the coating material at the time of weighing is known, the mean film thickness can then be calculated. The overall uncertainty of the method is around ± 5 %; films ranging from under 1 µm up to 1 mm in thickness can be measured.

These methods have some drawbacks that make them more difficult to use and, in terms of accuracy and speed, severely limit their application in determining the scattering coefficient, absorption coefficient, and hiding power. They are integral methods yielding only a mean film thickness over a large known surface area: Thickness variations cannot be detected and characterized; in other words, these methods do not meet the test that the thickness should be measured where the reflection measurement is done. What is more, the limited relative sensitivity of commercial analytical balances means that the accuracy is strongly dependent on the weight of the substrate. For example, the balance must have a relative sensitivity better than $2 \cdot 10^5$ (e.g., 1 mg in 200 g) if 1 µm variations are to be determined to 10 % accuracy on glass substrates. Another problem with these methods is that the coating density must be exactly determined or calculated; in pigmented coatings, this usually means that the PVC, the density of the pigment, and the density of the medium must be known.

In the *wedge cut method*, a sharp cutting tool is used to make a scratch in the film, creating a wedge-shaped cut with a certain inclination angle. The specimen is placed under a measuring microscope or profile measuring microscope and the horizontal dimension of the cut is measured. From this measurement and the known section angle, the film thickness is calculated. The uncertainty is around ± 10 %, and films from 10 µm to 2 mm thick can be measured. In coating systems, it is also possible to measure the thickness layer by layer, provided the layers can be distinguished by color or structure.

Off-the-shelf instruments – a cutting device and a measuring microscope – can be employed. The cutter has a hard-metal blade (usually replaceable) having a known edge angle and a guidance system that positions the blade in a consistent way relative to the film surface. A suitable measuring microscope will have a magnification of 50× and a scale with 100 divisions. Commercial profile measuring microscopes have 200–400× magnification. When a measuring microscope is used to determine the projection b of the section flank, the film thickness h is calculated from b and the section angle α (see Fig. 7-1). The film thickness h is read directly off the profile cut under the profile measuring microscope.

A number of factors can falsify the results given by these methods:

- Surface roughness requires more individual measurements if a representative average is to be obtained. The rougher the surface, the more uncertain the measurement
- Deformation of the coating occurs frequently when none-too-strong coatings are cut; as a result, the method cannot be used for such films
- Errors can also result from breaking of the cut edges, especially with extra-brittle paint coatings

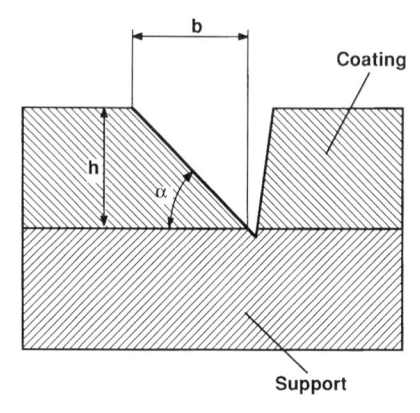

Fig. 7–1. Wedge cut (schematic).

Even though the cut requires only a relatively small area, this method cannot really be classed as nondestructive, because the cut has to be made quite near the place where the reflection measurement is to be carried out.

The *dial-gauge/micrometer method* involves measuring the height of the coating surface relative to a fixed datum plane and then measuring the height of the substrate after removal of the film. The difference in height gives the film thickness. This method has been reported to have ± 20% uncertainty. Below 20 µm, it has an absolute uncertainty of at least ± 3.6 µm. Coatings from 3 µm up to several mm thick can be measured. A commercial standard instrument comprises a measuring table, a fixture for the specimen, and a dial gauge mounted so that it can be moved over the table. Any mechanically contacting instruments or suitable micrometers can be employed, regardless of whether measured values are acquired mechanically, optically, or electronically. The measuring area of the dial gauge is a sphere or spherical cap whose radius, for metrological reasons, should be as large as possible (at least 1.5 mm); the force applied must not exceed a maximum value (1.5 N). The requirements are even more stringent for the measurement of organic coatings, such as the ones of concern here: The radius must be ca. 30 mm, the force between 0.1 and 0.5 N. Ordinary varnish solvents or paint removers are used to remove the coating from the measurement area.

Aside from not being nondestructive, this method has the drawback that the measuring device exerts a mechanical reaction on the object being measured, and the pressure is great enough to produce errors through local deformation of the coating even when the force applied is relatively slight. There are limits on how much the measuring sphere of the dial gauge can be enlarged, because otherwise a sphere with too large a diameter will measure only maximum values on rough surfaces, not the desired mean values over small specimen areas. In order to extrapolate to zero applied force, extensive series of measurements and costly calculations would have to be done.

7.1.3 Pneumatic Method

The pneumatic method is really a variety of the dial-gauge method that almost ideally satisfies all the requirements of Section 7.1.1. It permits the nondestructive, non-contact, and also very accurate point-by-point determination of film thickness at locations where the film reflectances are also measured. The method is independent of the binder and pigment content and, in the case of rough surfaces, permits the measurement of the mean coating thickness in small regions, which is the quantity of interest in practice.

Pneumatic distance measurement techniques are based on the fact that the volume of an air stream is governed by the narrowest cross section of the flow channel. Any change in the smallest cross section will alter the quantity of air flowing through the channel per unit time. In the pneumatic method of measuring film thickness, the air stream from a measuring nozzle is directed against the surface of the test film in such a way that the spacing between the film and the nozzle outlet is the most constricted part of the channel. With the substrate fixed in place, differences in coating thickness alter the smallest flow cross section. Two kinds of measurement are made on this principle, volume and pressure. The method described here employs the pressure principle, because the components available on the market for this method are more resistant to wear and less susceptible to malfunction, over a wider linear range, than the ones available for the volume method. In the pressure method, the pressure drop is measured within the air stream ahead of the exit from the nozzle. The pressure drop is a function of the flow rate and thus of the slit size, that is, the clearance between the nozzle and the coating surface (Fig. 7-2). Off-the-shelf instruments (for other purposes) offer transformation ratios of up to $2 \cdot 10^4$; that is, a difference of 1 µm results in a 20 mm shift in the manometer water level. Special models have accuracies down to 0.025 µm.

Figure 7-3 is a diagram of a pneumatic instrument for measurements of film thickness. The names of the parts are indicative of their functions. The ball-bearing posts (2) insure plane-parallel movement of the nozzle guide plate (3) over the baseplate (1), exact to ± 1 µm. The column assembly and the measur-

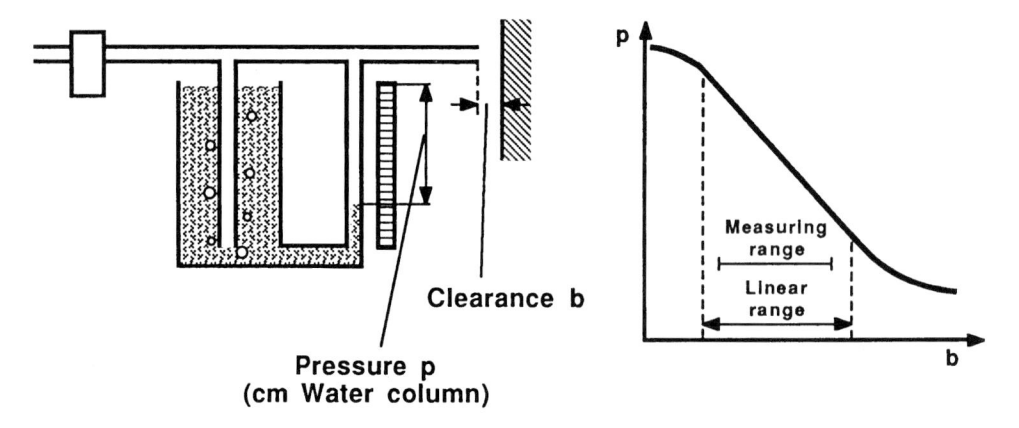

Fig. 7–2. Pneumatic measurement of film thickness (schematic diagram of pressure method).

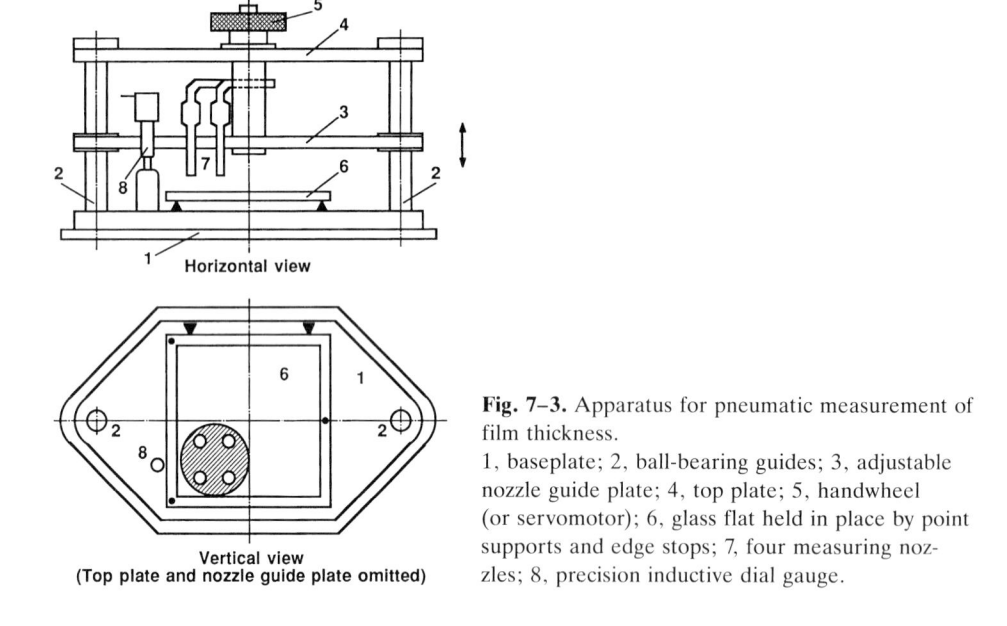

Fig. 7–3. Apparatus for pneumatic measurement of film thickness.
1, baseplate; 2, ball-bearing guides; 3, adjustable nozzle guide plate; 4, top plate; 5, handwheel (or servomotor); 6, glass flat held in place by point supports and edge stops; 7, four measuring nozzles; 8, precision inductive dial gauge.

ing nozzles are available commercially (see [26] in Section 3.7); the apparatus as a whole must be shop-fabricated.

To determine the scattering and absorption coefficients or the hiding power and transparency of paint films, black or white plane-parallel glass plates (100 mm × 100 mm × 5 mm) are used. The reflectance of the applied coating is measured on each of four circular regions; the thickness of the paint film is then measured at four fixed points in the same regions, with the four nozzles of the pneumatic apparatus. The glass plate rests on three points in the apparatus and is fixed in place by three edge stops.

Because the linear range of the nozzles (Fig. 7-2) is not wide enough to give the needed high sensitivity at the coating thicknesses encountered in practice, the pneumatic nozzles are used solely as null indicators. The nozzle guide plate is lowered to a constant, specified distance away from the coating surface. The actual clearance measurement is done separately, with a precision inductive dial gauge to the side of the glass plate; over a linear range of up to 2 mm, this guarantees a maximum error no greater than 0.02 µm.

When the measurement is performed semiautomatically, the nozzle guide plate is lowered by hand and the value for the uncoated glass flat is recorded. In the subsequent measurement on the coated plate, this value is input by means of a digital setpoint adjuster. In the fully automatic device, the plate is raised and lowered by a servomotor controlled by the nozzles. The absolute film thickness is output in analog or digital form (with the aid of an analog-digital converter).

High accuracies can be achieved with the pneumatic method. The absolute error of measurement can be held under 0.3 µm, which corresponds to coefficients of variation smaller than 1% for a film 30 µm thick. The accuracy is higher than that of the dial gauge method because no mechanical pressure is exerted on the specimen.

7.2 Scattering and Absorption Coefficients

7.2.1 Kubelka-Munk Coefficients S and K of White Pigments

Methods for determining the scattering coefficient S are set forth in standards.*

What these methods yield is not the absolute S but a relative (i.e., referred to a reference pigment) scattering coefficient per unit mass. Such results can be meaningfully compared only if the measurements are performed with equal masses of pigment or equal PVC values. Therefore, the S values obtained (m^2/L) of the pigment/binder mixtures are then referred to the pigment mass concentration PMC, i.e., to the mass of pigment (m^2/g).

The exclusive use of relative values has to do with results from round-robin tests performed by DIN, which showed that absolute scattering coefficients could be reproduced comparatively well in the same laboratory but no ade-

---

* See "Scattering Power, Relative" in Appendix 1.

quate agreement was possible from one laboratory to another. In contrast, relative scattering powers showed quite satisfactory agreement. Those results are no longer considered entirely representative for present-day conditions, because essential technical elements of specimen preparation (such as dispersing equipment and methods) and color measurement (such as the calibration of instruments) have been substantially improved since they were reported. There are, accordingly, virtually no objections to absolute measurement today, even when the same techniques are employed.

The standards distinguish between

– Black-ground method
– Gray-paste method

(Because of its intimate link to the determination of lightening power in the Beer's law region, the gray-paste method will be described in Section 8.3.2.)

Black-ground method

The black-ground method* is based on evaluating the Kubelka-Munk equations over a black substrate, as described in Section 3.3.3. The German standard initially leaves preparation methods (medium used, pigment concentration, dispersing conditions, coating application conditions) open – these are to be agreed on between testing laboratories – and limits itself to specifying the measurement and evaluation procedure. The pigment for test and the reference pigment are dispersed in the agreed-on medium. From the resulting pigmented media, a non-hiding coat over black and an optically infinitely thick coating are prepared. Instead of the reflectances in the middle of the visible spectrum, the CIE tristimulus value Y is measured for each coating. The calculation employs this Y value, plus either the pigment mass concentration (β) and film thickness (h) or the pigment mass per unit area (m_F) of the non-hiding coat**; in either case, the methods of determination must be agreed on. This point shows a second reason that relative coefficients are attractive: the possibility of doing away with the little-loved measurement of film thickness. (In other words, if the pigment mass per unit area m_F can be obtained without it, e.g., from the formulation, then the film thickness does not have to be measured.) An example of this procedure will be described further on.

The colorimeter should be set up to measure the CIE tristimulus value Y. The parties must agree as to whether method A or B will be used for the measurement. Black glass plates having a plane surface and CIE tristimulus values $Y < 5$ are to be used as substrates; these are available commercially. The test

---

* See "Scattering Power, Relative, Black-ground Method" in Appendix 1.
** Not to be confused with the pigment surface concentration (see Section 4.1.4).

7.2 Scattering and Absorption Coefficients

cannot be done with pastes (no well-defined film thickness), only with solid films or plastic films or plastic wafers. In the last two of these cases, a contact liquid (e. g., immersion oil) must be used to establish optical contact between the specimen and the substrate.

Measurement method: Measure CIE tristimulus values Y for the coatings under standard illuminant D65 or C at several points of the coating. The number of points to be measured should be dictated by the size of the sample port and the size of the surface regions on which the film thickness and pigment mass per unit area are determined. Determine Y values to four decimal places and average the measured values on each coating to obtain Y_∞ (hiding coat) and Y_b (non-hiding coat). Select the thickness of the non-hiding coat such that the difference between Y_∞ and Y_b is at least two units.

Evaluation method: Apply surface correction (eq. 3.11) to measured CIE tristimulus values Y, setting $Y/100 = \varrho$ (or R, depending on configuration) in each case. When measuring high-gloss specimens by method B of Section 6.1.2, add 0.04 to the ϱ values before applying the correction. From the corrected reflectances ϱ^*, calculate auxiliary variables a and b with equations (3.118) and (3.119). From a, b, and the corrected reflectance ϱ_b^*, calculate to four decimal places the product Sh from equation (3.124):

$$Sh = \frac{1}{b} \text{Arcoth } x = \frac{1}{2b} \ln \frac{x+1}{x-1} \tag{7.1}$$

$$\text{where } x = \frac{1 - a\varrho_b^*}{b\varrho_b^*}$$

Here ϱ_b^* is the corrected reflectance over black, and a and b are the auxiliary variables in the Kubelka-Munk theory. Now, with Sh, β, and h, or from Sh and m_F, first calculate the scattering coefficient per unit PMC β, $S_p = S/\beta$:

$$S_p = \frac{S}{\beta} = \frac{Sh}{\beta h} = \frac{Sh}{m_F} \tag{7.2}$$

Here β is the pigment mass concentration (PMC) and m_F is the pigment mass per unit area, g/cm². The formula to be used from (7.2) will depend on whether β and h or m_F is determined experimentally. The test and reference values of β or m_F should be as nearly equal as possible. From the values $S_{p,T}$ (test pigment) and $S_{p,R}$ (reference pigment), calculate the relative scattering power per unit pigment mass as

$$S_r = \frac{S_{p,T}}{S_{p,R}} \tag{7.3}$$

Annexes of the German standard give very detailed examples of preparation, measurement, and calculation procedures. Annex 1 cites as test medium a long-oil alkyd resin modified with linseed oil; as dispersing equipment, high-speed impeller mills and single-roll and triple-roll mills. The method used to determine pigment mass per unit area in this example is to detach 100 cm² of the film with the help of a solvent, then combust and ignite the film.

Annex 2 contains another example with PVC-P (plasticized polyvinyl chloride) plasticizer containing materials. The dispersing equipment is a mixing roll; the film is brought into contact with the black glass plates by an immersion oil used as contact liquid. The pigment mass per unit area is determined gravimetrically, with the aid of the pigment mass portion known from the preparation.

The following description of a test is based on this example and also relates to PVC-P materials.

Preparation method:
Test medium: PVC-P materials (basic mixture A*)
Dispersing equipment: mixing roll with heating; speed 30 min⁻¹, roll width 350 mm, roll diameter 150 mm
Film preparation device: multi-platen press with heating
Colorimeter for method A
Substrate: black glass plate with CIE tristimulus value $Y \leq 4$
Contact liquid: immersion oil, refractive index 1.515
Formula: 1 g TiO₂ pigment + 99 g basic mixture A for test and reference
In double-roll mill, premix quantity of basic mixture at 130 °C until plastic. Uniformly add pigment and mix (friction 1:1.2, total mixing time 8 min). Prepare two mixtures each from reference and test. Using spacers (1 mm thick) in the multi-platen press, press films of the pigmented media at 140 °C (pressing time 5 min, pressure 1500 N/cm²). Cut films into pieces 50 mm × 40 mm.

The CIE tristimulus values Y are measured in the way already described. The condition $2 \leq Y_\infty - Y_b \leq 6$ is virtually fulfilled. For the layer of infinite thickness, pieces (at least four) were stacked on the black glass plate together with immersion oil until the CIE tristimulus value Y remained constant. The pigment mass per unit area was determined as $m_F = m_f w/F$, where w is the pigment mass portion, $w = m/(m + m_M)$, F is the area, m is the weight of pigment, m_M the weight of basic mixture A, and m_f the weight of the film.

By way of example, Figure 7-4 shows a printout of the results from one such test.

Note that the quantity determined according to the standard and in the above examples is the relative scattering coefficient *per unit pigment mass*. Theoretically, this is identical to the zero-th approximation of the lightening power $P_W^{(0)}$ (Section 4.7.1). In comparing two pigments, it is often more infor-

---

* See "PVC, Plasticized, Basic Mixtures" in Appendix 1.

7.2 Scattering and Absorption Coefficients

mative to compare the relative scattering coefficient S_v *per unit pigment volume*, which is obtained by multiplying the mass coefficient by the density ratio: $S_v = (\varrho_T/\varrho_R)S_r$ (see Section 8.3.2 for more detail). Thus, for pigments of equal density, $S_v = S_r$.

As has been shown, the method described can be used without change for determining the absolute S and K values as well. Of course, the film thickness h must be measured, but once this is done, the scattering coefficient S is obtained by simply dividing the experimental product Sh (eq. 7.1) by h. The absorption coefficient is then given by equation (3.121): $K = S(a - 1)$. If h is measured in mm, S and K will have units mm^{-1} = m^2/L.

Finally, it should be recalled that the scattering coefficient is a nonlinear function of the pigment concentration. The product Sh is then also a nonlinear

```
Program "S(REL)"
Relative scatterung power of white pigments (Black ground method),
according to ISO 787-24******************************************
Date: ..-..-..
Binder: PVC-P                           Apparatus: Double-roll mill
Colorimeter: N.N., d/8, C/2                        Multi-platen press
Manufacturing method:                              Black glass plates
   Method C1.1.5, recipe X                         Contact liquid
   Dispersing method: C

Input:  ***********************************************************
        | Reference is TiO2-1                | Test is   TiO2-2
        |           PVC:  .312 %             |           PVC:   .328%
        |           PMP: 1.00 %              |           PMP:  1.00 %
        |                                    |
Film  | Area      Mass    Y(inf)  Y(b)     | Area      Mass     Y(inf) Y(b)
No    | cm2       g                        | cm2       g
------+------------------------------------+------------------------------
1     | 20.44     3.606   82.7    80.8     | 21.22     4.461    86.5   83.1
2     | 19.71     3.671   83.2    81.3     | 21.32     4.173    86.5   83.2
3     | 19.33     3.443   82.6    80.5     | 20.71     3.825    85.9   82.6
4     | 19.03     3.447   82.8    80.9     | 20.28     3.673    86.0   82.8
      |                                    |
Output:************************************************************
      | Reference                           | Test
Film  | Pigment mass   Scatt. coeff.        | Pigment mass   Scatt. coeff.
      | per area       per unit mass        | per area       per unit mass
      | kg/m2          m2/kg                | kg/m2          m2/kg
------+-------------------------------------+------------------------------
1     | .01764         897.4                | .02102         765.5
2     | .01957         938.3                | .01957         833.9
3     | .01781         844.5                | .01847         814.8
4     | .01811         890.7                | .01811         852.6

Means:                  892.7                                816.7
Relative scattering power: 91.5%
```

Fig. 7–4. Relative scattering powers of white pigments (computer output).

function of pigment concentration, so that doubling the pigment mass per unit area and simultaneously halving the film thickness will not necessarily preserve the same value for *Sh*. The product will, however, be constant if both the concentration and the doubled concentration lie in the Beer's law region. The standard takes this point into account by prescribing that the PVC be reported. Outside the Beer's law range, it is not possible to calculate the lightening power (the mass ratio at which the scattering power of the test is equal to that of the reference; see Chapter 4) from the relative scattering power, because of the nonlinear relationship between scattering power and pigment content. In general, then, there is no expectation that relative scattering powers will be in agreement with lightening powers for given test and reference pigments.

7.2.2 Kubelka-Munk Coefficients *S* and *K* of Black and Colored Pigments

In principle, the method described in the preceding section is also applicable to determining the Kubelka-Munk coefficients of black and colored pigments. All that changes is that white glass plates can be used instead of black ones with dark pigments. It is still better to use commercially available black/white glass plates right from the outset. These measure 10 cm × 10 cm and consist of a black part and a white part cemented together. The crucial point for a consistent film thickness over the black and white parts is that there must be no step at the transition. The CIE tristimulus value *Y* of the white part should be 80 ± 2, while that of the black part should be < 5. It is also desirable to use these black/white plates for the hiding coat; reflection measurements will quickly tell whether the *Y* difference between the coat over black and the coat over white is less than a significance threshold (e.g., 0.002). A point to remember is that the determination of the coefficients from a non-hiding coat over black or white, together with a hiding coat, is generally more accurate than the determination based on a film of equal thickness over black and white (see Section 3.7). The reason for this gain in accuracy is that the determination of the auxiliary variable *a* requires just one measurement in the first case (ϱ_∞, eq. 3.120) but four in the second case (ϱ_b, ϱ_w, ϱ_{ob}, ϱ_{ow}, eq. 3.122), so that the error of the reflection measurement acts on the error of *a* four times. The measurement of ϱ_∞ should be dispensed with only when special difficulties are encountered in applying the hiding coat, for example with poorly hiding pigments. For a program, it is advantageous to ask first whether a measurement of ϱ_∞ is available; if it is, *a* is calculated from ϱ_∞.

A further gain in accuracy comes about because either ϱ_b, ϱ_{ob} or ϱ_w, ϱ_{ow} can be substituted in equation (3.120) for *S*, provided that all four measurements have been done. It is desirable to take the pair that gives the smaller error in *S*, as was shown in Section 3.6.2 (eqs. 3.145 to 3.149). With this improvement, the program automatically looks for the best way to calculate *S* and *K*. As an aid to programming, the calculations are listed in full in Table 7-1.

7.2 Scattering and Absorption Coefficients

Table 7–1. Formulas for calculating the error of S. Primed symbols denote partial derivatives $(\partial/\partial \xi)$ with respect to the variable ξ identified by the subscript.

$$y = \frac{1 - a\,(\varrho^* + \varrho_o^*) + \varrho^* \varrho_o^*}{b\,(\varrho^* - \varrho_o^*)} \tag{3.121}$$

$$S = \frac{1}{bh} \cdot \text{Arcoth}\, y \qquad \text{(same as eq. 3.124)}$$

$$z = \frac{1}{b^2\,h\,(y^2 - 1)} \tag{3.148, 3.149}$$

where
1) $s = s_b \quad \varrho^* = \varrho_b^* \quad \varrho_o^* = \varrho_{ob}^* \quad y = y_b \quad z = z_b$
2) $ s_w \quad \varrho_w^* \quad \varrho_{ow}^* \quad y_w \quad z_w$

$$\acute{s}_{\varrho^*} = -z\,\frac{1 - 2a\varrho_o^* + \varrho_o^{*2}}{(\varrho^* - \varrho_o^*)^2}$$

$$\acute{s}_{\varrho_o^*} = z\,\frac{1 - 2a\varrho^* + \varrho^{*2}}{(\varrho^* - \varrho_o^*)^2} \tag{3.147}$$

where

1) $\varrho^* = \varrho_b^* \quad \varrho_o^* = \varrho_{ob}^* \quad z = z_b \quad \acute{s}_{\varrho^*} = \acute{s}_{\varrho_b^*} \quad \acute{s}_{\varrho_o^*} = \acute{s}_{\varrho_{ob}^*}$
2) $ \varrho_w^* \quad \varrho_{ow}^* \quad z_w \quad \phantom{\acute{s}_{\varrho^*} =\ } \acute{s}_{\varrho_w^*} \quad \phantom{\acute{s}_{\varrho_o^*} =\ } \acute{s}_{\varrho_{ow}^*}$

$$\acute{\varrho}^* = \frac{d\varrho^*}{d\varrho} = \frac{1}{\dfrac{d\varrho}{d\varrho^*}} = \frac{1}{\acute{\varrho}} = \frac{(1 - r_2\varrho^*)^2}{(1 - r_o)(1 - r_2)} \tag{3.146}$$

(derivative of Saunderson correction)

where
1) $\acute{\varrho}^* = \acute{\varrho}_{ob}^* \quad \varrho^* = \varrho_{ob}^*$
2) $\phantom{\acute{\varrho}^* =\ } \acute{\varrho}_{ow}^* \quad \varrho_{ow}^*$
3) $\phantom{\acute{\varrho}^* =\ } \acute{\varrho}_b^* \quad \varrho_b^*$
4) $\phantom{\acute{\varrho}^* =\ } \acute{\varrho}_w^* \quad \varrho_w^*$

$$dS^2 = \acute{s}_{\varrho^*}^2 \cdot \acute{\varrho}^{*2} + \acute{s}_{\varrho_o^*}^2\,\acute{\varrho}_o^{*2} \tag{3.145}$$

where

1) $dS = dS_b \quad \acute{s}_{\varrho^*} = \acute{s}_{\varrho_b^*} \quad \acute{s}_{\varrho_o^*} = \acute{s}_{\varrho_{ob}^*} \quad \acute{\varrho}^* = \acute{\varrho}_b^* \quad \acute{\varrho}_o^* = \acute{\varrho}_{ob}^*$
2) $ dS_w \quad \phantom{\acute{s}_{\varrho^*} =\ } \acute{s}_{\varrho_w^*} \quad \phantom{\acute{s}_{\varrho_o^*} =\ } \acute{s}_{\varrho_{ow}^*} \quad \phantom{\acute{\varrho}^* =\ } \acute{\varrho}_w^* \quad \phantom{\acute{\varrho}_o^* =\ } \acute{\varrho}_{ow}^*$

If

$$dS_b^2 < dS_w^2$$

then it is preferable to calculate S from ϱ_{ob}, ϱ_b; otherwise, from ϱ_{ow}, ϱ_w.

Accuracy problems in the determination of S and K can arise if the pigment has a very high absorption. In order to differentiate ϱ and ϱ_o, the films must then be made very thin or the pigment concentration must be greatly reduced. There are, however, limits to how far either change can be carried: Usable films less than 30 μm thick are difficult to make, while excessive dilution of the pigment means that not all incident photons hit a pigment particle ("sieve effect"; see Section 3.7), so that the premises of the Kubelka-Munk theory may not be fulfilled. In such cases, there is nothing to do but use other methods to determine S (e. g., in white or black pastes; see Chapter 8). If S and K are used in determining the hiding power or transparency, the strongly absorbing regions of the spectrum can be neglected entirely in the evaluation in certain cases since the hiding layer practically depends upon the weakly absorbing regions (see example below).

Figure 7-5 presents a flowchart for the determination of S and K. In a long loop, the wavelength-dependent reflectances are input and checked for logic (e.g., ϱ must always lie between ϱ_o and ϱ_∞, regardless of the substrate) and accuracy (the difference between ϱ and ϱ_o or between ϱ and ϱ_∞ must be at least 0.02), and the Saunderson correction is applied to all values. Next comes a conditional jump based on whether ϱ_∞ is present or not. If all the other four reflectances are present (regardless of whether ϱ_∞ is), the error calculation is done by two methods: over black substrate and over white substrate. The S (and K) values with the smallest error are output.

An example will illustrate the method. Consider the determination of S and K for a paint containing a yellow iron oxide pigment. The paint was prepared in accordance with the "drying system" method (see Section 6.2.3); the vehicle was Alkydal® F48 and the PVC was equal to 1%. The measured reflection spectra ϱ_b, ϱ_w, ϱ_{ob}, ϱ_{ow} are shown in Figure 7-6. A printout of the results appears in Figure 7-7, while Figure 7-8 is a plot of the spectral S and K values. This is a highly instructive example, showing that the required accuracy makes it impossible to determine S and K at short wavelengths, at least for the selected concentration and coating thickness. The program has generated a message ("Pigment mass per unit area too large") that implies a suggestion to reduce the coating thickness and/or the pigment concentration. Where S is calculable, the last column in Figure 7-7 states whether the calculation was based on ϱ_b or ϱ_w.

As in Section 7.2.1, the gray-paste method of ISO 787–24 should also be mentioned. Absolute scattering and absorption coefficients can also be determined by this method. Because of the close interrelationship with the relative tinting strength or lightening power, this method is discussed in Chapter 8.

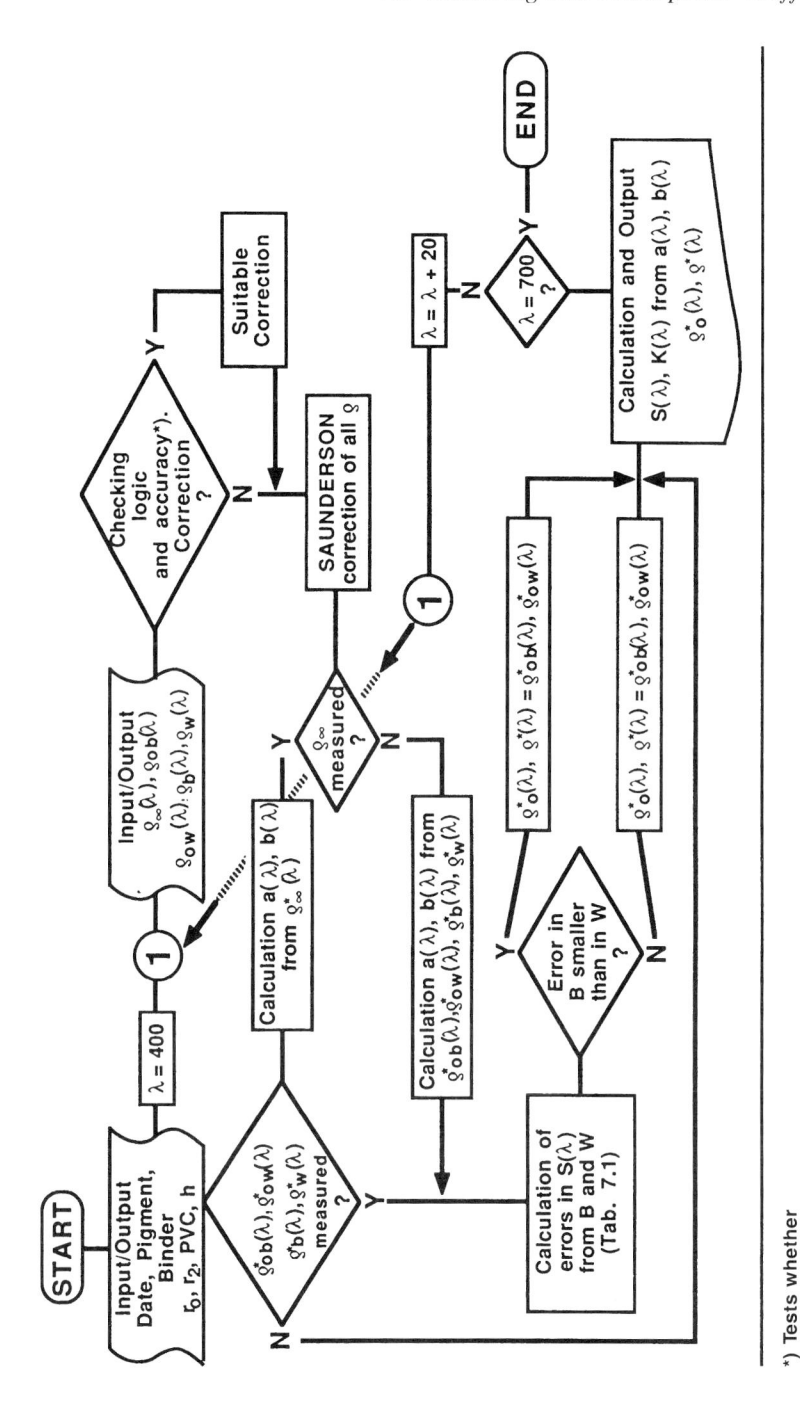

7.2 Scattering and Absorption Coefficients

Fig. 7-5. Flowchart for calculation of $S(\lambda)$ and $K(\lambda)$.

*) Tests whether
 – ϱ always lies between ϱ_0 and ϱ_∞
 – difference between ϱ and ϱ_0 resp. ϱ_∞ is at least 0.02
 If NO then
 – small differences between measured values in correct sequence
 – printout contains message warning of limited accuracy

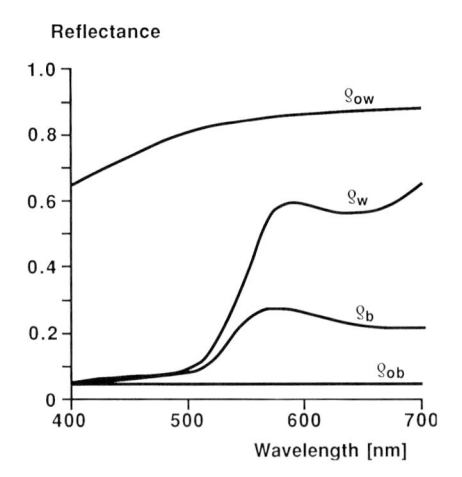

Fig. 7-6. Reflection spectra of a yellow iron oxide pigment in a alkyd resin paint (reflectances are ϱ_b over black, ϱ_w over white, ϱ_{ob} for black substrate, ϱ_{ow} for white substrate).

7.2.3 Four-Flux Coefficients s^+, s^-, k'

The computational methods described in Chapter 3 fully cover the region of technical interest with regard to accuracy and complexity. The two-flux (Kubelka-Munk) and four-flux calculations in "explicit" form (Sections 3.2 and 3.3) are simple enough that they can be performed on a desktop computer (beyond fundamental operations, only exponential and log functions are required). There is, however, a mild constraint on reliability. On the other hand, "implicit" many-flux calculations, that is, based on numerical integration of the differential equations, are best carried out on a medium-sized or large digital computer; the results are as reliable as the best experimental data. Even if many-flux calculations are mathematically rather complicated and tedious, they have taken on a standard form since the renowned publications of MUDGETT and RICHARDS, so that most of the work can now be done with readily available subroutines. It is therefore most remarkable that only a few publications have dealt with even the experimental aspects of many-flux theories and the four-flux theory in particular.

Our first example is taken directly from MUDGETT and RICHARDS ([17] in Section 3.9). The authors report how values of the coefficients s^+, s^- and k' can be obtained from S and K with the calculable coefficients of the Legendre polynomials. This means that values for these coefficients can be derived from the Kubelka-Munk coefficients with the aid of approximation formulas. What is more, however, the authors cite experimental data based on measurements on plastic wafers. The specimens in the first series were colored with a TiO_2 (rutile) pigment; those in the second series, with a red dye. As a consequence, the scattering and absorption coefficients were measurable separately.

The measured transmittances and reflectances were compared with the values obtained by the authors from scattering and absorption coefficients that

7.2 Scattering and Absorption Coefficients

```
Program "S AND K"
Scattering and absorption coefficients (Black ground method),
according to DIN 53 164************************************************
Date:..-..-..                    Pigment: Yellow iron oxide A
Binder: Alkydal F 48             PVC: 1%
Preparation: Instruction 00      Film thickness measurement: gravimetric
Colorimeter, Standard observer: N.N., d/8(incl)
Input:    ******************************************************************
Film thickness h:  .0476 mm
```

Lambda nm	Rho(inf)	Rho(b)	Rho(w)	Rho(0,b)	Rho(0,w)
400	---	.053	.053	.048	.671
420	---	.056	.056	.048	.721
440	---	.063	.063	.048	.760
460	---	.068	.068	.048	.784
480	---	.073	.073	.048	.801
500	---	.092	.092	.048	.817
520	---	.129	.154	.048	.838
540	---	.192	.300	.048	.848
560	---	.258	.504	.048	.857
580	---	.260	.579	.048	.864
600	---	.250	.583	.048	.872
620	---	.236	.568	.048	.883
640	---	.229	.568	.048	.893
660	---	.219	.579	.048	.901
680	---	.217	.608	.048	.905
700	---	.218	.643	.048	.914

```
Output:   *******************************************************
KUBELKA-MUNK Coefficients
```

Lambda nm	a	b	S 1/mm	K 1/mm	Calculated from:
400	Pigment mass per unit area too large				
420	Pigment mass per unit area too large				
440	Pigment mass per unit area too large				
460	Pigment mass per unit area too large				
480	Pigment mass per unit area too large				
500	Pigment mass per unit area too large				
520	2.462	2.250	14.23	20.81	Rho(w)
540	1.538	1.169	16.48	8.87	Rho(w)
560	1.184	.633	19.49	3.500	Rho(w)
580	1.137	.540	18.25	2.492	Rho(w)
600	1.144	.556	17.07	2.461	Rho(w)
620	1.170	.607	15.77	2.680	Rho(w)
640	1.180	.627	15.05	2.715	Rho(w)
660	1.187	.640	13.88	2.598	Rho(w)
680	1.170	.607	13.38	2.276	Rho(w)
700	1.148	.564	13.19	1.952	Rho(w)

Fig. 7-7. Determination of spectral S and K values (computer output).

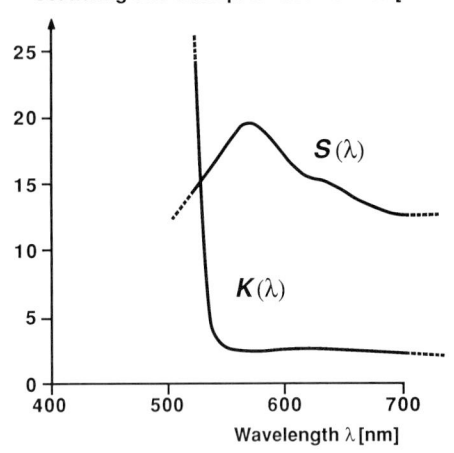

7 Determination of Hiding Power and Transparency

Fig. 7-8. Scattering coefficient S and absorption coefficient K for a yellow iron oxide pigment.

had been calculated by the method just mentioned, that is, from series of Legendre polynomials. The agreement was excellent. The coefficients for the surface reflection were also found close to the known values $r_0 = 0.04$ and $r_2 = 0.6$.

The coefficients referred to S and K had the following values:

$$s^+/S = 2.306, \quad s^-/S = 0.464, \quad k'/K = 0.5$$

These figures provide strong confirmation of Kubelka's conjecture that the absorption coefficient in the parallel flux (k') would be half as large as in the diffuse flux (K) (see Section 3.2.2). This hypothesis also holds to a good approximation for the backscattering coefficients (s^-, S). As expected, the coefficient s^+ is much larger than S – more than twice as large.

With these results, equation (3.19) can now be used to calculate

$$\mu/S = s^+/S + s^-/S + k'/S = 2.306 + 0.464 + (a-1)/2 = 2.270 + a/2$$

which will be needed for the calculation of p and q with equations (3.30) and (3.31). Equation (3.90) with the specified values of $\varrho_{\text{dif},\infty}$ (or $\varrho^*_{\text{dif},\infty}$ after the Saunderson correction) then yields the reflectance $\varrho^*_{g,\infty}$ for directional irradiation. We have used these data in calculating, for the first time, the (uncorrected) reflectances $\varrho_{g,\infty}$ over the reflectance range of 0.05 to 1. There is no problem with expanding the scope in this way, because mixtures of TiO_2 and the red dye can be prepared experimentally for any K/S. The results appear in Table 7-2, which shows the actually expected values (i.e., after Saunderson back-correction); as expected, only small differences (< 0.05) are seen between the reflectances for the directional and diffuse cases, with the latter values being somewhat higher throughout the series. This is also in agreement with

Table 7–2. Reflectances for diffuse and directional irradiation, calculated from data of MUDGETT and RICHARDS.

$\varrho_{\text{dif},\infty}$	$\varrho_{\text{g},\infty}$
0.05	0.046
0.10	0.083
0.15	0.125
0.20	0.168
0.25	0.213
0.30	0.259
0.35	0.306
0.40	0.354
0.45	0.403
0.50	0.452
0.55	0.503
0.60	0.554
0.65	0.607
0.70	0.660
0.75	0.714
0.80	0.769
0.85	0.826
0.90	0.883
0.95	0.941
1.00	1.000

expectations, for there is less scattering per unit path length for the directionally incident flux than for the diffuse one. The transfer of a photon from the forward to the backward direction thus requires, on average, a longer path length than in the diffuse case. What is more, experiment has not yet provided sufficiently sure confirmation for these differences between diffuse and directional irradiation; they are too large to be excused as "uncertainty of measurement" or "poor linearity at moderate reflectances." The reason cannot be identified with exactness; undoubtedly it is a fact that the conditions "directional" and "diffuse" cannot be realized precisely enough. Conventional instruments allow measurements only with conical, quasi-parallel, or diffuse irradiation, and the d/d geometry cannot be implemented neatly.

Other implications of the MUDGETT and RICHARDS work include the following: The phase function for a scattering medium can be obtained from Mie equations if the scattering is due to approximately spherical particles, or from pure scattering experiments if optically thin layers can be fabricated. Similar methods can be used for determining the absorption coefficient, insuring that the scattering and absorbing characteristics of the medium are described adequately for use in many-flux methods. For a simpler but less accurate calculation, the two-flux or four-flux method can be employed along with the empirical relations stated above for the effective scattering and absorption coefficients.

Two problems that must be allowed for in two-and four-flux calculations relate to the angular distribution of the radiant flux in the interior of the medium. The internal surface reflection coefficients react directly to changes in this angular distribution, as do the effective scattering and absorption coefficients. What causes the errors in the two-flux calculation is that the radiant flux departs more from the approximate diffuse distribution as the K/S ratio increases. Such deviations occur in layers with low pigment concentration (e. g., translucent plastics) or very thin layers (e. g., printing inks). It might be possible to counter these effects with relatively simple empirical algorithms in many practical cases, for example in order to achieve better and faster convergence for color formulations.

The second application example appears in a publication of ATKINS and BILLMEYER (see [23] in Section 3.9). The authors compared the results of four-flux calculations with experimental results on TiO_2 specimens. They find a value of $s^+/(s^+ + s^-) = 0.84$ for the forward scattering as a fraction of total scattering of a directional radiant flux. In the MUDGETT and RICHARDS example, the fraction was $2.306/(2.306 + 0.464) = 0.832$; this represents a good result, although it must be kept in mind that the two results are not for the same TiO_2 pigment.

Finally, the literature does not report any test to determine s^+, s^-, and k' using the formulas of Section 3.2.5. The computational derivation of these coefficients would greatly simplify the problems that arise with translucent specimens and in color formulations. Once coefficients were determined, the laborious procedure involving implicit many-flux theories could be rendered superfluous.

7.3 Transparency

7.3.1 Single-Point and Multi-Point Methods

Transparency is significant for a wide variety of applications (transparent lacquers, printing inks, etc.). That uniform testing standards* have nonetheless been devised can be attributed to the fact that such applications have in common one physical phenomenon, transparence. What is more, in view of the broad range of applications, no preparation method has been prescribed, so this point must be covered by an agreement between the parties. It should be noted that transparent pigments have to be dispersed with special care, because

---

* See "Transparency" in Appendix 1.

their small particle size means that they exhibit extremely strong adhesion forces. A preparation method will be set forth in Section 7.3.2.

The standards therefore stop with general provisions that must always be considered (geometry used in measurements, corrections to be applied, calculation of results). What is determined is the transparency number of a medium, which can be unpigmented or pigmented. If the specimen preparation conditions (vehicle system, application method, film thickness, pigment concentration) are held constant, transparent pigments can also be subjected to comparative evaluation by this method.

The definition of the transparency number T is expanded somewhat from the definition stated in Section 3.5.1. In Figure 3-10, the film thickness (abscissa) could be replaced by the mass of the coating (at constant pigment concentration) or by the pigment concentration (at constant film thickness) without changing the nature of the curves. The h in equation (3.139) could then be replaced by the mass of the system per unit area (e.g., the "ink load" in the testing of printing inks) or the pigment concentration. It goes without saying that the units of the transparency number T will vary from case to case. If, for example, the pigment mass per unit area (unit g/m^2, especially suitable for testing of transparent pigments) is taken as the pigment portion, the transparency number will have units of g/m^2 too, and T_m then tells how much pigment is required to coat 1 m² of substrate so that the color difference is $\Delta E^*_{ab} = 1$.

The test is performed by coating a black/white substrate and using a colorimeter to measure the color differences over the black substrate relative to ideal black. As a rule, the measured values have to be corrected before the color difference is calculated, because the coating filters the light incident on the substrate and reflected from it. This correction, which should take care of the departure of the real substrate from an ideally black substrate, is applied by preparing a drawdown of the same thickness on the white substrate. The transparency number is obtained from the color difference at one or more film thicknesses (if the pigment concentrations are equal) or at one or more concentrations (if the film thicknesses are equal).

The black/white glass plates must fulfill the same requirements as in Section 7.2.2. With regard to the geometry of the colorimeter, care must be taken to minimize the effect of specimen surface characteristics on the measured value. On principle, method A of Section 6.1.2 can be used if the value of surface reflection is subtracted. In the case of high-gloss plane specimens, however, the measurement can be carried out by method B of Section 6.1.2 (45/0 or 8/d(ex)). For non-high-gloss specimens, overlacquering is recommended; this step is essential in the case of bronzing specimens, particularly such as occur with printing inks.

The needed correction of the measured values to eliminate the effect of the real black substrate is based on the filter action of the coating, mentioned above. This is found as the ratio of the reflectometer value R' ($R' = \varrho$ or $R' = R$) of the coating over white (R'_w) to the value of the white substrate (R'_{ow}). The reflectometer value for the black substrate (R'_{ob}) is multiplied by this ratio and the product is subtracted from the specimen reflectometer value R'_b:

$$R'_{corr} = R'_b - R'_{ob} \frac{R'_w}{R'_{ow}} \tag{7.4}$$

If the measurement is done by the tristimulus method, the same computational scheme is applied to the reflectometer values R'_x, R'_y, R'_z or the CIE tristimulus values X, Y, Z. The CIELAB color differences are calculated by the method of Section 6.2.2.

Because the transparency number so defined is a function of h, its determination requires at least two specimens differing in film thickness (*multi-point method*); the transparency number obtained in this way is valid in the range between the two thicknesses h_1 and h_2, which should differ by at least ten times the standard deviation of the h measurement. With the color differences ΔE_1 and ΔE_2 measured on the two films, the transparency number becomes

$$T = \frac{h}{\frac{h - h_1}{h_2 - h_1}(\Delta E_2 - \Delta E_1) + \Delta E_1} \tag{7.5}$$

where h is an arbitrary film thickness or pigment mass per unit area lying between h_1 and h_2.

If the two specimens differing in film thickness have roughly equal transparency numbers, that is, when the increase in color difference is approximately linear for small film thicknesses or pigment concentrations, the transparency number can be determined in a simpler way: The multi-point method reduces to the *single-point method*. The formula then simplifies to a form identical to equation (3.139):

$$T = \frac{h}{\Delta E^*_{ab}(h)} \tag{7.6}$$

These methods of determination have the drawback that the coloring power or tinting strength cannot be taken into the evaluation of two transparent pigments without more details. The standards therefore suggest determining transparency numbers after tinting strength matching for the respective film thicknesses or pigment concentrations. This problem will be treated more fully in the method (see the next section) based on the principle of spectral evaluation and using the transparency-coloring power and transparency-coloring degree defined in Section 3.5.

An example of the two-point method follows. The test system is a transparent yellow iron oxide pigment in an alkyd resin, 1% PVC. The following were measured:

Film thicknesses
$h_1 = 0.035$ mm
$h_2 = 0.052$ mm

Color differences relative to ideal black
$\Delta E_1 = 12.83$
$\Delta E_2 = 18.87$

7.3 Transparency

Now the two-point formula gives

$$T = \frac{h}{\frac{h-0.035}{0.052-0.035}(18.87-12.83)+12.83} = \frac{h}{355.3 \cdot h + 0.39}$$

This implies, for a film thickness of, say, $h = 0.045$ mm, a transparency number of $T = 0.002748$ L/m² $= 2.748$ mL/m². If the only measured values available were those for the first point, h_1 and ΔE_1, the single-point formula (eq. 7.6) would yield $T = 0.002727$ L/m² $= 2.727$ mL/m². Multiplying this value by the PVC of $\sigma = 0.01$ and the pigment density of $\varrho = 4.08$ g/mL yields a transparency number related to the pigment mass of $T_m = 0.1113$ g/m². This value is also obtained if the pigment mass per unit area is determined instead of the film thickness and substituted for h.

7.3.2 Method Based on Principle of Spectral Evaluation

The calculation employs the equations of Section 3.6.3. To make the programming easier, a flowchart is presented in Figure 7-9 and the required formulas in closed form are summarized in Table 7-3. In addition to the equations needed for calculating the transparency number, those for the coloring power Φ, the transparency-coloring power L, and the transparency-coloring degree G are also listed.

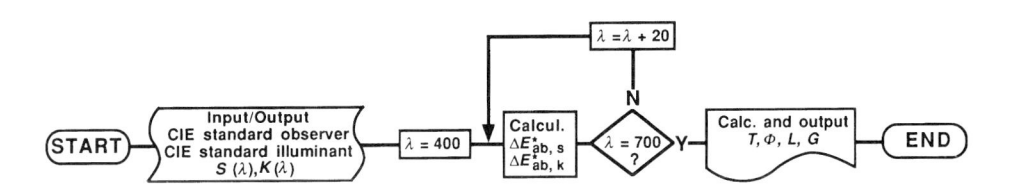

Fig. 7-9. Flowchart for determination of transparency by the principle of spectral evaluation.

The determination of transparency will now be demonstrated for an example. A transparent red iron oxide pigment is prepared in an alkyd resin vehicle, and the preparations are applied in

– a non-hiding coat over white
– an opaque coat

The reflection spectra are then measured. After the film thickness has been determined for the non-hiding coat, the Kubelka-Munk coefficients S and K

7 Determination of Hiding Power and Transparency

Table 7–3. Calculation of transparency number, coloring power, transparency-coloring power, and transparency-coloring degree.

Primes denote application of the operator $(\partial/\partial h)_{h \to 0}$ to the respective unprimed quantities.

$$\acute{R}_s = \int \bar{r}_\lambda (1 - r_o)(1 - r_2) S_\lambda \, d\lambda \qquad (3.158)$$

$$\acute{R}_k = 2 \int \bar{r}_\lambda \cdot \frac{1 - r_o}{1 - r_2} \cdot K_\lambda \, d\lambda \qquad (3.159)$$

where 1) $\acute{R}_s = \acute{X}_s$ | $\acute{R}_k = \acute{X}_k$ | $\bar{r}_\lambda = \bar{x}(\lambda)$ (CIE color matching functions)
2) $\acute{Y}_s$ | $\acute{Y}_k$ | $\bar{y}(\lambda)$
3) $\acute{Z}_s$ | $\acute{Z}_k$ | $\bar{z}(\lambda)$

$$\acute{R}_s^* = 7.787 \cdot \frac{\acute{R}_s}{R_n} \qquad (3.156)$$

$$\acute{R}_k^* = \frac{1}{3} \cdot \frac{\acute{R}_k}{R_n} \qquad (3.157)$$

where 1) $\acute{R}_s^* = \acute{X}_s^*$ | $\acute{R}_k^* = \acute{X}_k^*$ | $\acute{R}_s = \acute{X}_s$ | $\acute{R}_k = \acute{X}_k$ | $R_n = X_n$ (CIE tristimulus values
2) $\acute{Y}_s^*$ | $\acute{Y}_k^*$ | $\acute{Y}_s$ | $\acute{Y}_k$ | Y_n of standard illuminant)
3) $\acute{Z}_s^*$ | $\acute{Z}_k^*$ | $\acute{Z}_s$ | $\acute{Z}_k$ | Z_n

$$\acute{a}^* = 500\,(\acute{X}^* - \acute{Y}^*); \quad \acute{b}^* = 200\,(\acute{Y}^* - \acute{Z}^*); \quad \acute{L}^* = 116 \cdot \acute{Y}^* \qquad (3.155)$$

where 1) $\acute{a}^* = \acute{a}_s^*$ | $\acute{b}^* = \acute{b}_s^*$ | $\acute{L}^* = \acute{L}_s^*$ | $\acute{X}^* = \acute{X}_s^*$ | $\acute{Y}^* = \acute{Y}_s^*$ | $\acute{Z}^* = \acute{Z}_s^*$
2) $\acute{a}_k^*$ | $\acute{b}_k^*$ | $\acute{L}_k^*$ | $\acute{X}_k^*$ | $\acute{Y}_k^*$ | $\acute{Z}_k^*$

$$\Delta \acute{E}_{ab}^* = \sqrt{\acute{a}^{*2} + \acute{b}^{*2} + \acute{L}^{*2}}$$

where 1) $\Delta \acute{E}_{ab}^* = \Delta \acute{E}_{ab,s}^*$ | $\acute{a}^* = \acute{a}_s^*$ | $\acute{b}^* = \acute{b}_s^*$ | $\acute{L}^* = \acute{L}_s^*$
2) $\Delta \acute{E}_{ab,k}^*$ | $\acute{a}_k^*$ | $\acute{b}_k^*$ | $\acute{L}_k^*$

Transparency number: $\qquad T = 1000/\Delta \acute{E}_{ab,s}^* \quad \text{mL} \cdot \text{m}^{-2}$ $\qquad$ (3.160)

Coloring power: $\qquad \Phi = \Delta \acute{E}_{ab,k}^*/1000 \quad \text{m}^2 \cdot \text{mL}^{-1}$

Transparency-coloring power: $\quad L = T \cdot \Phi$ $\qquad$ (3.141)

Transparency-coloring degree: $\quad G = \sqrt{L(L+2)} - L$ $\qquad$ (3.142)

can be determined from it and the two spectra (calculation outlined in Section 7.2.2). Then, by using a program running on a desktop computer, the transparency number, coloring power, transparency-coloring power, and transparency-coloring degree are calculated.

Preparation method:
Precondition the test pigment at 105°; determine density* and loss on ignition** (needed for analytical determination of PVC of finished paint). Use Alkydal® F481 in the following preparation:
87.5 g Alkydal, 52% in white spirit
0.57 g Ca octoate, 4%
10.3 g white spirit
0.9 g Ascinine® R55
0.74 g Bayer OL silicone oil®, 5% in xylene
Add dryer (to 95 g clear paint):
0.36 g Co octoate, 6% in xylene
0.91 g Pb octoate, 24% in xylene
Calculate PVC from pigment density and loss on ignition for pigment and paint (both by titrimetric determination of iron content). At 7.5% PVC, disperse in agate vessels on planetary mill. Terminate dispersion operation only when coarseness < 10 μm***. To prepare non-hiding coatings, dilute paint to 1% PVC and apply by a spinning film spreader on (black/)white glass plates according to Section 7.2.2. After 24 h drying, coatings should have thicknesses between 30 and 70 μm. Measure film thicknesses with pneumatic device (method of Section 7.1.3). For hiding coat, dilute starting paint to 5% PVC (because hiding is difficult to achieve with 1% PVC); apply in several coats, allowing to dry between coats (ca. 15 min required). Place in fresh-air cabinet dryer for at least 72 h to dry completely.

Measurement and evaluation method:
Measure reflection spectra with spectrophotometer (method A of Section 6.1.2). Determine scattering and absorption coefficients in 20 nm intervals as in Section 7.2.2. Use CIE color matching functions from Stearns' tables (Table 7-4; see also Section 3.6.7 and [32] in Section 3.9).

Next, the S and K values obtained are input to the calculation of the transparency number and the other transparency quantities. Figure 7-10 is a printout of such a calculation.

---

* See "Density of Pigments" in Appendix 1.
** See "Loss on Ignition" in Appendix 1.
*** See "Fineness of Grind" in Appendix 1.

Table 7-4. CIE color matching functions in 20 nm intervals (after STEARNS).

λ nm	C/2° $\bar{x}(\lambda)$	C/2° $\bar{y}(\lambda)$	C/2° $\bar{z}(\lambda)$	D65/2° $\bar{x}(\lambda)$	D65/2° $\bar{y}(\lambda)$	D65/2° $\bar{z}(\lambda)$
400	0.044	−0.001	0.187	0.136	0.002	0.628
420	2.926	0.085	14.064	2.548	0.070	12.221
440	7.680	0.513	38.643	6.680	0.455	33.666
460	6.633	1.383	38.087	6.364	1.322	36.507
480	2.345	3.210	19.464	2.190	2.939	18.157
500	0.069	6.884	5.725	0.051	6.822	5.515
520	1.193	12.882	1.450	1.342	14.121	1.621
540	5.588	18.268	0.365	5.778	19.013	0.389
560	11.751	19.606	0.074	11.290	18.861	0.069
580	16.801	15.989	0.026	16.260	15.462	0.025
600	17.896	10.684	0.012	17.922	10.673	0.012
620	14.031	6.264	0.003	14.047	6.288	0.003
640	7.437	2.897	0	7.029	2.734	0
660	2.728	1.003	0	2.507	0.922	0
680	0.749	0.271	0	0.707	0.256	0
700	0.175	0.063	0	0.166	0.060	0
Totals	98.044	100.000	118.102	95.017	100.000	108.813

7.4 Hiding Power

7.4.1 General (Graphical Method)

According to a German standard*, a "hiding power value" is used to characterize the hiding power of pigmented media such as paints and plastic films. The standard does not contain instructions for preparation, and it leaves open such questions as the method used to measure the film thickness. It is general in application (i.e., for both chromatic and achromatic coatings) and even offers a way of obtaining hiding power values on arbitrary contrast substrates.**

The hiding power value is defined (consistently with Section 3.4.1) as the area of a well-defined contrast substrate such that a hiding layer can be created

---

* See "Hiding Power, Pigmented Media" in Appendix 1.
** For other standards see "Hiding Power" in Appendix 1.

7.4 Hiding Power 311

```
Program "TRANSP"
Transparency number, Color strength, Transparency Power, Degree of Transparency,
according to VÖLZ, Industrial Color Testing****************************************
Date: ...-..-..                          Pigment: Red iron oxide (transp.) B
Binder: Alkydal F 481
Preparation: Instruction  N.N.
S and K evaluation: PVC 1%, gravimetr. film thickness measurement: .0473 mm
Input:     **********************************************************************
```

Lambda n m	S 1/mm	K 1/mm
400	2.187	133.9
420	2.474	146.7
440	2.268	126.5
460	2.057	102.5
480	1.557	84.3
500	1.556	71.8
520	1.311	59.0
540	1.357	46.4
560	1.193	28.54
580	1.038	15.62
600	1.003	8.77
620	.963	5.91
640	.921	4.38
660	.887	3.424
680	.869	2.747
700	.840	2.160

```
Output:   **********************************************************************
TRANSPARENCY NUMBER                      T:    1.518 ml/m2
COLORING POWER                           Phi: 24.86  m2/ml
TRANSPARENCY-COLORING POWER              L:    37.74
TRANSPARENCY-COLORING DEGREE             G:    98.7%
```

Fig. 7–10. Determination of transparency number, coloring power, transparency-coloring power, and transparency-coloring degree of a transparent red iron oxide pigment based on the principle of spectral evaluation (computer output).

with a unit volume or unit mass of the pigmented medium. "Hiding layer" means that a stated hiding criterion – here, an agreed-on color difference between the two fields of the contrast substrate supporting the layer – is satisfied. In comparison with the definition in Section 3.4, however, the present definition of hiding power value is extended in the same way that we treated the transparency number (Section 7.3.1). There are, accordingly, two hiding power values:

– D_v in m²/L, the area such that a hiding layer can be created with 1 L of the pigmented medium

– D_m in m²/kg, the area such that a hiding layer can be created with 1 kg of the pigmented medium

If the pigment concentration is held constant, the hiding power values of pigments can be compared. Up to ca. 10 % PVC, the hiding power value can actually be corrected for the nominal concentration by linear interpolation (see Chapter 4, "Beer's Law").

In order to determine the hiding power value, coatings made of the pigmented medium and having various thicknesses are applied to the contrast substrate. After the films have dried, the color differences between the two contrasting fields are determined. A plot of color difference (ordinate) versus the reciprocal of the film thickness or of the mass per unit area (abscissa) is prepared. The hiding power value is obtained by going to the ordinate value of the agreed-on hiding power criterion and reading off the corresponding abscissa. The reason for using the reciprocals is that a quasilinear plot results; what is more, the abscissa, found as just described, gives the hiding power value directly with no further calculation.

As a rule, the contrast substrates are the black/white glass plates described in Section 7.2.2, with nominal Y values of $Y < 5$ for the black and $Y = 80 \pm 2$ for the white parts. The standard also permits the use of black/white contrast cards, but these can no longer be recommended because they amplify the errors of measurement to an unacceptable degree and also because pneumatic measurement of the film thickness is not possible with cards. A gray/brown contrast substrate is also permitted; the CIE tristimulus values are specified as

- Gray field: $X = 30 \pm 2$, $Y = 32 \pm 2$, $Z = 33 \pm 2$
- Brown field: $X = 9 \pm 1$, $Y = 9 \pm 1$, $Z = 8.5 \pm 1$

If a hiding power value is to be determined on a free film, that is, a film that has no optical contact to a substrate, a hollow light trap is to be used for the black substrate (special black standard, see Section 6.2.2) and a $BaSO_4$ white standard for the white (see Section 2.1.2). The mass per unit area of a free film can be determined very accurately by weighing a piece whose area has been measured. Most commonly, however, it is possible to produce an optical contact with the substrates described above. The film together with a suitable contact liquid (having approximately the same refractive index as the specimen) is placed on the contrast substrate, and a rubber roller is used to press it against the substrate and expel any included air.

Just as described in Section 6.2.2 for full-shade determinations on colored media, the colorimetric measurement on the contrasting fields is done by method A of Section 6.1.2. The ΔE_{ab}^* values are determined on at least four distinct films. One of the methods of Section 7.1 is used to measure the film thickness. If D_m is desired, the mass of the medium per unit area must be determined; if no free film is available for a direct determination, the formula $m_F = \varrho_M h$ makes it possible to calculate m_F from the density ϱ_M of the medium. Now ΔE_{ab}^* is plotted versus the reciprocal of h or m_F, and a curve is fitted to the points. From the hiding power criterion ($\Delta E_{ab}^* = 1$ is preferred), a vertical is dropped to the horizontal axis and the hiding power value is read off.

Figure 7-11 presents an example of the evaluation method (yellow iron oxide

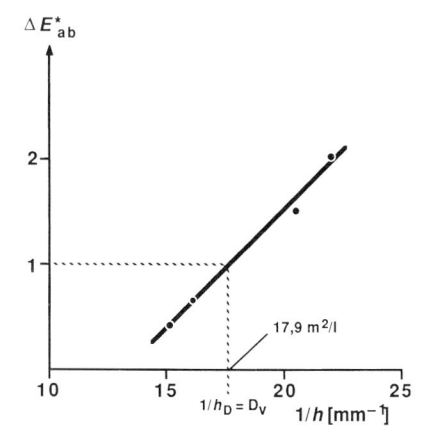

Fig. 7–11. Graphical determination of hiding power value D_v for a yellow iron oxide pigment.

pigment, 5 % PVC in alkyd resin paint). The hiding power value obtained is 17.9 m²/L.

The hiding power value is strongly influenced by the substrate contrast, increasing with decreasing contrast. On the other hand, the final result is also affected by what hiding criterion is selected; as the critical ΔE^*_{ab} decreases, the hiding power value also falls off. For this reason, the commonly used criterion $\Delta E^*_{ab} = 1$ represents a good compromise between the weaker contrast substrates often found in practice and the need for a lower critical color difference due to the lower perceptual threshold of the human eye.

7.4.2 Achromatic Case

Another German standard* also defines a hiding power value as in the preceding section. As before, the pigment concentration, test medium, and dispersing must be agreed on between the parties. The hiding criterion is now the lightness difference $\Delta L^* = 1$, because the small color difference between the coating over white and black can be set equal to the lightness difference in the ideal case (i.e., if the coating really is achromatic). Hiding and non-hiding coats of the pigmented medium for test are applied and the CIE tristimulus values Y are measured. From these results and the film thickness or the mass of medium per unit area (measured on the non-hiding coat), the hiding power value D_v or D_m is determined from tables, with formulas, or by an iterative calculation. The following method is an extension of the one found in the standard, specifically in two points (see Section 3.4.2):

---

* See "Hiding Power, White and Light Gray Media" in Appendix 1.

314   7 Determination of Hiding Power and Transparency

- Inclusion of gloss in the computer-simulated matching
- Calculation of scattering and absorption components of the hiding power value

The discussion of the preceding section can be adapted to cover the preparation of the pigmented layers and the measurement of the CIE tristimulus values Y and film thickness or mass per unit area.

The calculation of hiding power values was described fully in Section 3.4.2; ϱ is now replaced by $Y/100$ as it was there. After application of the Saunderson correction to the measured values and calculation of the auxiliary variables a and b, the scattering coefficient and the factor α are calculated. The three approaches mentioned above can now be used:

1) The α table (see [24] in Section 3.7). Because α depends only on the CIE tristimulus value Y_∞ of the hiding coat and its degree of gloss g, the table is entered with two parameters, ϱ_∞^* and g, and α is either read off directly or found by linear interpolation. The table is restricted to the range $0.7 < \varrho_\infty^* < 1$, corresponding to $50 < Y_\infty < 100$. The standard also gives a table starting directly from the measured Y_∞ (i.e., bypassing the Saunderson correction), but this leaves g out of account; the α table in the standard has only the values for the $g = 1$ column. The table need not be reproduced here, because the second approach now to be described will be more convenient and offer sufficient accuracy.

2) Nonlinear regression equations. They are

$$y_o = (a_{02}\, \varrho_\infty^* + a_{01})\, \varrho_\infty^* + a_{00} \tag{7.7}$$

$$y_1 = (a_{12}\, \varrho_\infty^* + a_{11})\, \varrho_\infty^* + a_{10} \tag{7.8}$$

Once y_o and y_1 are known, α is calculated from their values and g:

$$\alpha = y_o + g(y_1 - y_o) \tag{7.9}$$

It has proved useful to split the region of interest ($0.7 < \varrho_\infty^* < 1$) into three parts:

Region I: $0.7 \leq \varrho_\infty^* < 0.9$
Region II: $0.9 \leq \varrho_\infty^* < 0.97$
Region III: $0.97 \leq \varrho_\infty^* < 1$

The coefficients a_{ik} needed for the calculation are listed in Table 7-5. The formulas are quite accurate, with error less than 0.2% in all cases.

3) Direct (iterative) calculation. The equations of Section 3.4.2 are used. Figure 7-12 gives a flowchart of the calculation along with all needed equations. In

Table 7–5. Coefficients for the calculation of $a(\varrho^*_\infty, g)$ with the regression formulas.

	Region I $0.7 \leq \varrho^*_\infty < 0.9$	Region II $0.9 \leq \varrho^*_\infty < 0.97$	Region III $0.97 \leq \varrho^*_\infty < 1$
a_{00}	0.88869	1.30598	5.53410
a_{01}	-1.32952	-2.23071	-10.96049
a_{02}	0.47622	0.96232	5.46863
a_{10}	0.86519	1.27680	5.53121
a_{11}	-1.28370	-2.17282	-10.95552
a_{12}	0.45315	0.93288	5.46579

principle, the hiding power values can also be calculated for black and dark gray coatings; the algorithm must then include tests for the selection of the value function ($\varrho - gr_o < 0.008856$?) as well as formulas valid for the lower range. If, however, a method based on the principle of spectral evaluation is available (see the next section), there is no need for these modifications because the method is applicable generally to chromatic and achromatic layers.

Finally, the hiding power values are calculated with equation (3.133).

In the iterative procedure (3) with the equations of Section 3.4.3, there is no difficulty in calculating the scattering and absorption components of the hiding power. When the a table (1) or the regression formulas (2) are employed, computational aids are called for. The tabular solutions use equations (3.133) and (3.136):

$$D_{\text{sca}} = \alpha_S S; \qquad D_{\text{abs}} = \beta_K K = \beta_K (a - 1) S$$

Here α_S and β_K depend only on the degree of gloss g; Table 7-6 gives their values. If formulas are preferred, the two values are given by

$$\alpha_S = 0.042298 - 0.000767 \, g \tag{7.10}$$

For β_K, two cases are distinguished:

$$g < 0.8063: \beta_K = -(0.05986 \, g + 0.04386)g + 0.42263$$
$$g > 0.8063: \beta_K = 0.34779 \qquad \text{(independent of } g\text{)} \tag{7.11}$$

The values from these equations differ by less than 0.3% from the tabulated values.

Parts 2 and 3 of the German standard give two very thoroughly worked-out examples of the determination of hiding power values for white specimens. The

316   7 *Determination of Hiding Power and Transparency*

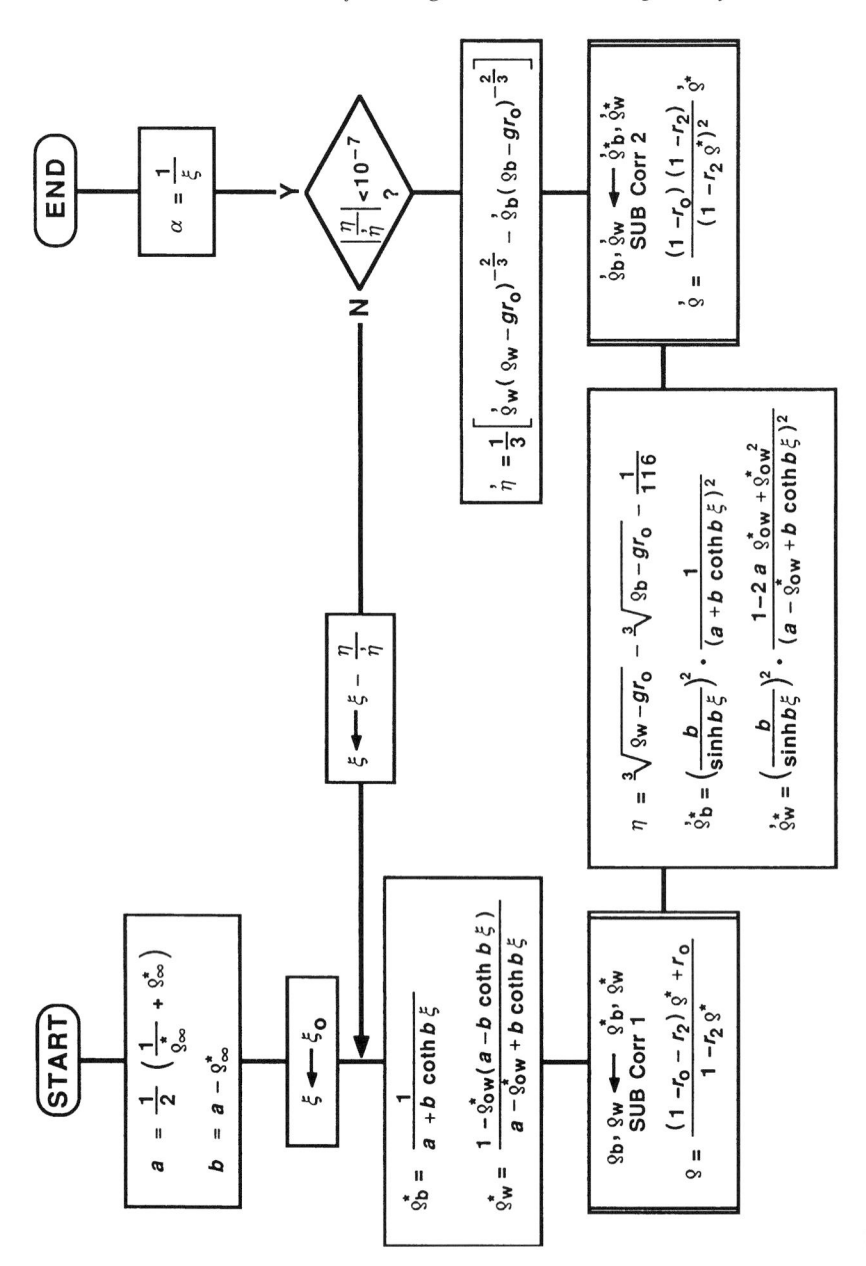

Fig. 7-12. Flowchart for iterative determination of hiding power value of white and light gray coatings.

Table 7–6. The functions $a_S(g)$ ("hiding power due to scattering alone") and $\beta_K(g)$ ("hiding power due to absorption alone").

g	a_S	β_K
0	0.042298	0.42342
0.1	0.042222	0.41732
0.2	0.042146	0.41071
0.3	0.042070	0.40347
0.4	0.041993	0.39542
0.5	0.041917	0.38630
0.6	0.041840	0.37571
0.7	0.041763	0.36290
0.8	0.041686	0.34841
0.9	0.041608	0.34779
1	0.041531	0.34779

pigmented medium in Part 2 is an air-drying alkyd resin paint with a TiO$_2$ pigment; in Part 3, a plasticized polyvinyl chloride (PVC-P) also with a TiO$_2$ pigment. What follows here is an earlier example (TiO$_2$ pigment in alkyd resin paint) with film thickness measurement by the gravimetric method; the computer printout appears in Figure 7-13. The hiding power value and its components are obtained with an iterative routine based on the flowchart of Figure 7-12. The calculation gives $\alpha = 0.05861$, the same result as found in the α table. Equation (7.9) for $0.9 \leq \varrho_\infty^* < 0.97$ (Region II) yields $\alpha = 0.05864$, which differs by 0.05 % from the calculated value. The hiding power value is then $D_v = 14.17$ m^2/L in both cases. For the scattering and absorption components, equations (3.133), (3.136), and (3.137) give the same results as the iterative calculation: $\delta_{sca} = 71.2\%$ and $\delta_{abs} = 1.1\%$.

One point in favor of using this method for white pigments is that the result is easily derived from tables or simple formulas. There is, however, another point grounded in the hiding power criterion. The hiding power value obtained by this method will be the same as in Section 7.4.1 only with extrapolation to $\Delta L^* = 1$ (instead of $\Delta E_{ab}^* = 1$) in the graphical evaluation. For a nearly white or nearly gray specimen, the chroma and hue differences between white and black should be of a higher order than the lightness difference, but this condition is not always adequately fulfilled. The effect is exaggerated because the CIELAB system underestimates lightness differences and overestimates chroma differences. As a result, the undertone in many cases becomes an effect of less than second order. This difference between the hiding power criteria manifests itself in the following way: Two specimens equal in lightness, one being a neutral white and the other exhibiting a slight yellow undertone, would have the same hiding power value over a black/white contrast substrate if the evaluation were based on $\Delta L^* = 1$, but if it were based on $\Delta E_{ab}^* = 1$ then the yellow-toned specimen would have a higher ΔE_{ab}^* between black and white than the neutral

```
Program "HIDING WHITE"
Hiding power of white and light gray media, according to DIN 55 984 (extended
method)      ****************************************************************
Date: ..-..-..-                        Pigment: TiO2-Z
Binder: Alkydal F 48                   PVC: 27.23%
Preparation: Instruction PP            Film thickness measurement: gravimetr.
Colorimeter, Standard Observer: NN, d/0 (incl.), C/2 degr
Input:        ****************************************************************
Film thickness h:       .0473 mm
Gloss degree g (of Rho(b)):            .764
Reflectances:  Rho(b):                 .8030
               Rho(inf)                .8688
Output:       ****************************************************************
Auxiliary quantities:   a = 1.0018854           b    =    .061435
Scattering coefficient:                         S(v)  =    241.73 1/mm
Hiding power value:                             D(v)  =    14.17 m2/l
Scattering component:                           Delta(sca) =  71.2%
Absorption component:                           Delta(abs) =  1.1%
Interaction component:                          Delta(int) =  27.7%
```

Fig. 7–13. Determination of hiding power value for a white paint containing a titanium dioxide pigment (computer output).

specimen would have a higher ΔE_{ab}^* between black and white than the neutral specimen, since it would have a chroma difference ΔC_{ab}^* contributing to the color difference ΔE_{ab}^*. To make ΔE_{ab}^* equal to 1, as specified by the criterion, the layer would have to be made thicker, and this would lead to a lower hiding power value. Accordingly, when comparing white pigments, it may sometimes be useful to determine the hiding power in accordance with the criterion $\Delta L^* = 1$.

In the evaluation with $\Delta E_{ab}^* = 1$, a yellow-toned specimen will therefore have a lower hiding power than a color-neutral specimen of the same lightness. The next section will show how this effect can show up in the scattering and absorption components of the hiding power.

A determination of the contrast ratio of white latex paints* is mentioned only peripherally. Such determinations tell nothing beyond whether a stated contrast over black and white is or is not attained at a given film thickness. They do not provide unambiguous information about what hiding power can be achieved or about the effectiveness of the paint.

---

* See "Emulsions Paints" in Appendix 1.

7.4.3 Based on Principle of Spectral Evaluation

The basis for determining the hiding power with the principle of spectral evaluation was treated in great depth in Section 3.6.2, which also presents the equations (repeated here in Table 7-7 as a programming aid). Figure 7-14 is a flowchart for the procedure including determinations of S and K (see Section 7.2.2). After the required data and the film thickness have been input, the program requests the needed reflectances as a function of wavelength in 20 nm steps. The CIE color matching functions are taken from STEARNS' table (Table 7-4; see Section 3.6.2 and [32] in Section 3.9). In each pass through this first loop, the scattering and absorption coefficients associated with this wavelength are calculated and output. When the loop terminates, the second loop is performed; it iterates h up to the hiding coat. A suitable iterative method must be used; that is, the change in film thickness must be derived from the residual difference between the actual color difference and the nominal value 1. When a color difference of 1 has been reached (to the required accuracy), the reciprocal of the hiding film thickness is output as the hiding power value.

In Section 3.6.2, Newton's method was also recommended for this calculation and the equations for the method were written out. Table 7-8 again lists these formulas as a programming aid; here the prime denotes a partial derivative with respect to h. The flowchart in Figure 7-15 shows, in the nested loop, the iterative calculation of the hiding power and its components in accordance with Table 7-8. An example (chromium oxide green) is given in Figure 7-16; the reflectances of the substrates, of the non-hiding coat over black and white, and of the hiding coat are shown. Figure 7-17 is a printout generated by the program for the same pigment; the calculations of transparency, coloring power, and so forth have now been included.

The program yields basically the same hiding power values as the procedure of Section 7.4.1. If, however, the procedure of Section 7.4.2 (achromatic layer) is used, departures may show up because of the $\Delta L^* = 1$ hiding power criterion. The effect of the hiding power criterion on the final result was discussed in the preceding section, where it was stated that white specimens showing an undertone may have lower hiding power values than neutral white specimens of equal lightness if $\Delta E_{ab}^* = 1$ is selected as the hiding power criterion. A further consequence of this behavior is that the "hiding power due to scattering alone," as calculated with the scheme described above, may be larger than the hiding power itself; in other words, the scattering component of the hiding power may be more than 100 %. The unavoidable conclusion is that the interaction component is negative. In such a case, absorption makes only a small positive direct contribution; on the other hand, the interaction makes a clearly larger negative contribution to the hiding power, since the undertone that reduces the hiding power comes about only through the interaction of scattering and absorption. This statement is illustrated by the example in Figure 7-18, where a white pigment with a slight yellow undertone was used.

7 Determination of Hiding Power and Transparency

Table 7-7. Formulas needed for determination of the hiding power.

Specified as measured values:

(I) ϱ_∞ Reflectance of hiding coat
      ϱ Reflectance of non-hiding coat
      ϱ_o Reflectance of substrate

OR

(II) ϱ_{ob} Reflectance of black substrate
      ϱ_{ow} Reflectance of white substrate
      ϱ_b Reflectance over black
      ϱ_w Reflectance over white

$$\varrho^* = \frac{\varrho - r_o}{1 - r_o - r_2(1-\varrho)} \tag{3.8}$$

(Saunderson correction: $r_o = 0.04$; $r_2 = 0.6$; or $r_o(\lambda)$, $r_2(\lambda)$ as inputs)

where   1) $\varrho^* = \varrho^*_\infty$ $\varrho = \varrho_\infty$
          2) ϱ^*_{ob} ϱ_{ob}
          3) ϱ^*_{ow} ϱ_{ow}
          4) ϱ^*_b ϱ_b
          5) ϱ^*_w ϱ_w

$$\text{(I)} \quad a = \frac{1}{2}\left(\frac{1}{\varrho^*_\infty} + \varrho^*_\infty\right) \tag{3.120}$$

or

$$\text{(II)} \quad a = \frac{(1 + \varrho^*_w \varrho^*_{ow})(\varrho^*_b - \varrho^*_{ob}) + (1 + \varrho^*_b \varrho^*_{ob})(\varrho^*_{ow} - \varrho^*_w)}{2(\varrho^*_b \varrho^*_{ow} - \varrho^*_w \varrho^*_{ob})} \tag{3.122}$$

$$b = \sqrt{a^2 - 1} \tag{3.120}$$

$$S = \frac{1}{bh} \cdot \text{Arcoth} \, \frac{1 - a(\varrho^* + \varrho^*_o) + \varrho^* \varrho^*_o}{b(\varrho^* - \varrho^*_o)} \quad \text{(Scattering coefficient; } \varrho, \varrho_o \text{ from I or II)} \tag{3.121}$$

$K = (a - 1) S$ (absorption coefficient)

Begin iteration loop: With $h = h_o$ as initial value

$$\varrho^* = \frac{1 - \varrho^*_o (a - b \cdot \coth bsh)}{a + b \cdot \coth bsh - \varrho^*_o} \tag{3.111}$$

where   1) $\varrho^* = \varrho^*_b$ $\varrho^*_o = \varrho^*_{ob} = 0$
          2) ϱ^*_w $\varrho^*_{ow} = 0.9048$ (from $\varrho_{ow} = 0.8$)

$$\varrho = \frac{(1 - r_o - r_2)\varrho^* + r_o}{1 - r_2 \varrho^*} - gr_o \tag{3.150}$$

Table 7–7. Continuation

(Saunderson back-correction with allowance for gloss)

where 1) $\varrho = \varrho_b$ | $\varrho^* = \varrho_b^*$
2) ϱ_w | ϱ_w^*

$R_b = \int \bar{r}_\lambda S_\lambda \varrho_{b,\lambda} \, d\lambda;$ $\qquad\qquad R_w = \int \bar{r}_\lambda S_\lambda \varrho_{w,\lambda} \, d\lambda$

$(S_\lambda = S(\lambda),$ standard illuminant) $\hfill (3.151)$

where 1) $R_b = X_b$ | $R_w = X_w$ | $\bar{r}_\lambda = \bar{x}(\lambda)$ (CIE color matching functions)
2) Y_b | Y_w | $\bar{y}(\lambda)$
3) Z_b | Z_w | $\bar{z}(\lambda)$

Branch in program:

A) $R/R_n > 0.008856$ | B) $R/R_n \leq 0.008856$

$R^* = \sqrt[3]{R/R_n}$ $\qquad\qquad R^* = 7.787 \cdot R/R_n + 0.138 \hfill (2.5)\,(2.6)$

where 1) $R^* = X_b^*$ | $R = X_b$ | $R_n = X_n$ (CIE tristimulus values of standard illuminant)
2) Y_b^* | Y_b | Y_n
3) Z_b^* | Z_b | Z_n
4) X_w^* | X_w | X_n
5) Y_w^* | Y_w | Y_n
6) Z_w^* | Z_w | Z_n

End of branch

$a^* = 500\,(X^* - Y^*)$

$b^* = 200\,(Y^* - Z^*)$

$L^* = 116 \cdot Y^* - 16 \hfill (2.7)$

where 1) $a^* = a_b^*$ | $b^* = b_b^*$ | $L^* = L_b^*$ | $X^* = X_b^*$ | $Y^* = Y_b^*$ | $Z^* = Z_b^*$
2) a_w^* | b_w^* | L_w^* | X_w^* | Y_w^* | Z_w^*

$\Delta r^* = r_w^* - r_b^* \hfill (2.11)$

where 1) $\Delta r^* = \Delta a^*$ | $r_w^* = a_w^*$ | $r_b^* = a_b^*$
2) Δb^* | b_w^* | b_b^*
3) ΔL^* | L_w^* | L_b^*

$\Delta E_{ab}^* = \sqrt{\Delta a^{*2} + \Delta b^{*2} + \Delta L^{*2}} \hfill (2.10)$

Return to beginning of loop (eq. 3.111) after applying correction to h; continue until $\Delta E_{ab}^* = 1$.

322   7 *Determination of Hiding Power and Transparency*

*) Starting value $h_0 = 0$

**) Δh is to be calculated from $\Delta E^*_{ab} = 1$ by means of a suitable iteration method

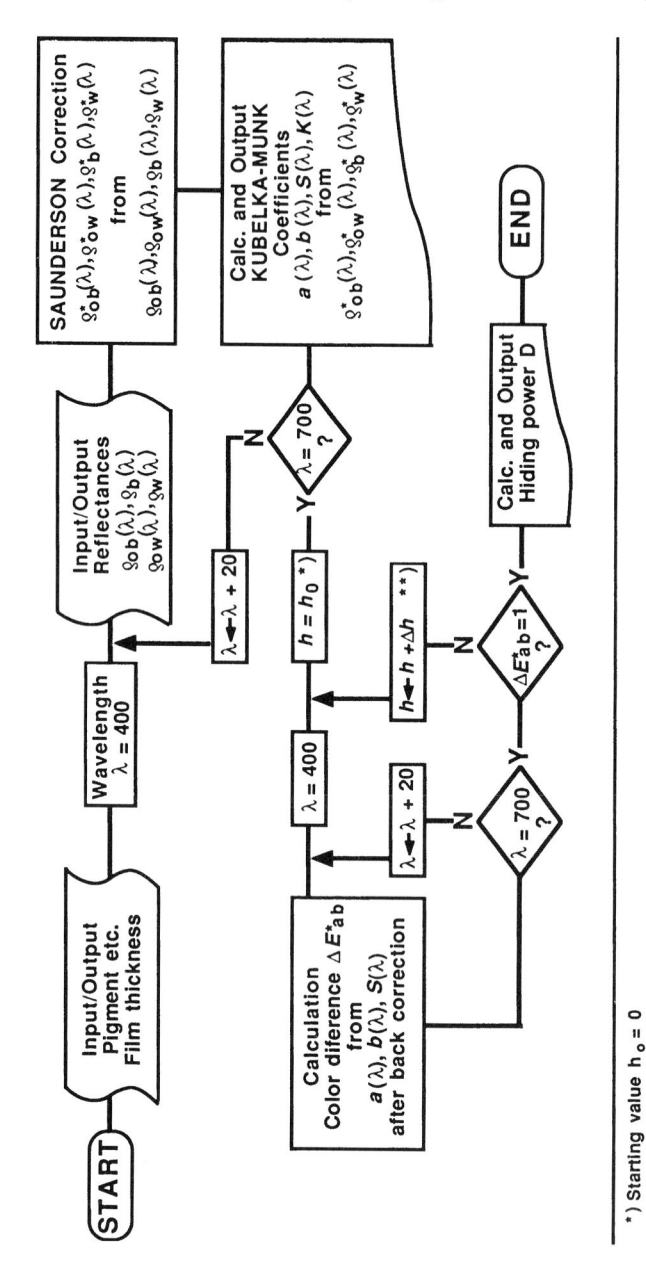

Fig. 7–14. Flowchart for determination of hiding power with the principle of spectral evaluation.

7.4 Hiding Power

Table 7–8. Formulas for h iteration (Newton's method).

Prime denotes partial derivative $\partial/\partial h$ of the unprimed quantity. See Table 7–7 for specified inputs.

After equation (3.111) of Table 7–7, insert:

$$\acute{\varrho}^* = \frac{bs^2\,[1 - \varrho_o^*(2a - \varrho_o^*)]}{[(a - \varrho_o^*)\sinh bsh + b \cdot \cosh bsh]^2} \tag{3.153}$$

where 1) $\acute{\varrho}^* = \acute{\varrho}_b^*$ | $\varrho_o^* = \varrho_{ob}^* = 0$
2) $\quad\quad\acute{\varrho}_w^*$ | $\varrho_{ow}^* = 0.9048$ (from $\varrho_{ow} = 0.8$)

On grounds of computational economy, it is desirable to calculate the three hyperbolic functions in equations (3.111) and (3.153) as follows:
$E = \exp(bsh)$, $\sinh bsh = x = (E - 1/E)/2$, $\cosh bsh = y = (E + 1/E)/2$, $\coth bsh = y/x$.

After equation (3.150) of Table 7–7, insert:

$$\acute{\varrho} = \frac{(1 - r_o)\,(1 - r_2)}{(1 - r_2\varrho^*)^2} \cdot \acute{\varrho}^*$$

where 1) $\acute{\varrho} = \acute{\varrho}_b$ | $\varrho^* = \varrho_b^*$ | $\acute{\varrho}^* = \acute{\varrho}_b^*$
2) $\quad\acute{\varrho}_w$ | $\quad\varrho_w^*$ | $\quad\acute{\varrho}_w^*$

After equation (3.151) of Table 7–7, append:

where 4) $R_b = \acute{X}_b$ | $R_w = \acute{X}_w$ | $\varrho_{b,\lambda} = \acute{\varrho}_b\,(\lambda)$ | $\varrho_{w,\lambda} = \acute{\varrho}_w\,(\lambda)$
5) $\quad\quad\acute{Y}_b$ | $\quad\acute{Y}_w$ | $\quad\acute{\varrho}_b\,(\lambda)$ | $\quad\acute{\varrho}_w\,(\lambda)$
6) $\quad\quad\acute{Z}_b$ | $\quad\acute{Z}_w$ | $\quad\acute{\varrho}_b\,(\lambda)$ | $\quad\acute{\varrho}_w\,(\lambda)$

After equations (2.5) and (2.6) of Table 7–7, insert:

A) $\acute{R}_b^* = \dfrac{R_b^* \acute{R}_b}{3\,R_b}$; $\acute{R}_w^* = \dfrac{R_w^* \acute{R}_w}{3\,R_w}$; B) $\acute{R}_b^* = 7.787 \cdot \dfrac{\acute{R}_b}{R_n}$; $\acute{R}_w^* = 7.787 \cdot \dfrac{\acute{R}_w}{R_n}$

where 1) $\acute{R}_b^* = \acute{X}_b^*$ | $\acute{R}_w^* = \acute{X}_w^*$ | $R_b = X_b$ | $R_w = X_w$ | $\acute{R}_b = \acute{X}_b$ | $\acute{R}_w = \acute{X}_w$
2) $\quad\quad\acute{Y}_b^*$ | $\quad\acute{Y}_w^*$ | $\quad Y_b$ | $\quad Y_w$ | $\quad\acute{Y}_b$ | $\quad\acute{Y}_w$
3) $\quad\quad\acute{Z}_b^*$ | $\quad\acute{Z}_w^*$ | $\quad Z_b$ | $\quad Z_w$ | $\quad\acute{Z}_b$ | $\quad\acute{Z}_w$

After equation (2.7) of Table 7–7, insert:

$\acute{a}^* = 500\,(\acute{X}^* - \acute{Y}^*);\quad\quad \acute{b}^* = 200\,(\acute{Y}^* - \acute{Z}^*);\quad\quad \acute{L}^* = 116 \cdot \acute{Y}^*$

where 1) $\acute{a}^* = \acute{a}_b^*$ | $\acute{b}^* = \acute{b}_b^*$ | $\acute{L}^* = \acute{L}_b^*$ | $\acute{X}^* = \acute{X}_b^*$ | $\acute{Y}^* = \acute{Y}_b^*$ | $\acute{Z}^* = \acute{Z}_b^*$
2) $\quad\acute{a}_w^*$ | $\quad\acute{b}_w^*$ | $\quad\acute{L}_w^*$ | $\quad\acute{X}_w^*$ | $\quad\acute{Y}_w^*$ | $\quad\acute{Z}_w^*$

7 Determination of Hiding Power and Transparency

Table 7-8. Continuation

After equation (2.11) of Table 7-7, append:

$$\text{where} \quad \begin{array}{l} 4) \ \Delta r^* = \Delta\acute{a}^* \\ 5) \quad \Delta\acute{b}^* \\ 6) \quad \Delta\acute{L}^* \end{array} \quad \begin{array}{l} r_w^* = \acute{a}_w^* \\ \acute{b}_w^* \\ \acute{L}_w^* \end{array} \quad \begin{array}{l} r_b^* = \acute{a}_b^* \\ \acute{b}_b^* \\ \acute{L}_b^* \end{array}$$

After equation (2.10) of Table 7-7, insert:

$$\eta = \ln \Delta E_{ab}^* \quad (\ln x \text{ converges to 0 faster than } x - 1)$$

$$\acute{\eta} = \frac{1}{\Delta E_{ab}^{*2}} (\Delta a^* \Delta\acute{a}^* + \Delta b^* \Delta\acute{b}^* + \Delta L^* \Delta\acute{L}^*)$$

$$h_{n+1} = h_n - \frac{\eta}{\acute{\eta}} \quad (\text{new value } h = h_{n+1}) \tag{3.152}$$

If

$$\frac{|h_{n+1} - h_n|}{h_n} < 10^{-6}, \text{ then } h = h_D \text{ and } D = \frac{1}{h_D}$$

Special cases

1. *"Due to scattering alone"*

Delete equation (3.111) of Table 7-7 and equation (3.153) of Table 7-8 and replace with:

$$\varrho^* = \frac{(1 - \varrho_o^*) Sh + \varrho_o^*}{(1 - \varrho_o^*) Sh + 1} \tag{3.113}$$

$$\acute{\varrho}^* = S \left[\frac{1 - \varrho_o^*}{(1 - \varrho_o^*) Sh + 1} \right]^2 \tag{3.154}$$

as equation (3.133) with

$$h = h_{D,\text{sca}}, \qquad D_{\text{sca}} = \frac{1}{h_{D,\text{sca}}}, \qquad \delta_{\text{sca}} = \frac{D_{\text{sca}}}{D} \qquad \text{(scattering component)}$$

Table 7-8. Continuation

2. *"Due to absorption alone"*

Delete equation (3.111) of Table 7–7 and equation (3.153) of Table 7–8 and replace with:

$$\varrho^* = \varrho_0^* \cdot e^{-2Kh}; \qquad \acute{\varrho}^* = -2K\varrho^* \qquad (3.112),\ (3.154)$$

as equation (3.133) with

$$h = h_{D,\text{abs}}, \qquad D_{\text{abs}} = \frac{1}{h_{D,\text{abs}}}, \qquad \delta_{\text{abs}} = \frac{D_{\text{abs}}}{D} \qquad \text{(absorption component)}$$

For each of cases 1 and 2, the same iteration loop can be used as in Table 7–7, supplemented by equations (3.153), (3.152), (1.133) (see above)

$$\delta_{\text{int}} = 1 - \delta_{\text{sca}} - \delta_{\text{abs}} \qquad \text{(interaction component)} \qquad (3.137),\ (3.138)$$

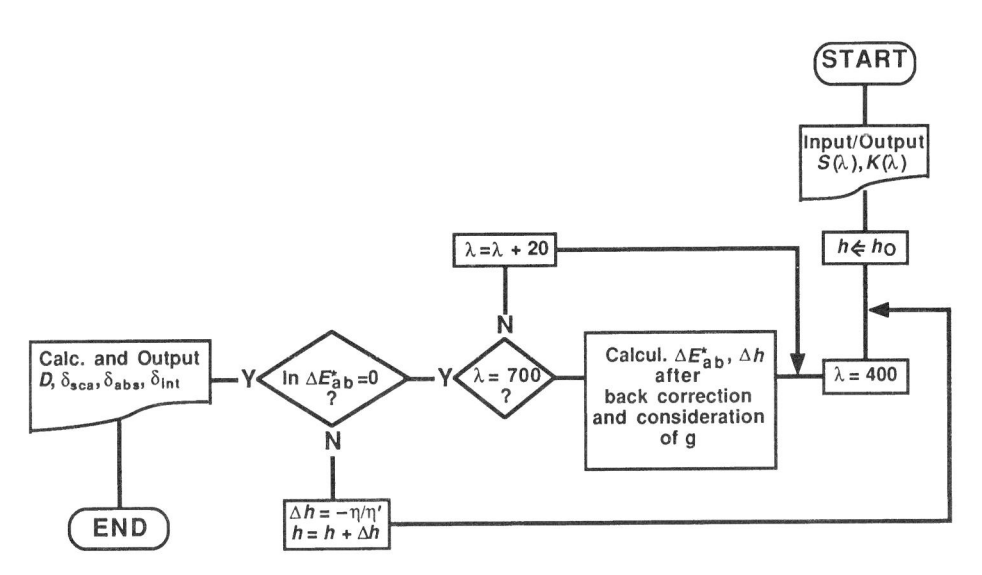

Fig. 7-15. Flowchart for determination of hiding power in accordance with the principle of spectral evaluation (NEWTON's method).

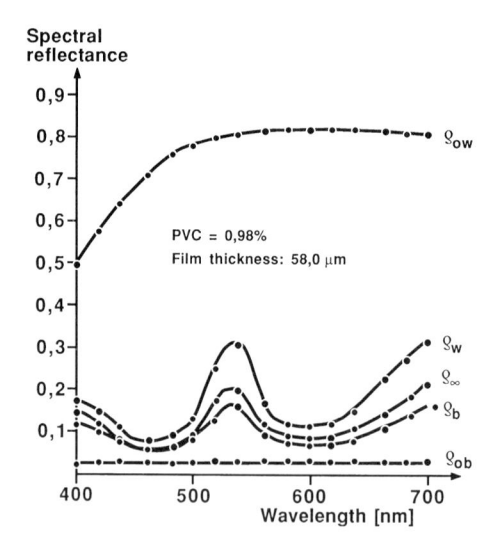

Fig. 7-16. Reflection spectra of a chromium oxide pigment in alkyd resin paint (reflectances: ϱ_b over black, ϱ_w over white; ϱ_∞, hiding coat; ϱ_{ob}, black substrate; ϱ_{ow}, white substrate).

The program described here seems to be a sort of final step in the determination of hiding power for arbitrary paints. The hiding power can be determined with a computer program that leaves virtually nothing to be desired. As has been shown, the determination of transparency (and the transparency quantities) has also been integrated into the program. At the same time, this program illustrates the universal applicability of the principle of spectral evaluation.

7.4.4 Economic Aspects

Any statement about hiding power, regardless of whether it is expressed as D_v in m²/L or as D_m in m²/kg, is a statement about effectiveness. This becomes particularly obvious if hiding power values are plotted versus the pigment volume concentration. The flowchart of Figure 7-19 makes clear what steps are required in order to plot the D versus σ curve (curve of hiding power value versus PVC).

Figure 7-20 shows D_v versus σ curves for two TiO₂ pigments (one anatase and one rutile). The plot demonstrates not only the well-known superiority of rutile pigments, but also the existence of a maximum in the D curves for both pigments, each maximum lying at a PVC value of around 30%. Further, the curves show the relationships between rutile and anatase with regard to a hiding coat (run a horizontal from one curve to the other). The maximum hiding power value of 21 m²/L for anatase, at 29% PVC, is already reached by the rutile pigment at a PVC of 13%. In other words, substantial amounts of pigment can be saved. On the other hand, more binder must be used; the

7.4 Hiding Power

```
Program "HIDING COLORED"
Hiding power and Transparency of colored pigments, according to the Principle of Spectral Evaluation
(VÖLZ, Industrial Color testing) ************************************************************
Date: . . - . . -.     Pigment: Chrome oxide green
Binder: Alkydal F 48       r0, r2: Standard values
Gloss degree: 94%          PVC : .98%
Film thickness: .05796 mm
Measuring conditions: d/8 (incl)
Rho(inf): measured
Rho(b): Input from keyboard     Rho(0b) : Input from keyboard
Rho(w): Input from keyboard     Rho(0w): Input from keyboard
Input:  ***********************************************************************
```

Lambda	Rho(inf)	Rho(0b)	Rho(0w)	Rho(b)	Rho(w)
400 nm	.1328	.0446	.5145	.1214	.1641
420	.1155	.0453	.5650	.1104	.1415
440	.0783	.0452	.6231	.0766	.0928
460	.0687	.0452	.6866	.0677	.0803
480	.0749	.0445	.7442	.0737	.0894
500	.0963	.0443	.7753	.0931	.1209
520	.1678	.0440	.7968	.1467	.2360
540	.2088	.0439	.8071	.1699	.3052
560	.1476	.0439	.8229	.1287	.2079
580	.1022	.0438	.8249	.0948	.1367
600	.0912	.0436	.8246	.0864	.1203
620	.0977	.0436	.8270	.0917	.1313
640	.1130	.0434	.8252	.1038	.1574
660	.1407	.0434	.8258	.1232	.2051
680	.1751	.0431	.8243	.1431	.2650
700	.2086	.0430	.8192	.1587	.3234

```
Output: ****************************************************************************
KUBELKA-MUNK Coefficients, related to PVC 1%
```

Lambda	a	b	K (1/mm)	S (1/mm)	Calculated from
400 nm	2.474	2.263	12.77	8.661	Rho(w)
420	2.931	2.755	15.42	7.984	Rho(w)
440	5.360	5.266	22.77	5.221	Rho(w)
460	7.026	6.954	25.88	4.295	Rho(w)
480	5.845	5.758	23.94	4.941	Rho(w)
500	3.778	3.643	18.62	6.704	Rho(w)
520	1.9411	1.664	9.371	9.958	Rho(w)
540	1.6114	1.264	6.722	10.99	Rho(w)
560	2.204	1.964	10.86	9.014	Rho(w)
580	3.461	3.313	16.31	6.630	Rho(w)
600	4.112	3.988	18.15	5.831	Rho(w)
620	3.696	3.559	16.78	6.222	Rho(w)
640	3.015	2.845	14.15	7.019	Rho(w)
660	2.320	2.093	10.80	8.182	Rho(w)
680	1.8664	1.576	8.004	9.238	Rho(w)
700	1.6125	1.265	6.089	9.940	Rho(w)

```
TRANSPARENCY
Standard Illuminant/Standard Observer: C/2 degr
Transparency number              T        .2602       ml/m2
Coloring power                   Phi      4.607       m2/ml
Transparency-coloring power      L        1.199
Transparency-coloring degree     G        75.9%
HIDING POWER
Hiding power value               D        6.171       m2/l
Scattering component             p(sca)   3.04 %
Absorption component             p(abs)   40.18 %
Interaction component            p(int)   56.78 %
*******************************************************************************
```

Fig. 7–17. Determination of hiding power, including transparency (computer output).

328   7 *Determination of Hiding Power and Transparency*

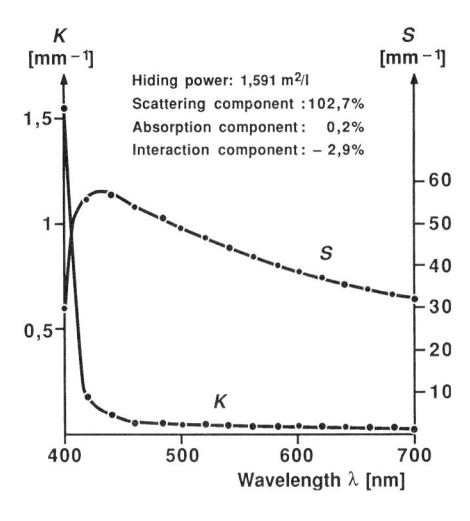

Fig. 7-18. Hiding power of a white pigment in alkyd resin paint (example of negative interaction component of hiding power).

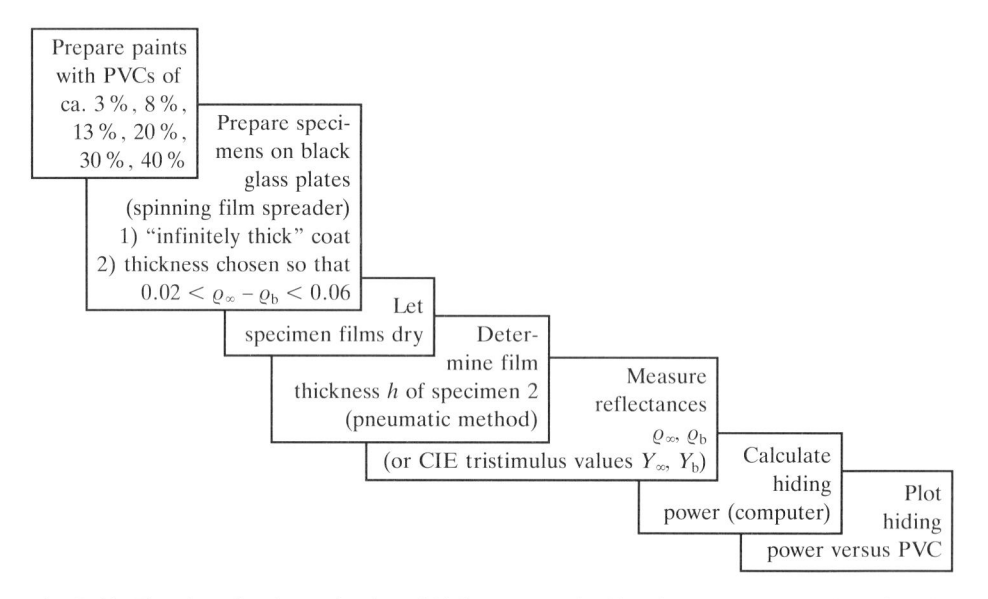

Fig. 7-19. Flowchart for determination of hiding power of white pigments as a function of PVC.

actual saving is not seen until the costs of pigment and binder for the hiding coats are added up.

A cost analysis based on the D_v versus σ plot is quite possible, as KEIFER has shown (see [27] in Section 3.9). The reciprocals of the ordinate values $1/D_v = h_D$ give the dry film thicknesses; the abscissa values σ and the densities of the pigment and medium give the mass fractions; and these are multiplied

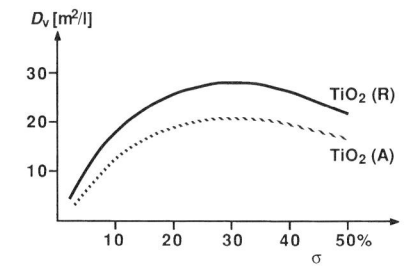

Fig. 7–20. Hiding power value D_v versus PVC (TiO$_2$ pigments in alkyd resin paint).

by the respective prices. The sum of the cost contributions for pigment and medium yields the desired total cost ϕ_L of the hiding coat of paint. The formulas are

$$\phi = h_D \, \sigma \varrho \Pi$$
$$\phi_M = h_D (1 - \sigma) \varrho_M \Pi_M \qquad (7.12)$$
$$\phi_L = \phi + \phi_M = h_D [(\varrho \Pi - \varrho_M \Pi_M) \sigma + \varrho_M \Pi_M]$$

where ϕ is the cost per unit area, US\$/m^2; ϱ is the density, g/cm^3 (= kg/L); σ is the PVC; Π is the price per unit weight, US\$/kg; h_D is the hiding film thickness, mm (= L/m^2); and the subscripts have the following meanings: no subscript, pigment; M, medium (unpigmented); L, pigmented paint. Figure 7-21 presents cost curves obtained in this way. The total cost (ϕ_L) curves show the expected distinct minima. From this plot it can be deduced that a hiding coat made with anatase pigment has a much higher total cost than one made with rutile pigment; the difference ranges from 25% to 30%, depending on the PVC. The component curves (ϕ and ϕ_M) show that not only the cost of the pigment but also the cost of the medium used to create a hiding coat are higher for anatase.

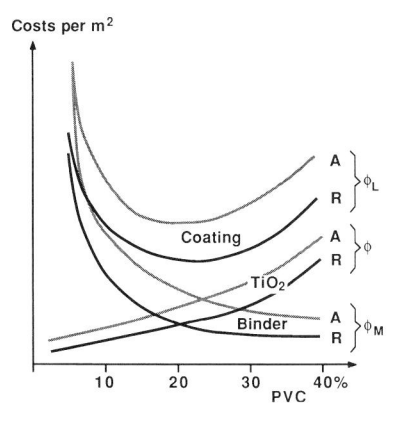

Fig. 7–21. Costs of hiding coats versus PVC.

The explanation of this result is that, when an optically less capable pigment is used at a given PVC, hiding requires a thicker film. This thicker coat, however, contains not just more pigment but also more of the medium. Hence it is easy to see that even small differences in the optical properties of pigments have major effects on the cost calculation.

The minimum in the cost curve of Figure 7-21 shows how much pigment must be used in order to obtain the least expensive hiding coat. The position of the cost minimum depends on the shape of the D_v versus σ curve, as has been shown. If the prices per unit volume for pigment and vehicle were equal ($\varrho \Pi = \varrho_M \Pi_M$), the price minimum in Figure 7-21 would coincide with the hiding power maximum in Figure 7-20, as can be inferred from equation (7.12). If the prices change, the minimum shifts and a new curve has to be plotted. For example, suppose the price of the pigment goes up; then the cost minimum moves toward lower PVC. This means that it has become more economical to use less pigment, even though the new pigmented medium hides less well and so must be applied in a thicker coat to hide.

The method described here makes it possible to arrive at correct assessments of pigments – with respect to optical performance and economics – on the basis of the hiding power versus PVC curve. Such results cannot be achieved through simple estimation.

8 Determination of Tinting Strength and Lightening Power

8.1 Content of Coloring Material

8.1.1 Dyes

Organic dyes can be quantitatively determined by methods based on physical optics or by chemical analysis. The recommended way to determine dye concentrations in solutions is through the concentration-dependent Kubelka-Munk function, which is discussed in Section 8.2 in connection with tinting-strength determination of dyes. For the test methods described here, the determination of concentration is not so important, because tests are commonly performed on preparations (paint, resin, aqueous solution) made up with a known quantity of dye in the first place.

This survey will be a brief one making no claim of completeness. When determination problems occur in relation to dyes, the extensive specialist literature must be consulted (see [3] in Section 1.4 and [32] in Section 4.10).

Chemical methods for determining the concentration of dyes can be employed only when the constitution of the dye, or at least its molecular weight, is known. The first technique to mention is elemental analysis, but only when the dye contains an element that the medium does not. Other methods are applicable to a dye dissolved in a liquid. The most important of such methods – because they are valid for nearly all groups of dyes – are reduction methods. Titanium(III) sulfate or titanium trichloride is used as reducing agent. Azo dyes in (acidic) aqueous solution can be titrated directly or indirectly; the same holds for a number of heterocyclic dyes. Azo and anthraquinone dyes can also be determined by oxidizing with chromic acid and measuring the volume of nitrogen evolved. Other basic dyes can also be determined by oxidimetric titration with cerium(IV) sulfate.

Precipitation methods are sometimes used for the quantitative determination of anionic dyes (e.g., with cetyl triammonium bromide in chloroform). In the case of basic dyes, anion-active compounds such as alkylsulfonates are suitable for precipitation titrimetry.

Some of the techniques listed are also appropriate when several dissolved dyes must be determined at the same time (dye mixtures).

8.1.2 Pigments

Again, in most test methods the quantity and hence the concentration of pigment are known from the outset. Even so, concentration determination may be useful, as for example in the determination of S by the black-ground method, already discussed. Procedures for the determination of pigment contents are set forth in standards:*

- Centrifuge methods, chiefly for production control by the producer and incoming inspection by the consumer (not suitable for separation of carbon black or finely divided SiO_2)
- Incineration methods (not suitable for determination of organic pigments)
- Filtration methods (also suitable for the separation of carbon black and finely divided SiO_2)

Of these groups, centrifuge and filtration methods are applicable chiefly to liquid paints; incineration, on the other hand, is also appropriate for determinations on surface coating finishes and other coatings as well as plastics. The residue generally contains all the non-combustible particles, that is, pigments, fillers, and other solid additives. Any of the three methods will require an analytical balance, a desiccator, and an oven.

Centrifuge Methods

The quantity determined is the mass fraction of solid particles insoluble in the selected solvent. A sample is diluted with a solvent or mixture of solvents, and the solid particles are separated by centrifugation. The solid material is dried and weighed, and the pigment content is calculated as a fraction. The following apparatus is required in addition to those listed above: a laboratory centrifuge with cups and taring device, and a ball-type stirrer. A low-boiling solvent should be chosen if possible.

Incineration Methods

The solvent is first evaporated off from the sample; the residue is then ashed at 450 °C if carbonates are present, at 700–900 °C otherwise. The pigment content is determined from the mass of the residue, either directly or after an analysis. Additional apparatus required includes porcelain crucible, crucible tongs, muffle furnace, or fast ashing device.

Filtration Methods

The sample is mixed with a suitable organic solvent or mixture of solvents, then with a solution of KOH in methanol. The mixture is filtered through a glass

---

* See "Pigment Content" in Appendix 1.

filter crucible with filter aid. The solid matter recovered is dried and then weighed. The pigment content is calculated from the mass of the solid component. The following additional equipment will be needed: glass filter crucible, safety pipet, filter aid, KOH solution in methanol (1 mol/L), analytical-grade methanol.

With many pigments and pigment mixtures, a simple determination of the residue is not enough; the residue must be subjected to analysis and examination. The analytical methods for inorganic pigments are contained in standards*.

8.2 Relative Tinting Strength

8.2.1 Dyes

A method for determining the relative tinting strength of dyes in solution is described in standards;** the procedure is a spectrophotometric one. On the basis of the relative tinting strength as defined for dyes (Section 4.6.1), the absorption coefficients (or extinction coefficients) are determined at the absorption maximum and the ratio is taken. The dye solutions must not scatter light, and the absorption curves must be highly similar in nature; that is, only dyes similar in hue may be compared. This kind of test applies to dyes such as are used in paints, inks, and so forth. The solution media (solvents but also binders) employed in preparing the solutions must satisfy the following requirements: The dyes must be completely soluble in the medium, the resulting solutions must be stable, and the results of the measurement must be such that they can be extended to lab practice.

Solutions of the test dye and reference are prepared; these should have concentrations as nearly equal as possible. In addition to the usual laboratory apparatus (pipets, volumetric flasks, analytical balance), a spectrophotometer or filter photometer with the proper glass cuvettes is necessary. No particular solution medium is specified, but the following are especially suitable:

– N-Methyl-2-pyrrolidone (particularly good for basic, oil-soluble, and metal complex dyes)
– Dimethylformamide
– Dimethylsulfoxide

---

* For a Listing, see [1] in Section 1.4.
** See "Tinting Strength, relative" in Appendix 1.

8 Determination of Tinting Strength and Lightening Power

When concentrated dye solutions must be further diluted, the above list can be supplemented with

- Toluene
- 2-Ethoxyethanol
- Ethanol
- Water

Binders and other additives (acids, bases, complexing agents, surfactants, antioxidants, etc.) can also be used as part of the medium.

Preparation method: Weigh out suitable amounts of solid dyes (e.g., 0.1 or 1 g) or dye solutions, with accuracy to 0.0001 g. Quantitatively transfer weighed samples to volumetric flasks (100 or 250 mL) and completely dissolve in solution medium. In the case of solid dyes, it is useful to predissolve the material (stirrer, heating). Fill volumetric flasks, homogenize contents, and dilute in one or more steps to a concentration for measurement such that the transmittance is 10–30% (corresponds to concentrations of 2 to 40 mg/L).

Measurement method: Within one hour after preparation (protect from light until measurement), measure transmission/extinction at the absorption maximum of the solutions; in the same cuvette, measure blank value (solution medium used). Take difference in extinction between measurement and blank (unless using a two-beam instrument). From it, calculate absorption coefficients k_β'' per unit mass concentration (specific extinction coefficients), using the formula $k_\beta'' = E/(h \cdot \beta)$, where E is extinction, h is length of path of the cuvette, and β is mass concentration of dye. The relative tinting strength is then found as the ratio of k_β'' values between test and reference dyes.

Figure 8-1 is a computer printout from a determination of relative tinting strength for dyes.

```
Program "P(REL), DYE"
Relative Tinting Strength of dyes in solution, according to DIN 55 978 ************
Date: ..-..-..          Colorimeter: Spectral ABC
                        Cuvette: 10 mm
Input:      *************************************************************
Test: Red dye T         Reference: Red dye R
Measured solution:      Instruction Dye/3   Measured wavelength: 505 nm
Mass concentration:     Test: 19 mg/l       Standard: 20 mg/l
Extinctions;            Test: 1.6094        Standard: 1.6607
Output:     *************************************************************
Relative Tinting Strength:   102 %
*************************************************************************
```

Fig. 8–1. Determination of relative tinting strength of dyes in solution (computer output).

Standard deviations obtained in round-robin tests between laboratories were less than 2%; in repetitions by the same laboratory, ca. 1%.

8.2.2 Inorganic Black and Colored Pigments (Lightness Matching)

Effective methods for determining the relative tinting strength and color difference after color reduction on inorganic black and colored pigments can be found in standards,* which set forth the conditions that must always be fulfilled if erroneous results are to be avoided. The German standard is concerned with some of the tinting strength criteria discussed in Section 4.6.2. The following criteria are permitted:

- Minimum (X, Y, Z) – the smallest of the CIE tristimulus values
- CIE tristimulus value Y (or lightness L^*)
- Color depth T_S

What is more, instructions for determining the color difference after color reduction are given.

The ISO standard includes the use of the zero-th approximation (see Section 4.6.4) to rationalize the determination of tinting strength with the aid of the Kubelka-Munk function. It describes an alkyd resin paste method with which the spectral absorption coefficients per unit mass and the relative tinting strength can be determined. The tinting strength criterion is $\varrho_{min}(\lambda)$, the reflectance at the absorption maximum (or else $\min(R'_x, R'_y, R'_z)$, i.e., the smallest of the reflectometer values).

Thus the prerequisites exist for an effective method in the case of inorganic pigments; the tinting strength criterion is the CIE tristimulus value Y (i.e., the lightness L^*); see Section 4.6.3.

The following equipment is needed:

- Triple-roll mill
- Automatic muller
- Paste film holder (see Section 6.2.3) or other suitable paste support
- Colorimeter (spectrophotometer or tristimulus instrument)

---

* See "Tinting Strength, Relative" in Appendix 1.

Preparation method:
White pigment paste with the following composition (by mass):
- 40 parts TiO$_2$ (type R2 of ISO 591; e.g., Bayertitan® R-KB2)
- 56 parts Alkydal® L 64 (Bayer AG)
- 4 parts Ca stearate

Thoroughly mix ingredients in order to wet the solids. Pass through triple-roll mill until Grindometer test* yields a value < 5 μm. Store paste in airtight container.

Place 3 g of white pigment paste on lower plate of muller; spread 0.12 g of inorganic pigment (test or reference) on it and mix with spatula. Distribute mixture in form of a circular ring. Mull in four stages of 25 revolutions each with a weight (1 kN) applied; after each stage, collect paste and redistribute.

Measurement method:
Measure CIE tristimulus value Y (also X and Z for color difference after color reduction) or reflectometer value R'_y (also R'_x and R'_z if necessary) of the reducing white on the paste film holder. Preferred procedure is method A of Section 6.1.2 (i.e., with surface reflection included).

Evaluation: Apply Saunderson corrections to all CIE tristimulus values (eq. 4.95). Use equation (4.61) to calculate Kubelka-Munk functions F, F_R; use equation (4.94) to calculate tinting strength. The tinting strength obtained in this way is identical to the relative absorption coefficient per unit mass, so the F values must first be divided by the pigment mass concentrations unless precisely 0.12 g of test and reference pigments was taken. To determine the color difference after color reduction, convert CIE tristimulus values X and Z of reference to matching concentration (with help of Kubelka-Munk functions and the relative tinting strength P just determined). From the CIE tristimulus values after Saunderson back-correction, calculate CIELAB coordinates by known method, and thence color differences. In the case of black pigments, it may be desirable to report the relative hue** as specified in Section 2.3.3 instead of the hue and chroma differences.

The flowchart for this determination appears in Figure 8-2; Table 8-1 lists the needed computational steps as an aid to programming. Here the reference is matched to the test pigment; this is recommended because the test is performed at a fixed PVC value corresponding to $\hat{m}$. If the test were matched to the reference, the pigment concentration would vary with the relative tinting strength of the test pigment. The summary of computational steps also shows how the measured CIE tristimulus values of the reference are converted to the CIE tristimulus values at matching.

---

* See "Fineness of Grind" in Appendix 1.
** See "Hue, Relative, of Near White Specimens" in Appendix 1.

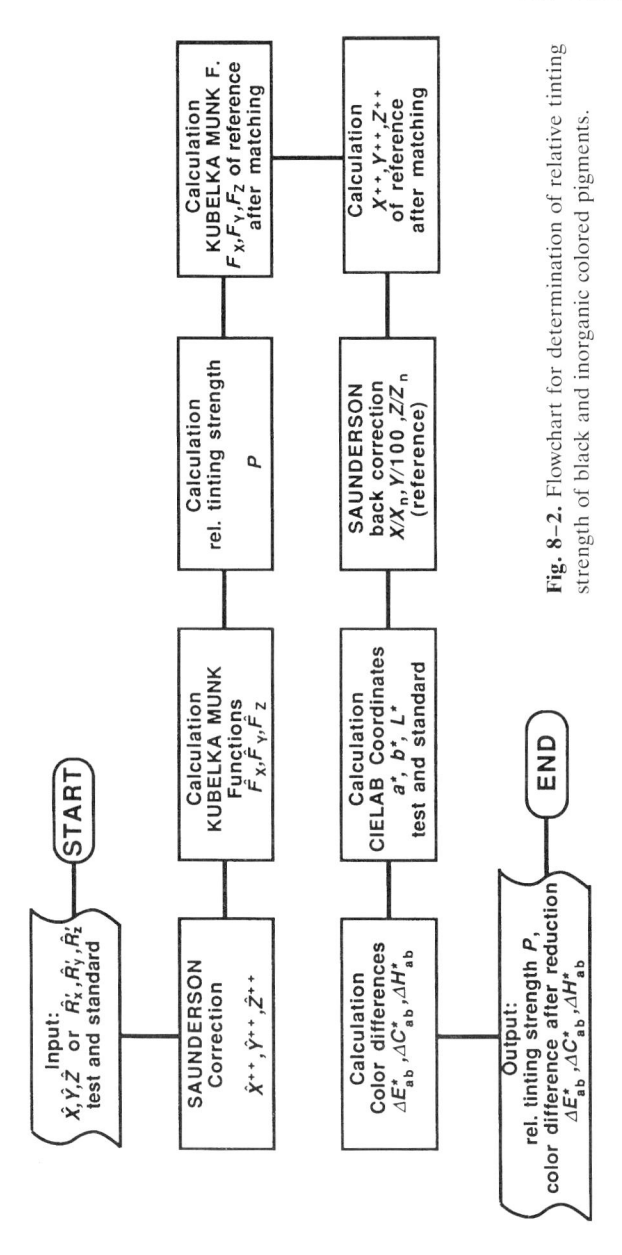

8.2 *Relative Tinting Strength* 337

Fig. 8-2. Flowchart for determination of relative tinting strength of black and inorganic colored pigments.

8 Determination of Tinting Strength and Lightening Power

Table 8–1. Formulas needed for calculating relative tinting strength and color difference after color reduction for inorganic colored and black pigments.

Given: $\hat{X}, \hat{Y}, \hat{Z}, \hat{X}_R, \hat{Y}_R, \hat{Z}_R,$ or $\hat{R}'_x, \hat{R}'_y, \hat{R}'_z, \hat{R}'_{x,R}, \hat{R}'_{y,R}, \hat{R}'_{z,R}$

$$\hat{R}^{++} = \frac{\hat{R} - 0.04}{0.36 + 0.6\,\hat{R}}; \qquad \hat{F} = \frac{(1 - \hat{R}^{++})^2}{2\,\hat{R}^{++}} \qquad \text{eqs. (4.95), (4.61)}$$

where
1) $\hat{R} = \hat{X}/X_n = (0.782\,\hat{R}'_x + 0.198\,\hat{R}'_z)/98$
2) $\hat{Y}/100 = \hat{R}'_y/100$
3) $\hat{Z}/Z_n = \hat{R}'_z/100$
4) $\hat{X}_R/X_n = (0.782\,\hat{R}'_{x,R} + 0.198\,\hat{R}'_{z,R})/98$
5) $\hat{Y}_R/100 = \hat{R}'_{y,R}/100$
6) $\hat{Z}_R/Z_n = \hat{R}'_{z,R}/100$

$\hat{R}^{++} = \hat{X}^{++}/X_n$	$\hat{F} = \hat{F}_X$
$\hat{Y}^{++}/100$	$\hat{F}_Y$
$\hat{Z}^{++}/Z_n$	$\hat{F}_Z$
$\hat{X}_R^{++}/X_n$	$\hat{F}_{X,R}$
$\hat{Y}_R^{++}/100$	$\hat{F}_{Y,R}$
$\hat{Z}_R^{++}/Z_n$	$\hat{F}_{Z,R}$

$$P^{(0)} = \frac{\hat{F}_Y}{\hat{F}_{Y,R}} \quad \text{or} \quad P^{(1)} = \hat{f}\,[1 + a\,(\hat{f} - 1)] \quad \text{with}\ \hat{f} = \frac{\hat{F}_Y - F_{BP}}{\hat{F}_{Y,B} - F_{BP}} \qquad \text{eqs. (4.92), (4.93), (4.94)}$$

$$R_R^{++} = 1 + \hat{F}_R P - \sqrt{\hat{F}_R P (\hat{F}_R P + 2)}; \qquad R_R = 0.04 + \frac{0.384\,R_R^{++}}{1 - 0.6\,R_R^{++}} \qquad \text{eqs. (4.111), (4.112)}$$

where
1) $\hat{F}_R = \hat{F}_{X,R}$
2) $\hat{F}_{Y,R}$
3) $\hat{F}_{Z,R}$

$R_R^{++} = X_R^{++}/X_n$	$R_R = X_R/X_n$
$Y_R^{++}/100$	$Y_R/100$
Z_R^{++}/Z_n	Z_R/Z_n

Two cases:

A) $R \leq 0.008856$ B) $R > 0.008856$

$$R^* = 7.787\,R + 0.138 \qquad R^* = \sqrt[3]{R} \qquad \text{eq. (2.5)}$$

where
1) $R = \hat{X}/X_n$ $R^* = X_T^*$
2) $\hat{Y}/100$ Y_T^*
3) $\hat{Z}/Z_n$ Z_T^*
4) X_R/X_n X_R^*
5) $Y_R/100$ Y_R^*
6) Z_R/Z_n Z_R^*

$$a^* = 500\,(X^* - Y^*); \qquad b^* = 200\,(Y^* - Z^*); \qquad L^* = 116\,Y^* - 16 \qquad \text{eq. (2.7)}$$

where
1) $X^* = X_T^*$ $Y^* = Y_T^*$ $Z^* = Z_T^*$ $a^* = a_T^*$ $b^* = b_T^*$ $L^* = L_T^*$
2) X_R^* Y_R^* Z_R^* a_R^* b_R^* L_R^*

$$\Delta a^* = a_T^* - a_R^*; \qquad \Delta b^* = b_T^* - b_R^*; \qquad \Delta L^* = L_T^* - L_R^*; \qquad \text{eq. (2.11)}$$

$$\Delta E_{ab}^* = \sqrt{\Delta a^{*2} + \Delta b^{*2} + \Delta L^{*2}} \qquad \text{eq. (2.10)}$$

$$\Delta C_{ab}^* = \sqrt{a_T^{*2} + b_T^{*2}} - \sqrt{a_R^{*2} + b_R^{*2}} \qquad \text{eq. (2.12)}$$

Table 8–1. Continuation

$$\Delta h_{ab} = \text{Arctan}\, \frac{b_T^*}{a_T^*} - \text{Arctan}\, \frac{b_R^*}{a_R^*} \qquad \text{eq. (2.14)}$$

$$\Delta H_{ab}^* = \sqrt{\Delta E_{ab}^{*2} - \Delta L^{*2} - \Delta C_{ab}^{*2}} \qquad \text{(same sign as } \Delta h_{ab}) \qquad \text{eq. (2.13)}$$

An example in the form of spectral reflection curves for reducing whites is presented in Figures 8-3 and 8-4. In the first case, the curves are for equal masses of test and reference; in the second, the curves are plotted after lightness matching. Finally, Figure 8-5 is a printout of a tinting strength determination.

For most purposes, the use of the zero-th approximation to obtain the tinting strength as described here is entirely adequate – for example, for production control. If there should be a need for a comparison between inorganic pigments with high light scattering power or a very great difference in tinting strength, so that more than ±20% departures from 100% relative tinting strength occur, then the first approximation is recommended (Section 4.6.4, eq. 4.92). The factor a in equation (4.93) depends on the reference alone and can easily be determined once for all.

8.2.3 Organic Pigments ("FIAF" Method Based on Principle of Spectral Evaluation)

The principles of this section were established by GALL (see [19], [23] in Section 4.10), whose work has become the basis for standards.* The German standard defines "standard color depths" and gives a procedure for the preparation of color depth standards (i.e., color samples of known standard color depth); these are used as controls in both visual and colorimetric matching. The same standard provides instructions for bringing about the standard color depth in color samples. We follow the German standards in determining the relative tinting strength of organic colored pigments and in describing Gall's "FIAF" method, a computer-aided method of colorimetric tinting strength determination by iterative color depth matching based on the principle of spectral evaluation.**

---

* See "Tinting Strength, Relative" and "Color Depth" in Appendix 1.
** The letters FIAF stand for the German phrase "*F*arbstärke durch *i*terativen *A*ngleich der *F*arbtiefen," with the meaning stated above.

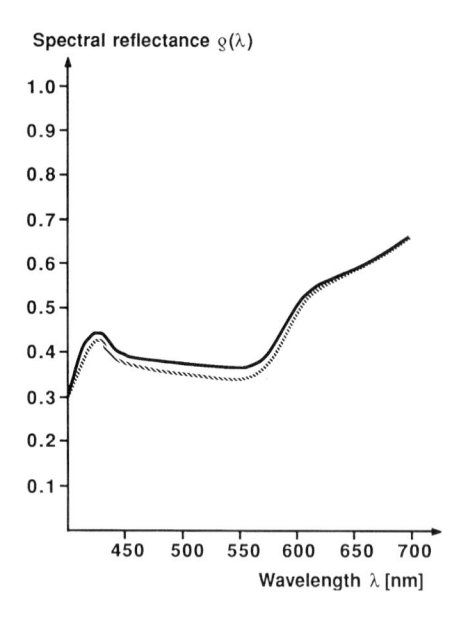

340   8 *Determination of Tinting Strength and Lightening Power*

Fig. 8-3. Spectral reflectance of two red iron oxide pigments in reducing white (equal added pigment masses).

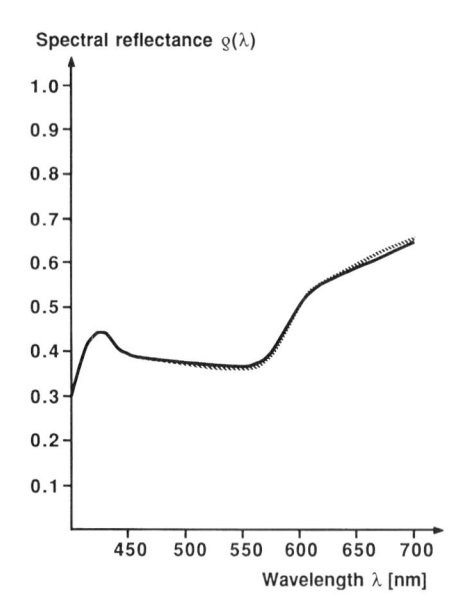

Fig. 8-4. Example from Fig. 8-3 after Y matching.

8.2 Relative Tinting Strength

The color depth standards are available commercially; they should be calibrated, that is, they should indicate the CIE tristimulus values and the color depth parameter $B = \Delta T_S$ (deviation from the color depth level), measured in the prescribed way. The visual adjustment of color samples to a given color depth level is of secondary importance in practice, so only the colorimetric technique needs explanation here. The color depth standards are necessary when results from different instruments or different laboratories are to be compared. Corrections to the measured CIE tristimulus values with respect to the nominal values of the calibrated standards take the form of ratios: $X_{corr} = (X_S/X_M) \cdot X$, where X_S is the nominal value for calibration and X_M is the value measured on the standard (and similarly for Y and Z).

Ordinary paint bases (white enamels) are used as test media (see Section 8.2.4 for examples); naturally, pastes can also be used as in Section 8.2.2.

GALL's method is contrasted to the method based on Y matching described in Section 8.2.2, where one Y value was specified and the second was adjusted; in the color depth matching method, both reducing whites (test and reference)

```
Program "P(REL)"
Relative Tinting Strength by means of lightness matching, according to ISO 787-24
(modified)       ****************************************************************
Date: ..-..-..
Standard is A          Test is B
Medium and white paste: according to ISO 787-24
Colorimeter, Standard Observer: N.N., C/2°
Measuring conditions: d/8 (incl)
Input:      ****************************************************************
Masses: White paste: 3 g   Pigment: .12 g
Reflectometer values:
         Reference:     Test:
R´(x)    47.74          47.02
R´(y)    41.03          39.93
R´(z)    39.68          38.18
Output:     ****************************************************************
Relative Tinting Strength:        P =      108 %
Tinting Equivalent:               1/P =    93 %
CIELAB Coordinates after matching:
         Reference:     Test;
a* =     15.0           15.7
b* –     1.7            2.2
L* =     69.4           69.4
Color difference after reduction:
Delta(E*(ab))      =      .9
Delta(L*)          =      .0
Delta(C*(ab))      =      .8
Delta(H*(ab))      =      .4
Delta(h(ab))       =      1.5°
```

Fig. 8–5. Relative tinting strength of a red iron oxide pigment by lightness matching.

must be adjusted to the specified color depth level.* The ratio of the masses of pigment required are input to the fundamental equation (4.75), which gives the relative tinting strength. The formulas needed for this procedure were already introduced in Section 4.5.6. The method results in the value $\Delta T_S = B$ of a selected standard color depth being made equal to zero for test and reference through variation of the amounts of pigment added to the white paste.

Now it is possible to describe Gall's FIAF method, which functions on the principle of spectral evaluation. A flowchart is shown in Figure 8-6. From the input reflection spectrum $\varrho(\lambda)$ at 16 chosen wavelengths (20 nm intervals), the computer calculates the CIE tristimulus values X, Y, Z, the CIE chromaticity coordinates x, y, and thence ΔT_S. A check is then made in order to see whether $\Delta T_S = 0$ to a sufficiently good approximation. If not, the mass-referred spectral Kubelka-Munk functions F/m are calculated from the Saunderson-corrected reflection spectrum, then multiplied by an appropriately modified mass of pigment. From the result, the new CIE tristimulus values and ΔT_S are calculated. This loop is repeated until $\Delta T_S = 0$. As a rule, two to four times around the loop will be required; five may be needed in quite rare cases. Once the target match for test and reference has been found, the tinting strength is output as the ratio of the two masses (along with the tinting strength equivalent if appropriate); in addition, the color difference after color reduction and the lightness, chroma, and hue differences are included in the output. Table 8-2 lists the needed equations as an aid to programming.

A problem is presented by the $a(\phi)$ factors (see Section 4.6.5): The German standard includes long tables for "manual" searching and calculating. GALL and FRIEDRICHSEN have written convenient formulas for computer calculations based on third-degree polynomials; these are excerpted here for 1/3, 1/9, and 1/25 ST and for illuminant/observer combinations C/2° and D65/10° (Tables 8-3 and 8-4), so that there is no need to reproduce the full tables. The formula for calculation is

$$a(\varphi) = a(\varphi_0) + K_1 W + K_2 W^2 + K_3 W^3$$

where $W = (\varphi - \varphi_0)/100$. A flowchart for the computer calculation is shown in Figure 8-7; the coefficients of the polynomials (tabulated values) must be stored in memory.

Figure 8-8 gives a printout by way of example. Two blue pigments were tested; the resulting value of the relative tinting strength is around 126%. The example is instructive in that a relative tinting strength based on lightness matching can also be determined from these data. It is only 117%; thus, less of the reference pigment would have to be added for matching. But this also

---

* SCHMELZER and co-workers propose the following departure from the standard: Test and reference are adjusted to the same (nonzero) value of B. Drawback: It is not possible to report the standard color depth level (SDS).

8.2 Relative Tinting Strength

Table 8–2. Formulas needed for calculating relative tinting strength and color difference after color reduction for organic colored pigments.

Given: $\varrho(\lambda)$, $\varrho_R(\lambda)$ with 16 chosen wavelengths spaced 20 nm apart

$$X = \int \bar{x}(\lambda)\, \varrho(\lambda)\, d\lambda$$

$$Y = \int \bar{y}(\lambda)\, \varrho(\lambda)\, d\lambda$$

$$Z = \int \bar{z}(\lambda)\, \varrho(\lambda)\, d\lambda \tag{2.3}$$

$$x = \frac{X}{X+Y+Z}; \quad y = \frac{Y}{X+Y+Z} \tag{2.4}$$

$$s = 10\sqrt{(x-x_n)^2 + (y-y_n)^2}$$

$$\varphi = \text{Arctan}\, \frac{y-y_n}{x-x_n} \tag{4.100}$$

$a(\varphi)$ from subroutine, Fig. 8–7, Tables 8–3 and 8–4

$$\Delta T_S = \sqrt{Y}\,(s \cdot a(\varphi) - 10) + b_i \tag{4.99}$$

$(b_{\frac{1}{3}} = 29,\ b_{\frac{1}{9}} = 41,\ b_{\frac{1}{25}} = 56)$

$$\varrho^*(\lambda) = \frac{\varrho(\lambda) - 0.04}{0.36 + 0.6\,\varrho(\lambda)} \tag{3.11}$$

$$F(\lambda) = \frac{(1 - \varrho^*(\lambda))^2}{2\,\varrho^*(\lambda)} \tag{3.117}$$

$$f(\lambda) = \frac{F(\lambda)}{m} \tag{4.64}$$

$$m_{i+1} = m_i\, 0.9^{\Delta T_S} \tag{4.101}$$

$$f(\lambda)\, m_{i+1} = F(\lambda) \tag{4.64}$$

$$\varrho^*(\lambda) = 1 + F(\lambda) - \sqrt{F(\lambda)\,(F(\lambda) + 2)} \tag{3.116}$$

$$\varrho(\lambda) = \frac{0.384\,\varrho^*(\lambda)}{1 - 0.6\,\varrho^*(\lambda)} + 0.04 \tag{3.10}$$

Iteration yields m_{end}; thence

$m_{\text{end}} \rightarrow F_{\text{end}}(\lambda) \rightarrow \varrho^*_{\text{end}}(\lambda) \rightarrow \varrho_{\text{end}}(\lambda) \rightarrow X, Y, Z.$

Thence a^*, b^*, L^* and a^*_R, b^*_R, L^*_R
and thence $\Delta E^*_{ab}, \Delta C^*_{ab}, \Delta h_{ab}, \Delta H^*_{ab}$ } as in Tab. 8–1

For black pigments: shade difference after color reduction as in Table 8–7.

344 8 *Determination of Tinting Strength and Lightening Power*

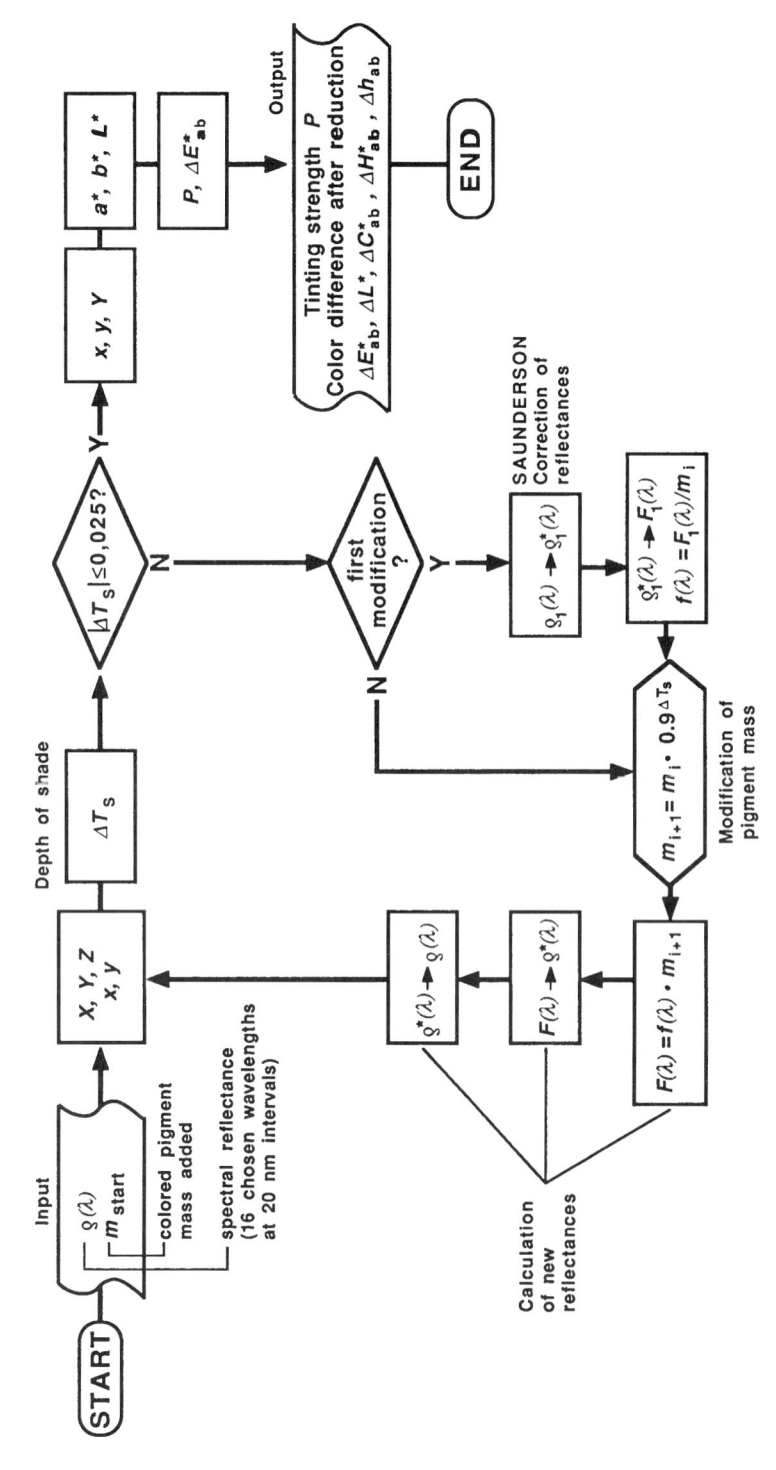

Fig. 8-6. Flowchart of FIAF program for calculating relative tinting strength and color difference after color reduction by Gall's method.

Table 8–3. Polynomial coefficients for calculating the $a(\varphi)$ factors in the determination of ΔT_S (after GALL and FRIEDRICHSEN). Standard illuminant C, 2° observer.

	Range	φ_0 (degrees)	$a(\varphi_0)$	K_1	K_2	K_3
$\frac{1}{3}$ SDS	1	0	1.971	1.88544	7.21387	9.80811
	2	52	3.523	0.569638	−0.97366	0.380866
	3	156	3.491	−0.369324	−5.51416	1.48145
	4	188	2.856	−2.81256	−2.37598	11.2539
	5	216	2.130	−2.74438	−25.4053	62.6152
	6	252	0.771	−0.421143	70.0625	−182.867
	7	276	2.177	2.79831	−12.0183	17.0195
	8	308	2.400	−0.492714	3.15607	−8.72205
	9	344	2.224	−2.95581	−5.12891	85.1523
	(10	360	1	1	1	1)
$\frac{1}{9}$ SDS	1	0	2.338	−0.899719	11.9614	−10.4897
	2	48	3.502	5.24835	−17.4316	19.165
	3	92	4.069	−0.36496	0.647095	−1.41742
	4	188	3.061	−1.78186	−3.7832	−4.89355
	5	224	1.701	−5.83112	14.4609	−12.9414
	6	244	1.010	4.65631	−0.212402	−5.78418
	7	288	2.525	1.51682	−1.57715	−0.630615
	8	344	2.769	−0.853577	−31.7139	125.77
	(9	360	1	1	1	1)
$\frac{1}{25}$ SDS	1	0	2.399	−3.06669	16.38	−10.9985
	2	24	2.454	6.05066	19.0391	−87.7031
	3	44	3.725	6.83469	−38.7412	69.4805
	4	72	4.126	−0.303894	5.60791	5.65527
	5	104	4.788	3.50299	−17.5785	20.2175
	6	144	4.671	−1.60059	−3.75781	2.90381
	7	196	3.231	−3.99182	−3.58398	−7.41406
	8	220	1.964	−8.67981	27.377	−17.6328
	9	236	1.204	3.00708	4.0166	−7.15625
	10	296	2.908	0.416885	−1.06287	−1.29846
	(11	360	1	1	1	1)

means that the color difference after color reduction will change. It is found that ΔE_{ab}^* is smaller than in the color depth matching method. The difference can be attributed primarily to a smaller chroma difference (saturation); the hue difference, on the other hand, is larger. These results imply that a greater difference in nuance is accepted when lightness matching is employed, even though in the case of color depth matching this larger difference would be perceived as interfering. This may indicate that questions of color formulation must be accorded more weight in color depth matching, while lightness differences are more significant in the case of Y matching.

8 Determination of Tinting Strength and Lightening Power

Table 8–4. Polynomial coefficients for calculating the $a(\varphi)$ factors in the determination of ΔT_S (after GALL and FRIEDRICHSEN). Standard illuminant D65, 10° observer.

	Range	φ_0 (degrees)	$a(\varphi_0)$	K_1	K_2	K_3
$\frac{1}{3}$ SDS	1	0	2.040	1.80164	9.15625	−12.6865
	2	52	3.669	1.44590	−3.59046	2.00006
	3	140	3.524	1.21893	−8.80359	7.20239
	4	196	2.710	−0.562195	−9.45264	1.99219
	5	236	1.101	−4.18735	16.9717	118.672
	6	252	1.351	7.98462	−9.83203	−14.2773
	7	276	2.504	1.54935	−3.91061	1.62125
	8	340	2.319	−2.57889	−14.03809	50.0312
	(9	360	1	1	1	1)
$\frac{1}{9}$ SDS	1	0	2.345	−0.634155	11.2192	−7.88379
	2	52	3.940	2.40570	−4.96634	2.47712
	3	148	3.864	−1.28906	1.00537	−3.89380
	4	224	1.756	−4.93951	−20.8896	188.117
	5	244	1.438	5.80115	−1.04980	−12.9629
	6	280	2.785	1.67851	−3.35052	1.46381
	7	336	2.932	−1.52710	−15.6045	49.1836
	(8	360	1	1	1	1)
$\frac{1}{25}$ SDS	1	0	2.360	−3.05167	11.0864	30.0801
	2	28	3.035	8.95835	−15.2949	−9.64844
	3	56	4.127	−0.28082	5.0835	−1.5742
	4	96	4.727	2.57528	−6.64651	2.61005
	5	148	4.636	−2.0872	0.12793	−3.27197
	6	224	1.687	−5.801	28.7262	27.2941
	7	244	1.895	6.0961	−8.30859	2.31348
	8	292	3.162	0.66899	−3.24585	0.796387

As Table 8-2 and equation 4.64 show, the zero-th approximation is employed in calculating the relative tinting strength. This is justified even for higher pigment concentrations, since organic pigments commonly have a lower refractive index (in the neighborhood of 1) and thus only slight light-scattering power. Thus the use of the zero-th approximation is proper even outside the Beer's law range, and therefore even at high relative tinting strength values differing by more than ±20% from 100% – as in the example.

For organic black pigments, it is desirable to employ Y matching; the procedure of Section 8.2.2 can thus be used.

8.2 Relative Tinting Strength

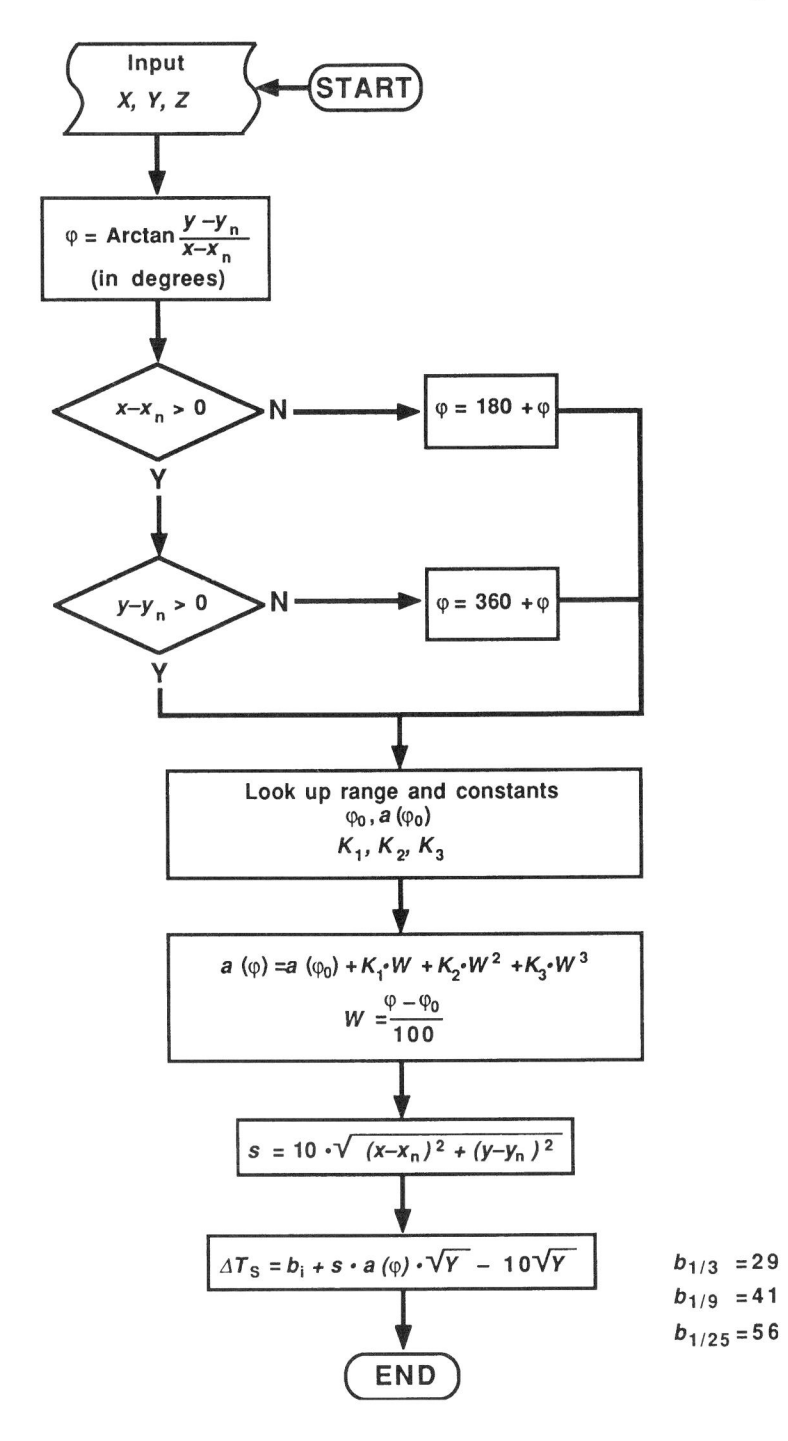

```
        ┌─────────┐
        │  Input  │◄───(START)
        │ X, Y, Z │
        └────┬────┘
             ▼
  ┌──────────────────────┐
  │ φ = Arctan (y−yₙ)/(x−xₙ) │
  │     (in degrees)     │
  └──────────┬───────────┘
             ▼
        ╱ x−xₙ > 0 ╲──N──► φ = 180 + φ ──┐
        ╲          ╱                      │
             Y                            │
             ▼◄───────────────────────────┘
        ╱ y−yₙ > 0 ╲──N──► φ = 360 + φ ──┐
        ╲          ╱                      │
             Y                            │
             ▼◄───────────────────────────┘
  ┌──────────────────────┐
  │ Look up range and    │
  │ constants            │
  │ φ₀, a(φ₀)            │
  │ K₁, K₂, K₃           │
  └──────────┬───────────┘
             ▼
  ┌──────────────────────────────────┐
  │ a(φ) = a(φ₀) + K₁·W + K₂·W² + K₃·W³ │
  │        W = (φ − φ₀)/100          │
  └──────────┬───────────────────────┘
             ▼
  ┌─────────────────────────────┐
  │ s = 10·√((x−xₙ)² + (y−yₙ)²) │
  └──────────┬──────────────────┘
             ▼
  ┌─────────────────────────────────┐
  │ ΔTₛ = bᵢ + s·a(φ)·√Y − 10√Y     │
  └──────────┬──────────────────────┘
             ▼
          (END)
```

$b_{1/3} = 29$
$b_{1/9} = 41$
$b_{1/25} = 56$

Fig. 8–7. Flowchart for computer calculation of ΔT_S (after GALL and FRIEDRICHSEN).

```
Program "FIAF"
Relative Tinting Strength by means of Color Depth matching, according to
DIN 53 235 and 55 986   ************************************************
Date: ..-..-..          Color Depth level: 1/3 ST
Standard is A           Test is B
Medium and white paste according to ISO 787-24
Colorimeter, Standard Observer: N.N., C/2°
Measurement Conditions: d/8 (incl)
Input:   *********************************************************************
Masses: White paste: 3 g         Pigment: .3 g
Spectral Reflectances:  on-line colorimeter
Output:  *********************************************************************
Relative Tinting Strength:       P =      126 %
Tinting Equivalent:              1/P =     80 %
CIELAB Coordinates of 1/3 ST:
         Reference:      Test:
a* =     - 10.5          - 11.2
b* =     - 27.8          - 28.2
L* =       72.4            73.0
Color Difference after reduction:
Delta(E*(ab))  =     1.1
Delta(L*)      =      .6
Delta(C*(ab))  =      .7
Delta(H*(ab))  =    - .6
Delta(h(ab))   =   - 1.1°
```

Fig. 8–8. Relative tinting strength of a blue pigment by color depth matching (FIAF method) (computer output).

8.2.4 Change in Tinting Strength

Test methods are set forth in the standard* for lacquer binding media and plastics. Test and reference pigments are dispersed in agreed-on vehicles in accordance with agreed-on conditions. Methods and vehicles listed in Section 6.2.1 are recommended. From the mill base, samples are taken after specified dispersion stages and mixed with a white paint (same vehicle) or with a specified white paste. The reflectances are then measured, and the Kubelka-Munk functions F of the test and reference mixtures are calculated from the results. The standard still prescribes $\varrho(\lambda)$ (min) as the tinting strength criterion; that is, the F value is to be determined at the reflection minimum (i.e., the absorption maximum). It goes without saying that different criteria (in particular the CIE tristimulus value Y) can be employed as agreed on.

Along with an analytical balance and porcelain dishes, the method requires the following:

---

* See "Change in Tinting Strength" in Appendix 1.

8.2 Relative Tinting Strength

- Bird applicator or paint spray gun
- Substrate material (glass plates, sheet metal, cardboard; also paste film holders as discussed in Section 6.2.3)
- Perforated cardboard template (for covering in measurements on pastes)
- Spatula
- Automatic muller
- Colorimeter (spectral or tristimulus)
- Oven
- White paint (20% TiO$_2$) or white paste (40% TiO$_2$) made up with same vehicle as used in mill base

The standard distinguishes between:

- Low-viscous mill bases
- High-viscous, pastelike mill bases

Preparation method 1 for low-viscous mill bases

Dispersing: generally with oscillatory shaking machine. Withdraw samples after fixed time intervals (initial sample after shortest dispersion time or stage, final sample after longest reasonable dispersion time or stage; tinting strength must reach at least 90% of potential final value).
Reducing white: Mix 3 g of ground mill base with such a quantity of white paint (generally 30 g) that a certain color quality as agreed on (an agreed-on color depth, according to the standard) is achieved.
Application: Apply reducing whites to substrates with Bird applicator or spray gun. Check hiding of the coat with contrast cards and rub-out test (see Section 6.2.1).

Preparation method 2 for high-viscous, pastelike mill bases

Dispersing: generally with automatic muller, otherwise same as method 1.
Reducing white: Mix 5 g of white paste with such a quantity of colored pigment paste (generally 0.1 to 0.5 g) that a certain color quality as agreed on (an agreed-on color depth according to the standard) is achieved. Disperse mixture in four stages of 25 revolutions in the automatic muller. After every stage, homogenize with spatula and collect toward center of plate.
Application: Apply reducing whites to substrates with Bird applicator. Check hiding with contrast cards and rub-out test (see Section 6.2.1).

Measurement and evaluation method

Measure CIE tristimulus values Y (in the standard, reflectance at the absorption maximum), apply Saunderson correction, and calculate Kubelka-Munk functions F. Use perforated template when measuring pastelike specimens. When rub-out effect is marked (i.e., $\Delta E_{ab}^* > 10$), measure the rubbed area.

8 Determination of Tinting Strength and Lightening Power

Plot F versus rubbing time or stages; also plot the reciprocals $1/F$ versus $1/t$ in a Pigenot diagram.

A practical example for the change in tinting strength was given in an earlier German standard.* Two copper phthalocyanine pigments were tested. The vehicles used were Alkydal® R35 (Bayer AG) and Maprenal® MF 800 (Hoechst AG); the white pigment was Bayertitan® R-KB-2. The composition is stated as follows:

Vehicle for dispersion (varnish), composition by mass:
  52% castor oil alkyd resin, 60% in xylene
  17% melamine resin, 55% in isobutanol
  19% xylene
  12% n-butanol

After sufficient stirring, the flow time** was determined by using a standard flow cup. If the flow time was > 25 s at 23°C, xylene and n-butanol were added to adjust.

White paint, composition by mass:
  52.6% castor oil alkyd resin, 60% in xylene
  17.2% melamine resin, 55% in isobutanol
  20.0% TiO_2 pigment
  5.0% n-butanol
  1.0% butylene glycol
  3.8% thickening paste (consisting of 87% white spirit, 10% Bentone® (Kronos Titan GmbH), 3% methanol)
  0.4% soybean lecithin

The TiO_2 pigment was ground in Alkydal (to a fineness*** of < 5 μm); only at this point was the melamine resin added. The flow time was supposed to be < 200 s (if not, an adjustment was made as in the varnish). Reducing whites were applied to contrast cards with a Bird applicator; the rest of the procedure was as in method 2 above. Measurement (of $\varrho_{min}(\lambda)$) and evaluation for pigments 1 and 2 yielded the diagrams of Figures 8-9 and 8-10.

Here an evaluation by the method of Section 4.7.3 should be included. The procedure is fairly simple if a conversion is first made from the curves of change in tinting strength to the functions $v(t)$. According to equations (4.94) and (4.116),

---

* DIN 53238, Annex 1 (withdrawn).
** See "Flow Time, Determination" in Appendix 1.
*** See "Fineness of Grind" in Appendix 1.

8.2 *Relative Tinting Strength* 351

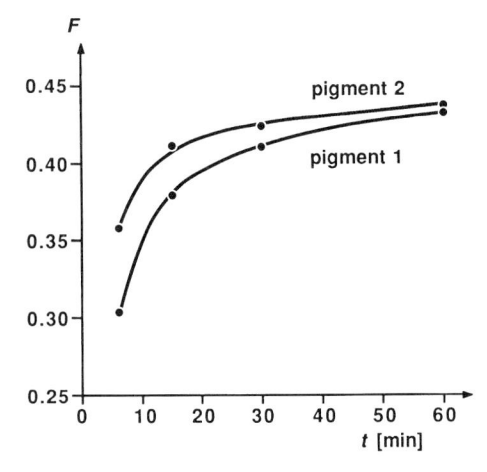

Fig. 8-9. Curves of change in tinting strength for copper phthalocyanine pigments, after DIN 53238 Annex 1.

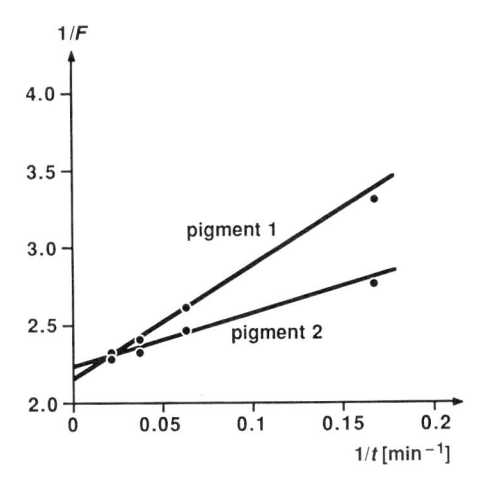

Fig. 8-10. Curves of change in tinting strength based on the curves of Fig. 8-9 (Pigenot diagram).

$$v = \frac{P - P_0}{1 - P_0} = \frac{F - F_0}{F_\infty - F_0} \tag{8.1}$$

where the values for F_∞ can be taken from the Pigenot diagram (Fig. 8-10). The following numerical examples are for pigment 1, so $1/F_\infty = 2.2$ and $F_\infty = 0.4545$. The initial values F_0 are the first measured values at $t = 6$ min, so a new time scale must be introduced: $t' = t - 6$. In order to determine the two constants A and B from equation (4.124), the measurements at $t_1 = 15$ min and $t_2 = 60$ min are used; thus

8 Determination of Tinting Strength and Lightening Power

$$t'_1 = 15 - 6 = 9,$$
$$t'_2 = 54,$$
$$v_1 = (0.3763 - 0.3021)/(0.4545 - 0.3021) = 0.487,$$
$$v_2 = 0.872.$$

Now A can be expressed in two ways:

$$\eta_1 = \left(\frac{1}{v_1} - 1\right)(e^{Bt'_1} - 1) = A$$

$$\eta_2 = \left(\frac{1}{v_2} - 1\right)(e^{Bt'_2} - 1) = A$$
(8.2)

Because η_1 must be equal to η_2, the function $\eta = \eta_2 - \eta_1$ must be made equal to zero so that a solution for B can be obtained. This is easily done with the aid of a diagram (η plotted versus B); Figure 8-11 shows that the zero crossing occurs at $B = 0.00765$ min^{-1} for this example. Naturally, a simple iteration program using the above equations could be written for a desktop computer. The value of A is then determined from η_1 or η_2 with equations (8.2); in the example,

$$A = (1/0.487 - 1)(\exp(0.00765 \cdot 9) - 1) = 0.0751.$$

For simplicity, we can set the rate constant $\varkappa_f$ equal to zero; then, from A and B, equations (4.124) and (4.125) yield the following rate constants and half-time for the example:

$$\varkappa_w = B = 0.00765 \text{ min}^{-1};$$

$$\varkappa_c = B\left(\frac{1}{A} - 1\right) = 0.00765\left(\frac{1}{0.0751} - 1\right) = 0.0941 \text{ min}^{-1}$$

$$t_{0.5} = \frac{A}{B}\left(1 - \frac{A}{2}\right) = \frac{0.0751}{0.00765}\left(1 - \frac{0.0751}{2}\right) = 9.45 \text{ min}$$

Further, the potential of tinting strength can be calculated with equation (4.126); in the example,

$$\Psi = F_\infty/F_0 - 1 = 0.4545/0.3021 - 1 = 50\%.$$

The results for pigments 1 and 2 are listed in Table 8-5.

Instead of Ψ, the quoted example gives the increase in strength IS (also listed in Table 8-5). When making the comparison between IS and Ψ, it must be kept in mind that the values of Ψ are more sharply defined (initial and final values), while in the case of IS it is not necessary to have F_∞; an F value obtained earlier, at least 90% of F_∞, can be used (here the F value for $t = 60$).

8.2 Relative Tinting Strength

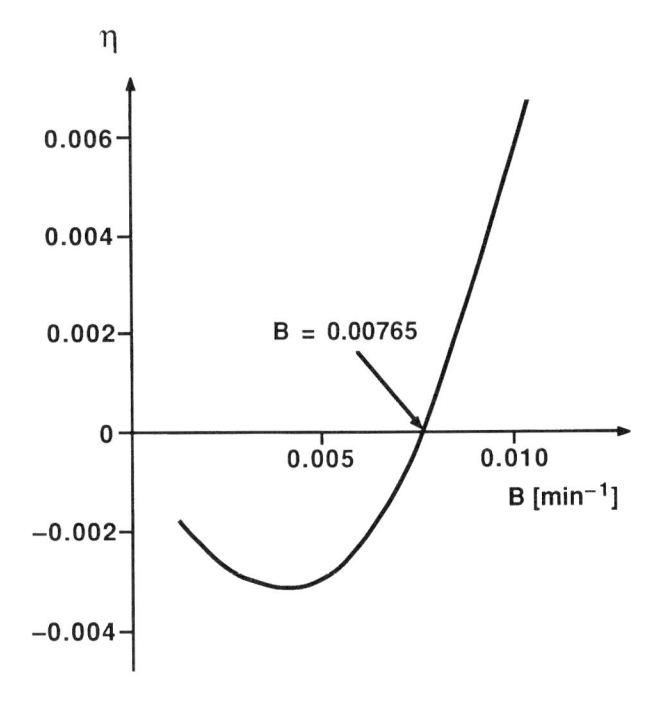

Fig. 8-11. Diagram for graphical determination of B.

(graph showing η vs B [min^{-1}], with $B = 0.00765$ marked)

Table 8-5. Evaluation of dispersing curves for two copper phthalocyanine pigments.

Quantity	Pigment 1	Pigment 2
$\varkappa_c$ in min^{-1}	0.0942	0.0716
$\varkappa_w$ in min^{-1}	0.00765	0.02325
$t_{0.5}$ in min	9.45	9.25
Ψ	50 %	24 %
IS	44	22

These examples show the following: If the half-times for both pigments are used to characterize their dispersing resistance, these are equal to within 2 %. If, on the other hand, the potential of strength Ψ (or the IS value) is considered, the results differ by a factor of two or more. It thus follows that the potential of strength Ψ is scarcely an appropriate measure of the dispersing behavior. At best, it does give some indication of what tinting strength can be attained.

Another noteworthy inference can be drawn from the figures in Table 8-5. A comparison of the $\varkappa_w$ values shows that the distribution on wall and grinding media for pigment 2 is better, by easily a factor of three, than for pigment 1. This indicates that pigment 2 can advantageously be dispersed with high grinding media fillage.

8.3 Lightening Power

8.3.1 Graphical Method

Instructions for the graphical method are provided in standards.* The determination is based on the lightening power as defined in equation (4.75): It is the mass of reference white pigment that, when added to a colored paste of stated composition, yields the same lightness as with the reference pigment. A calibration curve, showing the CIE tristimulus value Y for the reference colored paste as a function of PVC, is used in the evaluation. In order to determine the lightening power, one disperses the test pigment, at an agreed-on concentration, in a certain quantity of the agreed-on colored paste. In this way, what is measured is the CIE tristimulus value Y for the mixture with colored or black base. From the calibration curve (measured Y values versus white pigment PVC of reference), the nominal PVC is read off; this is the value that would give the same CIE tristimulus value Y. The lightening power is then obtained as the ratio of masses calculated from the two PVC values. It is useful to plot, over the calibration curve, the associated curve from which the lightening power can be read off directly.

The method can be further simplified by plotting Y directly against the pigment mass (instead of the PVC). The lightening power curve then becomes a straight line.

The standards present examples of lightening power determination by a gray paste method and a blue paste method. The composition of the black paste is chosen such that the concentration of the black pigment in the gray pastes remains constant for all PVC values; this corresponds to the constant volume method. Here we go through the example, but using the adding pigment method for simplicity; in this technique, the Y values are plotted directly against the added masses of white pigment, so that only one black paste has to be prepared.

Aside from the analytical balance, spatula, doctor blade, and so forth, the following apparatus will be needed:

- Triple-roll mill
- Automatic muller
- Colorimeter
- Measuring setup for measurements on pastes, such as the paste film holder of Section 6.2.3

---

* See "Lightening Power of White Pigments" and "Scattering Power, Gray Paste method" in Appendix 1.

Preparation method:
First prepare *master carbon black paste:* Carefully stir by hand
18.7 g Flammruss® 101 (Degussa)
81.3 g Alkydal® L64 (Bayer)
and pass through the triple-roll mill six times.
Composition of the *black paste* for, say, a PVC of 15% of the reference (density $\varrho = 4.1$ g/cm³):
- 91.72 g Alkydal® L64
- 5.11 g Aerosil® 200 (Degussa)
- 3.17 g master carbon black paste (see above)

For other black paste compositions for other test PVC values or other pigment densities, see Table 8-6.

In order to generate the calibration curve of the reference, add increasing masses of white pigment to equal masses of black paste and place on the lower

Table 8–6. Compositions of black pastes.

Pigment volume concentration (PVC)	Composition of black paste, g			Volume of white pigment per 2.50 g black paste (cm³)
	®Alkydal L 64 g	®Aerosil 200 g	Carbon black paste (Section 8.3.1)	
5%	92.04	5.12	2.84	0.123
6%	92.02	5.12	2.87	0.149
7%	91.98	5.12	2.90	0.176
8%	91.95	5.12	2.93	0.203
9%	91.92	5.12	2.96	0.251
10%	91.89	5.12	2.99	0.259
11%	91.86	5.12	3.03	0.288
12%	91.82	5.12	3.06	0.318
13%	91.79	5.12	3.10	0.349
14%	91.75	5.11	3.13	0.380
15%	91.72	5.11	3.17	0.412
16%	91.68	5.11	3.21	0.444
17%	91.64	5.11	3.25	0.478
18%	91.60	5.11	3.29	0.512
19%	91.56	5.11	3.33	0.547
20%	91.52	5.11	3.37	0.583
21%	91.48	5.11	3.41	0.620
22%	91.43	5.11	3.45	0.658
23%	91.39	5.11	3.50	0.697
24%	91.34	5.11	3.54	0.737
25%	91.30	5.11	3.59	0.777

8 Determination of Tinting Strength and Lightening Power

plate of the automatic muller. After mixing, distribute in the form of a circular ring. Mull in four stages of 25 revolutions each, under pressure of 1.5 N/cm². After each stage, rehomogenize and redistribute. Prepare and measure each calibration paste twice.

Testing method:
Mix together 2.5 g of black paste and 1.684 g of test pigment, using the method described above. Measure the CIE tristimulus value Y of the resulting gray paste. From the calibration curve, read off the mass to be added such that the reference will exhibit the same Y value as the test.

The calibration diagram for the example is shown in Figure 8-12 along with the lightening power line. From the curve it can be inferred that a reference quantity of 1.250 g corresponds to the Y value of the test, so that the lightening power is $P_w = 1.250/1.684 = 74\%$. This value is also read directly off on the upper line. Figure 8-12 further contains the CIE tristimulus values X and Z of the reference, measured at the same time. With their help, the X and Z values of the reference at matching can be read off; these can be further analyzed together with the measured test tristimulus value in order to obtain the shade difference after color reduction (see Section 2.3.3).

If the calibration curve were plotted as Y versus the logarithm of the added mass (instead of versus the mass itself), a straight line would again be obtained

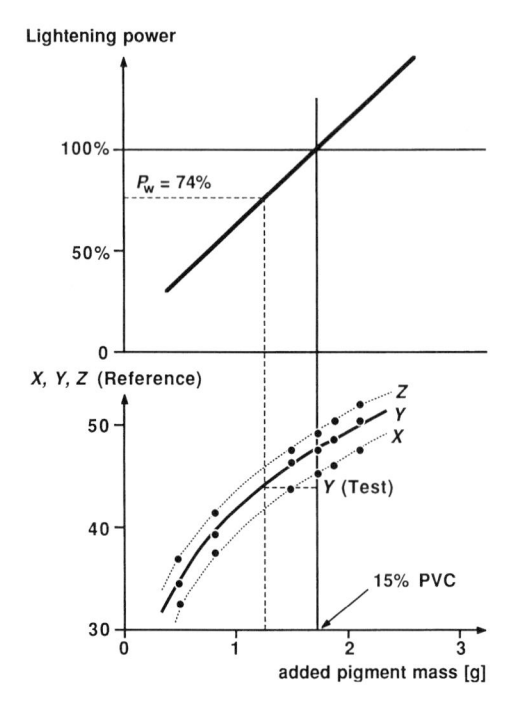

Fig. 8–12. (Bottom) Calibration curve for Y with CIE tristimulus values X and Z. (Top) Lightening power versus added pigment mass of reference.

in the Beer's law range of PVC, according to equations (3.118) and (4.49). Such a calibration line is the basis for GRASSMANN and CLAUSEN's method of determining the lightening power (see Section 4.8).

It has been shown that the increase in the scattering interaction outside the Beer's law range (see Fig. 4-20) means no linear variation of lightening power with PVC can be expected. The advantage of the standard method is that it permits lightening power (and shade difference after color reduction) to be determined as a function of the PVC of the reference pigment. For this purpose, a special black paste as characterized in Table 8-6 must be prepared from the master carbon black paste for every new PVC. The procedure described above – recording of a calibration curve, etc. – must then be repeated for every PVC. It is clear that the method of determination needs a good deal of rationalization. This is the subject of the sections that follow.

8.3.2 Rationalized Method

The method is described in the standards*. The Kubelka-Munk function is used for rationalization. As in Section 8.2.1, gray pastes with equal added masses of white pigment are prepared and their CIE tristimulus values Y are measured. After Saunderson correction of the Y values (eq. 4.112), the Kubelka-Munk functions F (eq. 3.117) are calculated for the reference and test. The ratio of scattering coefficients per unit pigment mass concentration then gives the relative scattering coefficient of the test pigment per unit mass (eq. 4.106). This is identical with lightening power, but only in the Beer's law range. The relative scattering coefficient itself is usually better stated in terms of volume; the conversion from the mass-referred quantity is easily given by the reference/test density ratio (eq. 4.103). When relative scattering coefficients obtained by this method are compared with those from the black-ground method of Section 7.2.1, the unlike nature of the methods means that complete agreement cannot always be expected.

The apparatus and materials for this test are as listed in Section 8.3.1, and the same master carbon black paste is again prepared. The ISO standard describes the analysis at a fixed test PVC of 17%; the German standard allows other PVC values, and it includes a table giving the required paste compositions for PVC values in 1% steps from 5% to 25% (see Table 8-6).

In parallel to the lightening power, the shade difference after lightening power matching can also be calculated from the pigment mass added to match. The formulas for this calculation are collected in Table 8-7 as a programming aid. A printout of results appears in Figure 8-13.

This method can be employed only in the Beer's law range; at higher pigment concentrations, the values contain errors, that is, they no longer agree

---

* See "Lightening Power of White Pigments" and "Scattering Power, Relative, Gray Paste Method" in Appendix 1.

8 Determination of Tinting Strength and Lightening Power

Table 8–7. Formulas needed to calculate lightening power and shade difference after lightening power matching in the Beer's law range.

Given: $\hat{X}, \hat{Y}, \hat{Z}$ and $\hat{X}_R, \hat{Y}_R, \hat{Z}_R$ for equal added pigment masses $\hat{m}$

$$\hat{R}^{++} = \frac{\hat{R} - 0.04}{0.36 + 0.6\,\hat{R}} \tag{4.112}$$

$$\hat{F} = \frac{(1 - \hat{R}^{++})^2}{2\,\hat{R}^{++}} \tag{3.117}$$

where
1) $\hat{R} = \hat{Y}/100 \qquad \hat{R}^{++} = \hat{Y}^{++}/100 \qquad \hat{F} = \hat{F}_Y$
2) $\hat{X}_R/X_n \qquad\qquad \hat{X}_R^{++}/X_n \qquad\qquad \hat{F}_{X,R}$
3) $\hat{Y}_R/100 \qquad\qquad \hat{Y}_R^{++}/100 \qquad\qquad \hat{F}_{Y,R}$
4) $\hat{Z}_R/Z_n \qquad\qquad \hat{Z}_R^{++}/Z_n \qquad\qquad \hat{F}_{Z,R}$

$$P_w = \frac{\hat{F}_{Y,R}}{\hat{F}_Y} \tag{4.106}$$

$$F_{X,R} = \hat{F}_{X,R} \cdot P_w; \qquad F_{Z,R} = \hat{F}_{Z,R} \cdot P_w \tag{4.67}$$

Then continue as in Table 8–1 (eqs. 4.111 ff.).

The CIELAB coordinates are substituted in the formulas for shade difference after color lightening power matching:

$$\Delta s = \sqrt{\Delta a^{*2} + \Delta b^{*2}}; \qquad \Delta\varphi = \operatorname{Arctan}\frac{\Delta b^*}{\Delta a^*} \tag{2.16}$$

From Table 2–3, $\Delta\varphi \to$ nature of undertone.

---

with the values obtained by the Y matching method of Section 8.3.1. In such cases, it is necessary to use the first approximation of the lightening power, equation (4.105). This means that a_w must be determined; it contains pigment data for the reference alone. The coefficient g_R of the scattering interaction of the reference white pigment occurs in it. This is expediently determined by using the dilution method (see section 4.4.1), or simply from two gray mixtures containing quantities of the reference that differ as widely as possible; in this case, two expressions are obtained from equation (4.48) (we now drop the subscript R):

$$\varphi_1 = \frac{1}{F_1} - \frac{1}{F_{BP}} = z_1\,\psi_1\,(1 - z_{\frac{5}{3}}\,\psi_1^{\frac{2}{3}})$$

$$\varphi_2 = \frac{1}{F_2} - \frac{1}{F_{BP}} = z_1\,\psi_2\,(1 - z_{\frac{5}{3}}\,\psi_2^{\frac{2}{3}}) \tag{8.3}$$

8.3 Lightening Power

```
Program "LP"
Lightening Power, according to ISO 787-24  *****************************************
Date: .. - .. - ..
Reference is A                              Test is B
Medium and black paste: according to ISO 787-24
Colorimeter/Standard Observer: N.N., C/2°
Measuring Conditions: d/8 (incl)
Input:     *********************************************************************
Masses: Black paste: 2.5 g                  Pigment: 2.0 g
Reflectometer values:
        Reference:                          Test:
R´(x)   46.88                               46.81
R´(y)   48.08                               48.79
R´(z)   52.50                               55.84
Output:    *********************************************************************
Lightening Power:                           105 %
Lightness:                                  74.9
Color Difference after matching:
        Type:                               Magnitude:
        blue                                Delta (s) 2.8
```

Fig. 8–13. Lightening power and shade difference after lightening power matching for a white pigment, obtained by the rationalized method (computer output).

and thence

$$g = z_{\frac{5}{3}} = \frac{\varphi_1 \psi_2 - \varphi_2 \psi_1}{\varphi_1 \psi_2^{\frac{5}{3}} - \varphi_2 \psi_1^{\frac{5}{3}}} \tag{8.4}$$

The desired factor a_w is calculated with equation (4.105):

$$a_w = g\,\hat{\psi}^{\frac{2}{3}} \text{ where } \hat{\psi} = \frac{\hat{m}}{\mu} \text{ and } \mu = m_{BP}\,\frac{\varrho}{\varrho_{BP}}$$

As a computational example, the values from Annex 1 of the German standard can be used again: $m_1 = 0.520$ g, $m_6 = 2.062$ g, $Y_1 = 32.6$, $Y_6 = 50.4$. The transformations made are, first, the Saunderson correction $(Y \to Y^{++})$ and, second, the calculation of the Kubelka-Munk functions $(Y^{++} \to F)$; see Table 8-7 for the formulas:

$$Y_1^{++}/100 = (32.6/100 - 0.04)/(0.36 + 0.6 \cdot 32.6/100) = 0.515$$

$$Y_6^{++}/100 = 0.700$$

$$1/F_1 = 2 \cdot 0.515/(1 - 0.515)^2 = 4.372$$

$$1/F_6 = 15.61.$$

Now, with $1/F_{BP} = 0.027$, equation (8.3) gives $\varphi_1 = 4.372 - 0.027 = 4.35$ and $\varphi_2 = 15.59$. Given $m_{BP} = 3$ g, $\varrho_{BP} = 1.285$ g/cm³, $\varrho = 4.1$ g/cm³, and $\mu = 3 \cdot 4.1/1.285 = 9.57$ g, the PMV values are found to be $\psi_1 = 0.520/9.57 = 0.0543$ and $\psi_6 = 0.2155$. By equation (8.4), then, $g = (4.35 \cdot 0.2155 - 15.59 \cdot 0.0543)/(4.35 \cdot 0.2155^{5/3} - 15.59 \cdot 0.0543^{5/3}) = 0.422$. Finally, with $\hat{m} = 1.689$ g, equation (4.105) yields $a_w = 0.422(11.689/9.57)^{2/3} = 0.1328$. In the DIN example (see Fig. 8-12), the mixture of the test gray paste has a CIE tristimulus value $Y = 44.0$. If the associated F value is substituted in equation (4.105) together with the value just calculated for a_w, the result is a lightening power of $P_w^{(1)} = 74\%$ – in extremely good agreement with the graphical solution (whereas the zero-th approximation would have given $P_w^{(0)} = 76\%$). Thus the example shows that the first approximation still supplies correct values far away from 100% lightening power.

If X and Z (or R'_x and R'_z) are measured at the same time, then from any pair of values (X_1, X_6 and Z_1, Z_6) the coefficients z_{X1}, $z_{X5/3}$ and z_{Z1}, $z_{Z5/3}$ of the Kubelka-Munk functions F_X and F_Z of the reference can be calculated with equation (4.109). Thus the CIE tristimulus values X_R and Z_R of the reference at matching are accessible, and from these together with X, Z of the test, the shade difference after lightening power matching can be determined.

Even with this new method, a black paste must be mixed for every new PVC, and new gray mixtures must be prepared. Section 8.3.3 will show how the entire curve of lightening power versus PVC can be obtained from a single gray mixture.

8.3.3 PVC Dependence from One Gray Mixture

This method requires only a single mixture of pigment with black paste but yields values for the lightening power throughout the PVC range of interest. The key to the method is a rather simple relation with which the dilution method (Section 4.4.1) can be used to obtain the coefficient $z_{w5/3} = g$ (eq. 4.52). Accordingly, one need only dilute the initial gray paste in a known ratio with additional vehicle – i.e., the black paste without black pigment – in order to calculate g from the two Kubelka-Munk functions. This test can be performed much more quickly than any of the determinations that have been discussed, because the pacing step is mixing in the automatic muller. All that is required for dilution is more stirring. One of the preparation methods set forth above can be employed for the preparation of the starting mixture.

The algorithm uses equations (4.51) and (4.52) to calculate g, then equation (4.48) to calculate the other coefficient z_1. If both coefficients are known for both test and reference, equation (4.102) can be used with $p_{w1} = 0$ to draw or plot the full curve of lightening power versus PVC. Table 8-8 lists out the formulas needed for programming, and Figure 8-14 shows a flowchart for the procedure.

The data needed for the dilution require some separate calculations; these are also described in Table 8-8. First, the volumetric dilution factor $q_D = \sigma/\sigma_A$

Table 8–8. Formulas needed for calculation of lightening power and shade difference after lightening power matching as functions of PVC.

Given: $\hat{\psi}_A, \hat{X}_A, \hat{Y}_A, \hat{Z}_A$ before dilution
$\hat{\psi}, \hat{X}, \hat{Y}, \hat{Z}$ after dilution $\Bigg\}$ of test and reference (subscript R)

Saunderson correction and calculation of Kubelka-Munk functions of all values with equations (4.112) and (3.117) (see Table 8–7):

$$g_Y = \frac{1}{\hat{F}_{Y,A}} \frac{\hat{F}_{Y,A} - \hat{F}_Y}{\sigma_A^{\frac{2}{3}} - \sigma^{\frac{2}{3}}} \quad \left(\text{where } \sigma = \frac{\psi}{1+\psi} \right) \quad (4.51), (4.52)$$

similarly g_X, g_Z of test and reference:

$$\varphi_A = \frac{1}{F_{Y,A}} - \frac{1}{F_{BP}}; \quad z_{Y1} = \frac{\varphi_A}{\psi_A (1 - g_Y \psi_A^{\frac{2}{3}})} \quad (4.48)$$

similarly z_{X1}, z_{Z1} of test and reference:

$$p_{wo} = \frac{z_{Y1}}{z_{Y1,R}}; \quad p_{w\frac{2}{3}}^2 = g_Y - p_{wo}^{\frac{2}{3}} g_{Y,R} \quad (p_{w1} = 0) \quad (4.104)$$

$$P_w^{(1)} = \frac{\varrho_R}{\varrho} p_{wo} \left[1 - p_{w\frac{2}{3}}^2 \left(\frac{\sigma}{1-\sigma} \right)^{\frac{2}{3}} \right] \quad (4.102)$$

$$\frac{1}{F_{X,A}} - \frac{1}{F_{BP}} = \bar{z}_{X1} P_w^{(1)} \frac{\sigma}{1-\sigma} \left[1 - g_X \left(P_w^{(1)} \frac{\sigma}{1-\sigma} \right)^{\frac{2}{3}} \right] \quad (4.109)$$

$$\frac{1}{F_{Z,A}} - \frac{1}{F_{BP}} = \bar{z}_{Z1} P_w^{(1)} \frac{\sigma}{1-\sigma} \left[1 - g_Z \left(P_w^{(1)} \frac{\sigma}{1-\sigma} \right)^{\frac{2}{3}} \right] \quad (4.109)$$

Then continue as in Table 8–1 (eqs. 4.111 ff.). The CIELAB coordinates are analyzed to give shade difference after lightening power matching, as in Table 8–7 (eq. 2.16 and Table 2–3). Lightening power and shade difference after lightening power matching can be plotted (manually or by computer) versus σ.

Dilution data:
If σ_A is to be lowered to σ by dilution, the volumetric dilution factor is $q_D = \sigma / \sigma_A$. The following must then be calculated:

$$\mu = m_{BP} \frac{\varrho}{\varrho_{BP}} \quad (4.4); \quad \psi = \frac{m}{\mu} \quad (4.6); \quad \sigma = \frac{\psi}{1+\psi} \quad (4.15)$$

$$w = \frac{m}{m + m_{BP}} \quad (4.1); \quad \varrho_{GP} = \frac{\varrho \sigma}{w} \quad (4.17); \quad m_M = m_{GP} \frac{\varrho_M}{\varrho_{GP}} \left(\frac{1}{q_D} - 1 \right) \quad (8.5)$$

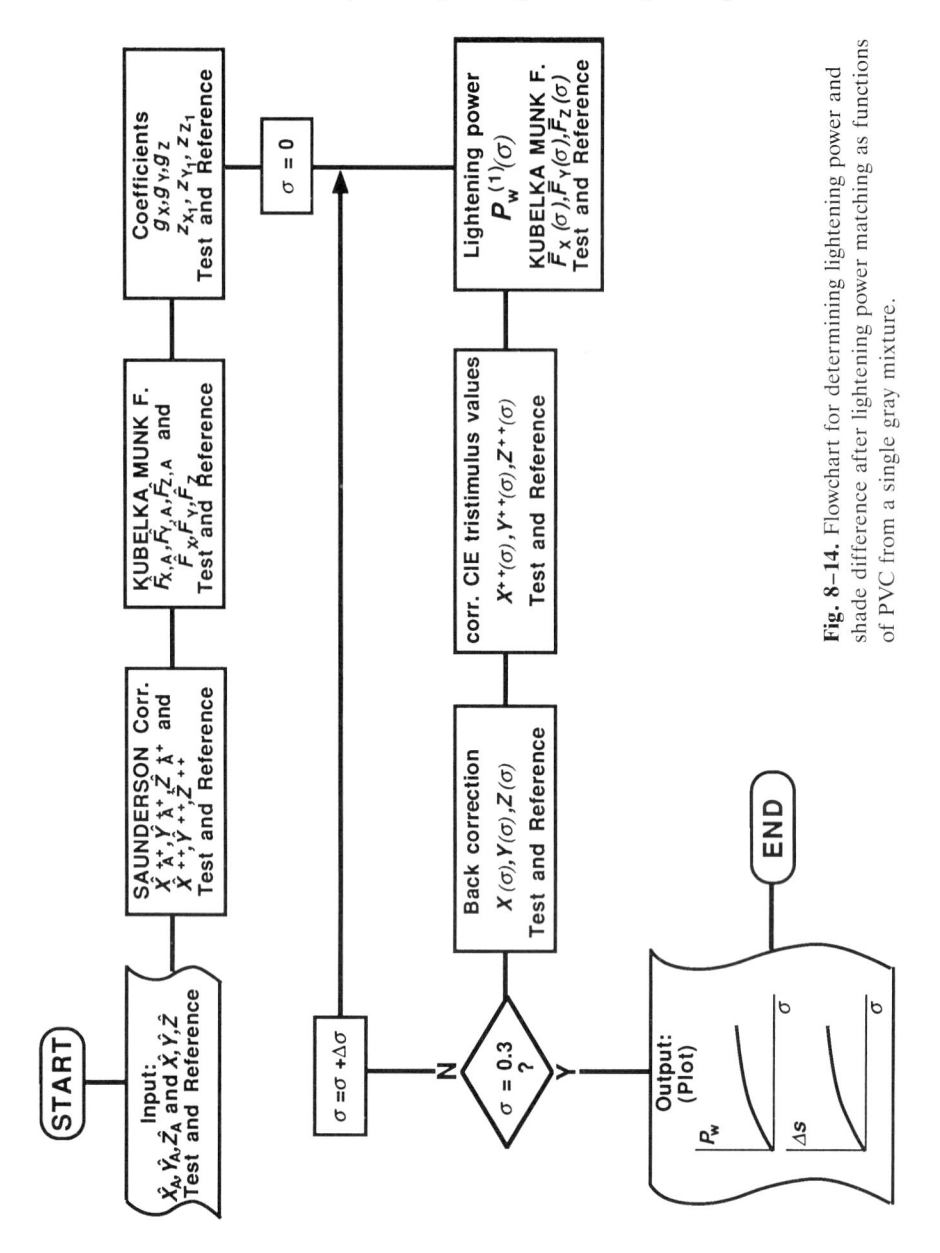

362   8 *Determination of Tinting Strength and Lightening Power*

Fig. 8–14. Flowchart for determining lightening power and shade difference after lightening power matching as functions of PVC from a single gray mixture.

is fixed. The volume ratio of white pigment/black paste mixture ("gray paste," subscript GP) V_{GP} to medium V_M to be added is then derived from $\sigma_A = V/V_{GP}$ and $\sigma = V/(V_{GP} + V_M)$, where V is the volume of white pigment:

$$\frac{V_M}{V_{GP}} = \frac{1}{q_D} - 1$$

and (8.5)

$$m_M = m_{GP} \frac{\varrho_M}{\varrho_{GP}} \left(\frac{1}{q_D} - 1 \right)$$

To convert volumes to masses, the density of the gray paste must be known. It can be found in the following way: First, the PVM μ (eq. 4.4) and thence, with the PMV ψ (eq. 4.6), the PVC σ in the gray paste are calculated (eq. 4.15). Together with the PMP w, equation (4.1), the density ϱ_{GP} of the gray paste is then obtained with equation (4.17).

The procedure can be illustrated by an example. The starting values are as follows: $m = 5$ g, $m_{BP} = 6.5$ g, $\varrho = 3.957$ g/cm³, $\varrho_{BP} = 1.286$ g/cm³. Then $\mu = 6.5 \cdot 3.957/1.286 = 20.00$ g, $\psi = 5/20 = 0.2500$, $\sigma = 0.25/(1 + 0.25) = 0.2000$, $w = 5/(5 + 6.5) = 0.435$, $\varrho_{GP} = 3.957 \cdot 0.2/0.435 = 1.819$ g/cm³. If the PVC is to be reduced from 20% to 0.5% by dilution, $q_D = 0.005/0.2 = 0.025$. The mass of medium required for 0.5 g of gray paste is then obtained from equation (8.5): $m_M = (0.5 \cdot 1.231/1.839) \cdot (1/0.025 - 1) = 13.05$ g.

Figures 8-15 and 8-16 give plotted curves of lightening power versus PVC for a total of six white pigments, all against the same reference. The curves show that pigments initially having a better relative lightening power may well

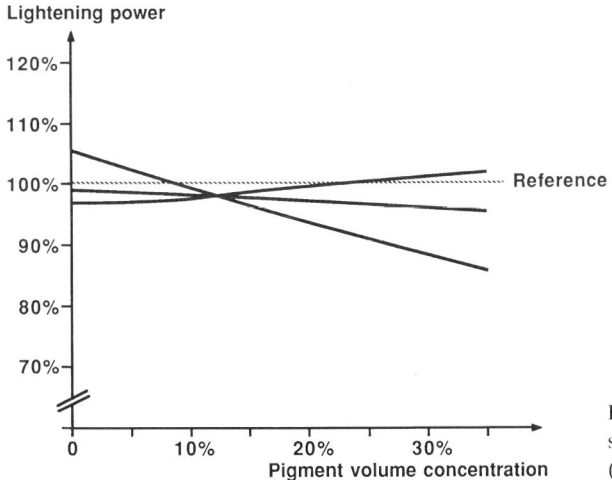

Fig. 8–15. Lightening power versus PVC for three rutile pigments (series 1).

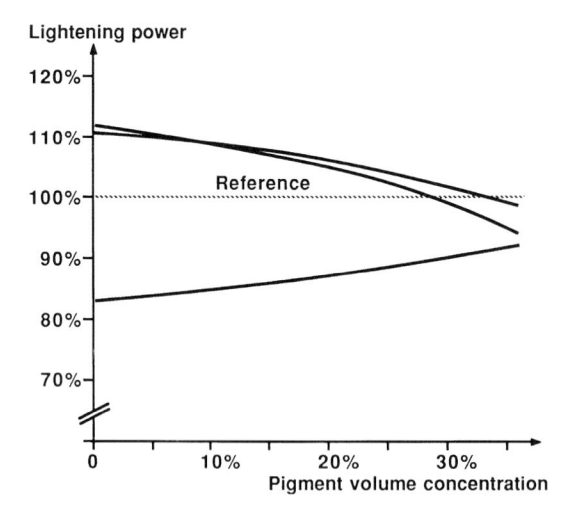

Fig. 8–16. Lightening power versus PVC for three rutile pigments (series 2).

become worse at higher PVC, and vice versa. It should be kept in mind that lightening power values for PVC $\sigma > 30\%$ do not necessarily represent reality, because of the possibility of a pigment concentration exceeding the critical PVC.

In calculating the shade difference after lightening power matching, one proceeds in an analogous way. Again, equations (4.52) and (4.109) are used to calculate g and z_1 from the Kubelka-Munk functions of the CIE tristimulus values X and Z, and the mass added at matching $\hat{m}P_w$ is substituted in the equations. Similarly to what was done in Tables 8-1 and 8-7, the strength and nature of the undertone are then inferred; they can also be plotted in suitable form as functions of the PVC.

This test can be performed in less than 20 minutes (up to plotting of the curves). At our current level of knowledge, an interim goal – a more rational test for lightening power – has thus been attained with the method just described.

Appendix 1

Listing of Standards

Key words	ISO	ASTM	BS	NF	JIS	DIN
Black value (Carbon black pigments)						55 979
Change in tinting strength (see Dispersion and PVC)						
Coating materials, terms and definitions	4618-1 to 4	D 16	2015	T 36-001 T 36-005	K 5500	55 945
Color depth						53 235 T 1 to 2
Color differences:						
CIELAB	7724-3	D 1729 D 2244 E 308	3900/D10	ISO 7724-3		6174
Conditions/evaluation of measurements Significance	7724-2		3900/D9	ISO 7724-2		53 236 55 600
Color in full-shade systems:						
Black pigments	787-25	D 3022				55 985 T 2
Colored pigments	787-25	D 3022				55 985
White pigments	787-25	D 2805				55 983
Color matching	3668		3900/D1	T 30-061	K 5400	6173 T 1
Color measurement (see Color differences)						53218
Color tolerances			4764			6175 T 1
Colorimetry	7724-1 7724-2	E 259 E 308	3900/D8 3900/D9	ISO 7724-1 ISO 7724-2		5033 T 1 to 9
Coloring materials:						
Classification				T 36-002		55 944
Terms and definitions	4618-1		2015	T 36-003		55 943
Density of pigments:						
Centrifuge method	787-23		3483/B9	T 31-201		
Pyknometer method	787-10	D 153	3483/B8	T 31-202	K 5101	ISO 787 T 1

Key words	ISO	ASTM	BS	NF	JIS	DIN
DIN Color System						6164 T 1 to 3
Dispersion, ease of:						
Alkyd resin and alkyd melamine system:						
Drying by oxidation:						53 238 T 30
						53 238 T 33
Stoving type						53 238 T 31
						53 238 T 32
Automatic muller	8780-5	D 387	3483/D5		K 5101	ISO 8780 T 5
Bead mill	8780-4		3483/D4			ISO 8780 T 4
Change in gloss	8781-3		3483/E3			ISO 8781 T 3
Change in tinting strength	8781-1		3483/E1			ISO 8781 T 1
Fineness of grind (see below)						
High speed impeller mill	8780-3		3483/D3			ISO 8780 T 3
Introduction	8780-1		3483/D1			ISO 8780 T 1
Oscillatory shaking machine	8780-2		3483/D2	T 30-067	K 5101	ISO 8780 T 2
Triple roll mill	8780-6		3483/D6			ISO 8780 T 6
Emulsion paints						53 778 T 3
Film thickness	2808		3900/C5	T 30-120 to 125	K 5400	50 982 T 1 to 3 ISO 2808
Fineness of grind	1524 8781-2	D 1210	EN 21524 3483/E2	EN 21524	K 5101	EN 21524 ISO 8781 T 2
Flow time, determination	2431		EN 535	EN 535	K 5400	EN 535
Gloss, measurement	2813	D 523 D 5307	3900/D5	T 30-064	K 5400	67 530
Hiding power:						
Pigmented media	6504-1	D 2805	3900/D7			55 987
White and light gray media		D 2805				55 984
Hue of near white specimens						55 980
Hue relative of near white specimens						55 981
Lightening power of white pigments	787-17	D 2745	3483/A5	T 30-023		55 982

Appendix 1

Key words	ISO	ASTM	BS	NF	JIS	DIN
Lightness:						
Emulsion paints (see above)						
White pigment powders						53 163
Loss on ignition (Red iron oxide pigments)	1248	D 3721				55913 T 1
Metamerism index						6172
Mixtures, Composition						1304
Particle size analysis:						1310
Basic terms:			3406/1 to 7			
Representation:						53 206 T 1
Basic terms	9276-1					66 141
Logarithmic normal diagram		D 1366				66 144
Power function grid						66 143
RRSB grid						66 145
Sedimentation method:						
Balance method					Z 8820	66 116 T 1
Basic standards		D 3360		T 30-044	Z 8822	66 111
Pipette method					Z 8821	66 115
Photometric properties of materials						5036 T 1 to 6
Pigment content						55 678
PVC, paste:						
Basic mixture						53 773 T 1
Test specimen preparation						53 773 T 2
PVC, plasticized:						
Basic mixture						53 775 T 1
Change in tinting strength						53 775 T 7
Test specimen preparation						53 775 T 3
PVC, unplasticized:						
Basic mixture						53 774 T 1
Test specimen preparation						53 774 T 2

Key words	ISO	ASTM	BS	NF	JIS	DIN
Radiometric properties of materials (see Photometric properties of materials)						
Reflection factor: paper, cardboard:						
Fluorescent			4432/1 to 3	Q 03-039		53 145 T 2
Nonfluorescent	2469			Q 03-038		53 145 T 1
Reflectometer (see gloss)						
Sampling	842	D 3925	4726	T 30-048	K 4001	53 242 T 1 to 4
					K 5101	
					K 5400	
Scattering power, relative:						
Gray paste method	787-24		3483/A6	T 30-021		ISO 787 T 24
Black ground method	6504-1					53 164
Significances (see color differences)						
Symbols for use in formulas	31		5775	X 02-001 etc. Z 8202		1301
						1303
						1304
Thermoplastics						53772
Tinting strength, relative:						
Change in	8781-1		3483/E1			ISO 8781 T 1
Dyes in solution		D 387				55 987
		D 3022				55 986
Photometric	787-24	D 4838	3483/A6	T 30-031		ISO 787 T 24
Visual	787-16		3483/A4	T 30-027		ISO 787 T 16
Transmission of optical radiation						1349
Transparency:						
Paper, cardboard	2469	D 589	4432/3	Q 03-006	P 8138	53 147
				Q 03-040		
Pigmented/unpigmented systems						55 988

Subject Index

A

absorbing power, principles   3–9, 67–127, 129–209, 211–35
– test method   239–41, 261, 285, 333–35
absorption coefficient   11, 67–127, 136, 137–39, 211–35
– four-flux theory   75–79, 137–39, 300–4
– Kubelka-Munk theory   75–77, 137–39, 290, 291–304
– – determination   291–304
– – – error   296–300
– – per unit mass   137, 336
– – – relative, determination   335
– – per unit volume   149–50, 195, 224–26
– many-flux theory   68–71
– test methods   259, 290, 291–304, 335–39
absorption cross-section   11, 224–26, 229–31
absorption index   11, 217–21, 229–31
absorption maximum   139, 167, 333, 335, 348
acceptability   16, 56–59, 248, 285; see also ellipsoid, acceptability
adding-pigment method   145–47, 336
agglomerate   130, 212, 243
aggregate   130
anisotropy, optical   220

B

Beer's law   137, 138
– region   139–42, 292, 346
black standard   see colorimeters
black value   258
brightness   see lightness

C

characteristic equation   45, 70, 264
chi-square distribution   43, 55
chroma, principles   29, 102, 170
– test methods   254–58, 317, 337
chroma difference, principles   33–34, 102, 170

– test methods   254–58, 317, 337
CIE chromaticity coordinates   23–26, 252, 342
CIE chromaticity diagram   25, 282
CIE color matching functions   23–26, 115–16, 159–60, 250–54, 308
CIE color notation system   20–26, 113, 187, 254
CIE standard colorimetric system   see CIE color notation system
CIE standard illuminants   17, 116, 159–62, 251, 293
CIE standard observer   see CIE color notation system
CIE tristimulus values, principles   23–26, 116, 158–59, 161–64
– test methods   250–54, 292, 335–37
CIELAB color space   31, 36–37, 168–71, 264–65, 281
CIELAB coordinates   28–37, 112–14, 168, 255, 257–58, 337
CIELAB system   28–37, 113, 115, 191–94, 252–54, 317–18
coating thickness   see film thickness
colloid chemistry   129, 131–36
color depth   170–71, 335, 341–42
– criteria   187–88, 341
– level   see standard depth of shade
– matching   187–89, 339, 341, 345
– standards   339, 341
color difference
– after color reduction   191–94, 335–39, 340, 341; see also shade difference
– principles   15, 101–2, 167, 170
– test methods   246–48, 305, 337
color formulation   15, 166–67, 180, 303, 345
color identification code   15
Color Index   39
color matching   172–89, 275–84
– trichromatic, theory   23

Subject Index

color measurement   10, 15–17, 286
– errors   53–56, 98, 266–70
– – ellipsoid   264–66
– – preparation   255, 266–70
– – – ellipsoid   267–70
– test errors   262–70
– – ellipsoid   262–70
color mixing   20
color, object   see object color
color order systems   37–39
color plane   29, 34–35
color rendering   15
color sample   37, 248–49, 341
color saturation   see saturation
color space, absolute   27, 38
– CIELAB   see CIELAB color space
color stimulus   9–12, 15, 27, 112, 260, 275
color systems   15, 26
– perceptual   26–37
– physiologically equidistant   27–28
color tolerances   see acceptability
colorimeters   17–19, 72–74, 244–48, 293, 335–36
– black standard   250, 258, 312
– calibration   250, 292, 341–42
– light sources   249–50, 277
– measuring conditions   19, 72–74, 244–48, 260, 293, 295, 336
– sample illumination   260
– spectrophotometer   see spectral method
– three-filter   see tristimulus method
– trichromatic matching   23, 27
– white standard   19, 250, 312
colorimetry   10–12, 15–48, 181, 241, 341
coloring material   3–9, 212, 247–48
see also dyes; pigments
coloring power   108–10, 118–20, 185–86, 285, 306
– condition (measurement)   see colorimeter
– determination   308
confidence interval, level   55, 254, 262–63, 271, 279
conformity (with standard)   see acceptability
constant-volume method   145–47, 354
contrast ratio   101–2, 318
contrast substrate   101–7, 143, 255, 296
corpuscular theory   67, 211–35
cosine distribution, Lambert's   72, 240–241
covariance   53–54, 262, 272
– matrix   54, 262
critical value   see confidence interval

D
daylight color   259
dilution method   145–47, 358
DIN Color System   39
dispersing (dispersion) of pigments
    see pigments, dispersing
dispersion (colloid chemistry)   129–36, 248–49
– density   136
dispersion, anomalous (optics)   230
distribution
– cumulative   213
– density   212
– width   213
see also normal distribution; lognormal distribution
dye
– content   67, 129–209, 261, 331–33
– principles   67, 129–36, 331–33
– test methods   247, 259–61, 275, 300–3, 331–33

E
effectiveness   101, 186, 326, 328–30
ellipse
– axes   216
– coefficients   41
– eccentricity   216
– standard deviation   41, 216–17
ellipsoid
– acceptability   56, 275–84
– coefficients   50–53, 262–70
– conditions   44, 49, 264
– normal form   45, 48, 55
– principal axes   44–47, 264–66
– – system   45, 266
– projections   47, 264–66
– radius   49
– standard deviation   42–43, 44–49, 54, 262–70
see also color measurement, errors
equi-energy stimulus   25
equivalent diameter   131
Eurocolor system   39

F
FIAF method   339, 341–48
fillers (extenders)   3, 131, 132, 220, 254, 332
film thickness
– principles   11, 75–77, 137
– test methods   254, 285–91, 325

– measurement 285–91
– – chemical, electrical, magnetic, optical methods 286
– – dial-gauge micrometer methods 288–89
– – errors 285
– – gravimetric methods 287, 312
– – pneumatic methods 289–91, 309, 312, 328
– – wedge cut method 287–88
floating 132, 269
flocculation 130, 132, 249
four-flux theory 67–68, 70, 75–93, 300
Fresnel equations 239, 243
Fresnel reflection 226, 228, 239
full shade 132, 246–47
– black 246, 254–58
– chromatic 254–58
– – determination 312
– white 254–58
full-shade system 249, 254, 275

G

gloss 102, 132, 240, 293
– degree of 105, 243, 247, 314–17, 325
– high 240, 247, 293, 305
– matte (mat) 240, 247
– measurement 239–44
gloss angle 243
gloss exclusion (trap) 246–47
gloss haze 132, 243, 247
gloss indicatrix 240, 241, 262
glossmeters 241–42, 261–62
goniophotometers see glossmeters
Grassmann's laws 20
gray scale 29

H

Helmholtz equation 222
hiding criterion 101–2, 311, 320–21
hiding power 101–7, 112, 113–18, 164–65, 285–330
– economic aspects 326, 328–30
hiding power value 101, 103–5, 105–7, 285, 310–30
– absorption component 105–6, 315–17, 325
– achromatic, determination 313–18
– graphical determination 310–13, 315, 317, 320–21
– interaction component 106–7, 318, 320–21, 325

– scattering component 105–6, 315–17, 324
histogram 213, 216
hue, principles 26–27, 52, 102, 167
– test methods 256, 317, 333
hue angle 29, 257
– difference 33, 52–53, 257, 273
hue difference, principles 33–34, 52, 102, 172
– test methods 256, 317, 337
hue (shade, undertone) 34–35, 191–94, 256–58, 336
– near-black specimens 256–58
– near-gray specimens 102, 256–58, 317
– near-white specimens 102, 256–58, 317

I

immersion method 221
incandescent light color 259
infinitely thick layer 89, 90, 292
integrating sphere 244–46, 255, 260

K

Kubelka-Munk
– function 95–98, 138, 331, 335–37
– – error of determination 98
– theory 11, 68, 144, 148–64, 224, 291–300

L

Lambert's law 68, 75, 82–83, 137
light absorption power see absorbing power
light scattering power see scattering power
lightening power
– determination 354–64
– – graphical 354–57
– – per unit PVC 360–64
– – rationalized method 357–60
– principles 129, 147, 165, 189–91
– test methods 292, 296, 298, 354–64
– – Reynolds 148
lightness 26–27, 168–69, 254–58, 319, 335, 337
– difference 33–34, 101–2, 254–56, 313
– matching 181–89, 335–39, 340, 341
likelihood see maximum likelihood method
lognormal distribution 213, 215, 217
– two-dimensional 215–17

M

many-flux theory 67–71, 300, 303–4
mass tone 254
matching criteria see tinting strength criteria

372 Subject Index

matrix inversion 262–63, 269
maximum likelihood method 56, 279
metamerism 23, 252, 258–59
– geometrical see Silking effect
– index 259
Mie minimum 227
Mie theory 11, 70, 211, 222–31, 303
Munsell system 28, 36, 38

N

Natural Color System 38
Newton's method (approximation) 104, 117–18, 279, 323–25
nonluminous perceived color see object color
non-self-radiating medium 17
normal distribution
– logarithmic see lognormal distribution
– one-dimensional 40, 213–14, 266
– three-dimensional 40–43, 266, 271, 277
– two-dimensional 41

O

object color 17, 19, 23, 239–84
one-flux theory see Lambert's law
optical brighteners 260
optical contact 293–95, 312
OSA-UCS (color order system) 39
Ostwald color system 39

P

partial oscillations 222–24, 226
particle size
– fractions 213
– parameters 131, 212–17
particle-size
– analysis 216–17
– distribution 131, 167, 211–17, 225, 230–31
– – analysis based on volume effects 217
– – class 213–14
– – light scattering analysis 217
particles
– anisometric 215, 224, 259
– counting 217
– diameter 131, 212, 220
– form 129–31, 215
– mass 131, 212
– number 212
– principles 11, 67, 129–36, 211–35
– size 11, 67, 129–31, 211–17, 305
– – analysis see particle-size analysis

– – distribution see particle-size distribution
– – optimal 227–28, 230–31
– surface area 131, 212, 220–21
– test methods 243, 259, 305
– volume 131, 213–14, 220, 225
paste film holder 255–56, 355–56
perceptibility 16, 277, 281
phenomenological theory 67–68, 211
Pigenot equation 195–96, 350–51
– diagram 351
pigment content
– determination, centrifuge method 332
– – filtration method 332–33
– – incineration method 332
– principles 11, 67, 129–36, 217
– test methods 243, 291, 332–33
pigment distribution see particle-size distribution
pigment mass 133
– concentration (PMC) 133–35, 291–93, 336, 357
– equivalent to the medium volume (PVM) 134–35, 182, 363
– equivalent to the medium volume of the mass portion (PMM) 134–35
– per unit area 292
– portion (PMP) 133–35, 291
pigment/medium volume ratio (PMV) 134–35, 357
pigment particle see particle
pigment/paste mixing 145–64
pigment surface area see particles, surface area
pigment surface concentration 136
pigment volume see particles, volume concentration (PVC) 113, 133–36, 243, 287
– critical 132, 139, 364
– portion (PVP) 136
pigments
– black 4, 5, 145, 150–54, 246–47, 296–300, 335–39
– colored, principles 4, 6–8, 105, 145, 157–64, 211–35
– – test methods 254–58, 263–64, 296–300, 335–39, 340, 341
– density 134–35, 287, 355
– dispersing (dispersion) 131–33, 212–17, 243, 292, 348–53
– – equipment 248, 292, 336, 349–50, 354–56
– – rate of 194–99, 352–53

Subject Index 373

– – resistance   199, 353
– fluorescent   4–5, 260
– interference   4, 262
– luminescent   4–5, 260–61
– luster   4
– metal-effect   4, 243, 261
– nacreous   4, 262
– phosphorescent   5, 260–61
– principles   3–9, 129–36, 211–35
– test methods   243, 291, 332–33
– transparent   304–7
– visual   23
– white, principles   4–5, 105–6, 145, 155–57, 211–35
– – test methods   254–58, 291–96
PMC   see pigment mass concentration
PMM   see pigment mass equivalent to the medium volume of the mass portion
PMP   see pigment mass portion
PMV   see pigment/medium volume ratio
polarized light   219, 222–23
power-law distribution   213
primary particle   130, 141
primary stimulus   20, 21, 24, 27
probability, cumulative   42, 216
probability density   42, 216
production monitoring (control)   15–17, 185, 255, 336
PVC   see pigment volume concentration
PVM   see pigment mass equivalent to the medium volume
PVP   see pigment volume portion

R

radiance factor   18
radiative transfer   68
RAL Color Register   38–39
Rayleigh scattering   226
reaction kinetics   195–99
reference stimulus   see primary stimulus
reflectance
– errors of measurement   98, 285, 296–300
– principles   10–12, 17–19, 67–127, 149–50, 230
– test methods   239–48, 285
reflection   see also reflectance
– coefficients   71–72, 119, 219, 239–48, 302
– factor   17–19, 71–74, 239–48, 285, 335
– law   219
– specular   83–88, 240, 261
reflectometer values   170, 242, 305, 335–36

refractive index   11, 67, 217–21, 239
– complex   218, 223
– mixing rules   221
retina   31
rhodopsins   23
RRSB distribution   213
rub-out test
– determination   349
– method   249
Ryde theory   67, 81

S

sampling   248
saturation   26–27, 38, 167, 176, 178, 179, 260
Saunderson correction   71, 150, 293, 336–37
scattering
– dependent (multiple)   see scattering interaction
scattering coefficient
– four-flux theory   75–79, 136, 300–4
– Kubelka-Munk theory   75–77, 141–42, 285, 290, 291–304
– – error of determination   114–15, 285, 296–300
– – per unit mass   291–300, 357–60
– – per unit volume   141, 149, 252, 295, 357
– – – relative   see scattering power, relative
– many-flux theory   67–71
– principles   11, 67–127, 138, 139–44, 211–35
– test methods   259, 290, 291–304
scattering cross-section   11, 224–26, 227–29
scattering interaction   11, 139–42, 358–60
scattering power, principles   4, 9–10, 67–127, 129–209, 211–35
– relative   285, 291–92
– – black-ground method   292–96, 357
– – determination, paste methods   357–60
– test methods   239–48, 285, 348, 349
SDS   see standard depth of shade
secular equation   see characteristic equation
sedimentation analysis   217
shade difference after color reduction   191–94, 341, 343–55, 357; see also hue (shade, undertone)
shape factor   215
sieve effect (Kubelka-Munk theory)   298
significance   16, 53–56, 254–55, 262–63, 270–75
– level   see confidence interval
– test   272–75

Silking effect   259
size reduction   see pigments, dispersing (dispersion)
Snell's law   219
special color differences   27–28, 116, 246–48, 313, 337
spectral colors   20–21
spectral evaluation, principle of   112, 113–15, 307–9, 339, 341–48
spectral locus   21, 26
spectral luminosity curve (spectral luminous efficiency function)   26
spectral method   118–19, 250, 307, 333–35
stabilization   see pigments, dispersing (dispersion)
standard depth of shade   176, 339, 344
standard deviation   40–41, 49–53, 264, 335
standard deviation ellipse   see ellipse, standard deviation
standard deviation ellipsoid   see ellipsoid, standard deviation
standard observer   see CIE color notation system
surface chemistry   220–21
surface correction   see Saunderson correction
surface reflection   71–74, 94, 239–48, 287, 336

T

tinting characteristic   168–71
tinting strength   164–89
– absolute   166, 185
– change in (increase in strength)   194–99, 243, 348–53
– criteria   167–72, 335
– matching   306, 337–39
– relative   112, 129, 243, 291–92, 331–64
– – of coloring materials, determination   333–53

– – of pigments, inorganic, determination   335–39, 341
– – of pigments, organic, determination   339, 341–48
tinting strength equivalent   166, 341, 342
transmission   see transmission factor; transmittance
transmission factor, principles   19, 79–83, 137
– test methods   239–48, 300, 302–3, 334
transmittance, principles   19, 70
– test methods   239–48, 300, 334
transparency   107–11, 118–20, 285–330
– number   107–11, 119–20, 285, 305–8
– – single-point method determination   306–7
– – two-point method   305–7
transparency-coloring degree   110, 119
– determination   306, 327
– detertransparency-coloring power   109, 119
– determination   306, 327
trichromatic principle   21–23
tristimulus method   170, 250–54, 306, 335–39
two-flux theory   see Kubelka-Munk theory

V

van de Hulst formula   226–29
van der Waals forces   131
vision, color   23
– cones   23, 31
see also pigments, visual
volume distribution   213, 217

W

wave equation   218, 222
wavelength, principles   11, 112, 139, 217–21
– test methods   217–21, 298
Weber-Fechner law   28
wetting   131
white locus   see equi-energy stimulus
white standard   see colorimeters

Name Index

A
Abramowitz, W.   65
Adams, E. Q.   62, 63, 65
Aitchinson, J.   235
Allen, T.   208
Atkins, J. T.   125, 127, 304

B
Beasley, J. K.   127
Beattle, H.   235
Beer, A.   129, 137–40 passim, 147, 151, 153, 156, 191, 207, 208, 292, 296, 312, 346, 357, 358
Berger, A.   65, 209
Berger-Schunn   see Berger
Bessel, F. W.   231
Billmeyer, F. W., Jr.   64, 65, 125, 127, 304
Blumer, H.   235
Booth, G.   13, 32
Born, M.   65
Bouguer, P.   68, 124
Brockes, A.   64, 65, 125, 126, 209, 229, 234, 235
Brossmann, R.   234, 235
Brown, J. A. C.   see Aitchinson
Brown, W. R. J.   63, 65
Büchel, K. H.   13
Büchner, W.   see Büchel
Buxbaum, G.   13

C
Chandrasekhar, S.   125, 127
Channon, H. J.   126
Clausen, H.   207, 208, 357
Cremer, M.   127

D
Descartes, R.   241, 264
Dettmar, H.-K.   234, 235
Duntley, S. Q.   124, 125, 126

E
Eitle, D.   125, 127
Euclid   38

F
Fechner, G. T.   28
Felder, B.   234, 235
Fisher, R. A.   64
Fresnel, A. J.   72, 219, 226, 228, 233, 234, 239, 243
Friedrichsen, K.   208, 342, 344–47
Fritzsche, E.   13
Fukshanski, L.   124, 126

G
Gall, L.   63, 65, 126, 127, 172, 176–180 passim, 187, 188, 207, 208, 339–47 passim
Gans, R.   235
Gardner, H. A.   208
Gauss, F. K. (C. F.)   16, 43, 64
Geissler, G.   208
Grassmann, H.   20, 63, 65, 207, 208, 357
Grassmann, W.   see Clausen
Guild, H.   63, 65
Gurevic, M.   124, 126

H
Hauser, P.   65, 234, 235
Hediger, H.   234, 235
Heine, H.   13
Heller, W.   235
Helmholtz, H. von   23, 61, 63, 65, 222
Hengsterberg, J.   65
Herbst, W.   13
Honerkamp, J.   235
Honigmann, B.   130, 206, 208, 234, 235
Howe, W. G.   65
Hubel, D. H.   65
Hund, F.   13
Hunger, K.   see Herbst
Hunt, R. W. G.   64

Name Index

J
Jaenicke, W.   208, 234
Joos, G.   235
Judd, D. B.   125, 127, 207, 208

K
Kaluza, U.   207, 208
Kämpf, G.   125, 127, 206, 208
Keifer, S.   125, 127, 132, 169, 172, 179, 206–9 passim, 328
Kerker, M.   234, 235
Klaeren, A.   208
Koch, O.   207, 208
Köhler, P.   see Buxbaum
Kortüm, G.   125, 127, 234
Koutnik, V.   235
Kreyszig, E.   65
Kroker, R.   208
Kronecker, L.   69
Kubelka, P.   11, 67, 68, 81, 83, 85, 90–129 passim, 138, 144–64 passim, 171, 181–89 passim, 207, 224, 225, 228, 233, 291, 293, 296–302 passim, 307, 327, 331, 335–49 passim, 357–60, 364

L
Lambert, J. H.   67, 68, 72, 75, 82, 83, 124, 137, 207, 240
Laplace, P. S.   218, 222
Legendre, A. M.   70, 300, 302
Leschonski, K.   208
Lowan, A. N.   234, 235

M
MacAdam, D. L.   63, 64
Madelung, E.   142
Maikowski, M. A.   234, 235
Maxwell, J. C.   218
McLaren, K.   see Booth
Meehan, E. J.   see Beattle
Metz, H. J.   235
Mie, G.   11, 70, 211, 212, 220–31 passim, 234, 235, 303
Mudgett, P. S.   125, 127, 225, 300, 303
Müller-Kaul, W.   see Geissler
Munk, F.   see Kubelka
Munsell, A. H.   28, 36, 38, 61, 63, 65
Murley, R. D.   235

N
Nassenstein, H.   234, 235
Newton, I.   63, 65, 104, 116, 117, 126, 279, 325
Nickerson, D.   62, 63, 65

O
Ostwald, W.   23, 39

P
Paffhausen, W.   65
Panek, P.   see Buxbaum
Pauli, H.   125, 127, 234, 235
Pigenot, D. von   196, 198, 206, 207, 208, 350, 351
Pluhar, P.   see Kroker

R
Rabe, P.   see Koch
Rayleigh, J. W.   226, 233, 234, 235
Rembrandt van Rijn, H.   240
Renwick, F. F.   124, 126
Reynolds, C. E.   148, 207
Rich, R. M.   see Howe
Richards, L. W.   see Mudgett
Richter, M.   63, 65
Robertson, A. R.   64, 65
Römer, H.   see Honerkamp
Ryde, J. W.   67, 81, 83, 88, 89, 124, 125, 126

S
Saunderson, J. L.   71, 80, 94, 99, 114, 115, 125, 126, 185, 202, 205, 297, 298, 299, 302, 314, 320–21, 336, 337, 342, 349, 357, 359, 362
Schliebs, R.   see Büchel
Schmelzer, H.   207, 208, 342
Schmitt, K.-H.   see Hengstenberg
Schmitz, O. J.   see Kroker
Schrödinger, E.   65
Schuster, A.   124, 126
Schweppe, H.   see Berger
Sharples, W. G.   see Booth
Silberstein, L.   65, 124, 126
Snell (Snellius van Roijen), W. S.   219
Stabenow, J.   130, 206, 208
Stange, K.   65
Stearns, E. I.   116, 126, 127, 254, 309
Stegun, I.   see Abramowitz
Steinbuch, K.   65
Stiles, W.   65

Stokes, G. G.   124, 126
Storr, B. V.   see Channon
Strocka, D.   65, 126, 127, 209
Stultz, K. F.   65
Sturm, B.   see Hengstenberg
Suzuki, S.   see Okuda
Sward, G. G.   see Gardner

T
Thiemann, J.   224

V
van de Hulst, H. C.   226, 228, 233, 234, 235
van der Waals, J. D.   131
Vial, F.   208
Völz, H. G.   13, 65, 126, 127, 179, 207, 208, 209, 235

W
Wald, G.   65
Walraven, J.   65
Weber, E. H.   see Fechner
Westwell, A.   see Booth
Winkler, O.   see Hengstenberg
Winter, G.   see Büchel
Woditsch, P.   see Buxbaum
Wright, W. D.   63, 65
Wyszecki, G.   65

Y
Young, T.   23, 63, 65

Z
Zollinger, H.   see Booth
Zorll, U.   208

Buxbaum, G. (ed.)

Industrial Inorganic Pigments

1993. XIII, 281 pages with 92 figures and 56 tables.
Hardcover. DM 188.00. ISBN 3-527-28624-1

The authors, who are renowned experts in their field, ensure a concise, up-to-date presentation of the chemistry, production, properties, applications and economic importance of industrial inorganic pigments.

This book is neither a list of commercially available products nor a compilation of company product specifications but instead provides the knowledge required for the optimal selection and use of inorganic pigments. It will prove to be an indispensable guide to all chemists, material scientists and practitioners in pigment-related fields.

Date of Information:
November 1994

VCH, P.O. Box 10 11 61,
D-69451 Weinheim,
Fax 0 62 01 - 60 61 84

VCH

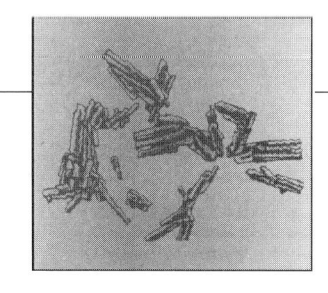

Herbst, W. /Hunger, K.

Industrial Organic Pigments
Production, Properties, Applications

1992. XIV, 630 pages with 95 figures, 6 in color and 38 tables. Hardcover. DM 296.00. ISBN 3-527-28161-4

Currently the most comprehensive source of information on synthetic organic pigments! It treats all aspects of applications of organic pigments from chemical and physical viewpoints. Relevant test methods are covered, and toxicological and ecological properties are outlined. Readers will find the book exceptionally useful as it considers the synthesis, properties and applications of organic pigments commercially available on the world market. They will appreciate the fact that standardized methods allow test results to be compared throughout the book.

From reviews of the German Edition:
"The volume can be recommended unreservedly to industrial and academic practitioners concerned in any way with the technological aspects of organic pigments. Presentation throughout is of the highest quality and the volume must now become the standard reference text in this important area of coloring matters."
Dyes and Pigments

"This wide-ranging reference work can be warmly recommended ..."
farbe + lack

Date of Information:
November 1994

VCH, P.O. Box 10 11 61,
D-69451 Weinheim,
Fax 0 62 01 - 60 61 84

VCH